ACCESS MANAGEMENT THEORIES AND PRACTICES

PROCEEDINGS OF THE SECOND INTERNATIONAL CONFERENCE ON ACCESS MANAGMENT

September 25-27, 2014
Shanghai, China

SPONSORED BY
Access Management Committee of the Transportation Research Board
Tongji University
Shanghai Jiaotong University
Research Institute of the Ministry of Public Security, PRC
Research Institute of Highway, PRC
Ministry of Transportation, PRC
Construction Institute of the American Society of Civil Engineers

EDITORS
Zhongyin Guo, Ph.D.
Baoshan Huang, Ph.D., P.E.
Zhongren Wang, Ph.D., P.E.

ASCE

CONSTRUCTION INSTITUTE

Published by the American Society of Civil Engineers

Library of Congress Cataloging-in-Publication Data

International Conference on Access Management (2nd : 2014 : Shanghai, China)
Access management theories and practices : proceedings of the Second International Conference on Access Management, September 25-27, 2014 / editors, Zhongin Guo., Ph. D., Bacshan Huang, Ph. D., P.E., Zhongren Wang, Ph.D., P.E. ; sponsored by Access Management Committee of the Transportatio Research Board, Tongji University, Shanghai Jiaotong University, Research Institute of the Ministry of Public Security, Research Institute of Highway, Ministry of Transportation, Construction Institute of th American Society of Civil Engineers.
pages cm
Includes bibliographical references.
ISBN 978-0-7844-1386-9 (print : alk. paper) 1. Roads--Interchanges and intersections--Congresses. Express highway interchanges--Congresses. I. Guo, Zhongyin. II. Huang, Bacshan. III. Wang, Zhongr (Transportation engineer) IV. National Research Council (U.S.). Transportation Research Board. Acces Management Committee. V. American Society of Civil Engineers. VI. Title.
TE176.I55 2015
625.7--dc23
2014041023

Published by American Society of Civil Engineers
1801 Alexander Bell Drive
Reston, Virginia, 20191-4382
www.asce.org/bookstore | ascelibrary.org

Errata: Errata, if any, can be found at http://dx.doi.org/10.1061/9780784413869

ISBN 978-0-7844-1386-9 (paper)
ISBN 978-0-7844-7898-1 (E-book PDF)
Manufactured in the United States of America.

Preface

A variety of topics in Access Management (AM) were covered in this special publication that contains 30 peer-reviewed technical papers. The selected papers were organized into four technical sections: (1) Access management policy and practices; (2) Safety analysis and improvement; (3) Freeway access management practices; and (4) Access management planning and design. The *Access management policy and practices* section contains development of access management, the impacts of AM on traffic or economic, and also practices of AM. The focuses and development of AM in different countries or districts and different periods are introduced. Some technical supporting researches influencing AM policy are also included in this section. The *safety analysis and improvement* section mainly aims to decrease or prevent accident and promotion traffic safety by using AM techniques. Some safety analysis or evaluation methods are introduced for different conditions, and corresponding suggestions or promotion methods are also proposed. The *Freeway AM practices* section focuses on freeway, the important part of access management. Auxiliary lanes, deceleration or acceleration lane, DDI are analyzed to find safe and effective solutions or designs. In addition, operation stability in freeway off-ramp or interchange is also discussed. The *AM planning and design* section aims to road planning and design by using AM techniques. Special cases such as roundabout, signalized or unsignalized intersections, as well as old city reconstruction are discussed. Corresponding planning, design schemes or parameters are also proposed.

Each paper was reviewed by two or more reviewers as well as the editors prior to being published in this ASCE Construction Institute Technical Publication. The authors were required to address the reviewers' comments until the paper met the satisfaction of the editors. All published papers are eligible for ASCE awards.

The papers collected in this publication were presented during the Second International Conference on Access Management (AM2014) held in Shanghai, China, September 25-27, 2014. The conference was chaired by Professor Marc Butorac and co-chaired by Professor Zhongyin Guo, Philip Demosthenes and Kristine Williams. The organizations that hosted this conference include Access Management Committee of the Transportation Research Board (TRB), Tongji University, Shanghai Jiaotong University, Research Institute of the Ministry of Public Security, PRC, Research Institute of Highway, and the Ministry of Transportation, PRC.

Zhongyin Guo, Ph.D., Tongji University
Baoshan Huang, Ph.D., P.E, University of Tennessee
Zhongren Wang, Ph.D., P.E, California Department of Transportation

Acknowledgments

The editors would like to appreciate the organizing committee providing technological support in administrating the paper submission. Thanks are extended to Laura Ciampa of ASCE, Marvin Oey, former director of ASCE Construction Institute (CI) and ASCE CI. Special thanks are given to the following individuals that provided timely and high-quality technical reviews for one or more papers submitted to this conference:

First Name	Last Name	Organization	Country
Tony	Abrahamson	A3 Transportation Engineering	South Africa
Majed	Al-Ghandour	North Carolina Department of Transportation	USA
Joe	Bared	Federal Highway Administration	USA
Anne	Benware	Capital District Transportation Commission	USA
Marc	Birnbaum	Caltrans	USA
Frank	Broen	Teach America	USA
Henry	Brown	University of Missouri	USA
Lacy	Brown	Texas A&M Transportation Institute	USA
Marc	Butorac	Kittelson & Associates	USA
Zach	Clark	Kittelson & Associates	USA
Rich	Cunard	Transportation Research Board	USA
Chris	Cunningham	Institute for Transportation Research and Education/North Carolina State University	USA
Barbara	De Ste. Croix	Washington State Department of Transportation	USA
Phil	Demosthenes	Transportation Consultant	USA
Karen	Dixon	Texas A&M Transportation Institute	USA
Bill	Eisele	Texas A&M Transportation Institute	USA
Eric	Fitzsimmons	University of Kansas	USA
Bill	Frawley	Texas A&M Transportation Institute	USA
Jim	Gattis	University of Arkansas	USA
Alan	Ginder	Port Authority of New York and New Jersey	USA
Jerry	Gluck	AECOM	USA
Tom	Hiles	HDR Engineering	USA
Robert	Hofrichter	Virginia Department of Transportation	USA
Christina	Hopes	Oregon Department of Transportation	USA
Chris	Huffman	Huffman Corridor Consultants	USA
Dane	Ismart	Louis Berger & Associates	USA
Reza	Jafari	RSTS, Inc.	USA
Yang Xiao	Kuan	Beijing University of Technology	China
Rick	Laughlin	HDR Engineering	USA
Robert	Layton	Oregon State University	USA
Pan	Liu	Southeast University China	China
Yifang	Liu	Illinois Institute of Technology	USA
Lionel	Lucien	Massachussetts Department of Transportation	USA

Jennifer	Ogle	Clemson University	USA
Basil	Psarianos	National Technical University of Athens	Greece
Yi	Qi	Texas Southern University	USA
Stuart	Samberg	RK&K	USA
Meeta	Saxena	Transportation Engineer	USA
Grant	Schultz	Brigham Young University	USA
Larry	Shannon	MS Consultants	USA
Jonathan	Soika	VHB	USA
Gary	Sokolow	Florida Department of Transportation	USA
Neil	Spiller	Federal Highway Administration	USA
Hein	Stander	AECOM	USA
Brad	Strader	LSL Planning	USA
Afia	Sultana	San Diego State University	USA
Daniel	Tran	University of Colorado	USA
Mike	Wahlstedt	TranSystems	USA
Haizhong	Wang	Oregon State University	USA
Xuesong	Wang	Tongji University	China
Zhongren	Wang	Caltrans	USA
Huaguo	Zhou	Auburn University	USA
Baoshan	Huang	The University of Tennessee	USA
Zhen	Yang	Tongji University	China
Junhua	Wang	Tongji University	China

Contents

Access Management Policy and Practices

Access Management Planning and Design

A Compendium Survey of International Access Management Practices and Concepts

Kristine Williams[1] and Basil Psarianos[2]

[1]Program Director, Planning and Corridor Management, University of South Florida, Center for Urban Transportation Research, Tampa, Florida. E-mail: kwilliams@cutr.usf.edu
[2]Professor, National Technical University of Athens, 9 HiroonPolytechniou St., GR-15780 Athens, Greece. E-mail: bpsarian@mail.ntua.gr

Abstract: The Transportation Research Board Access Management Committee initiated an effort to promote international understanding and integration of roadway access management into the transportation planning and design decisions of nations around the world. This paper summarizes findings of the first step of this effort, which is to identify the international state of the practice. Issues addressed include historical context, legal framework, processes for addressing roadway function, access control and enforcement schemes, associated land use and development regulation processes, pertinent policies and standards, transportation modes addressed, and related issues. This paper presents selected highlights from the following countries that responded to the survey: United States (U.S.), South Africa, China, South Korea, Greece, Germany, Poland, and the United Kingdom (U.K.).

INTRODUCTION

Transportation research and practice has spurred numerous advancements in recent years in roadway safety, operations, "completeness" (e.g., modal options), livability, economic development and the environment. Numerous international and national organizations, like the United Nations, the World Bank, the World Health Organization, the U.S. Transportation Research Board (TRB), the U.K. Transport Research Laboratory, are actively working to convey advances in transportation to other nations, particularly in the areas of roadway safety and economy.However, the benefits of these advancements are not understood or shared equally around the world. Reasons for this disparity range from a shortage of financial and human resources to a failure of policy makers to prioritize and take the necessary actions to strengthen the legal, economic and technical decision-making framework of a country.

The concept of access management is one of these important developments in transportation planning and engineering. Access management has the potential to significantly improve the safety and operation of the transportation system at relatively low cost. It is the careful consideration of the location, type and design of access to a roadway and adjacent land development and involves a range of strategies to reduce conflicts among the various facility users.From a technical standpoint, the benefits of implementing an access management scheme are well known and clearly documented. Nonetheless, a substantial disparity exists between the available technical documentation and guidance on the topic, and current long range transportation planning, urban planning, and roadway design practices. Such practices range from a systematic application based on technical engineering and planning guidance, to a complete absence or even adverse use of access control strategies in many parts of the world (see Fig. 1,Fig. 2).

In recognition of this issue, the TRB Access Management Committee has initiated an action plan to promote and support the acceptance and integration of access management in the transportation planning and design decision processes of nations around the world. One component of the action plan was a survey of the international state-of-the-practice in access management. The full results of the survey will be published in an international access management primer (TR Circular). This paper addresses selected findings and observations.

Fig. 1. Pedestrian crossing on a divided high-speed highway in Africa.
Source: German Research Assoc. for Road and Traffic Engineering (2011)

Fig.2.Arterial access problems in Papua, New Guinea.
(Source: Transport Research Laboratory, 1994)

METHODOLOGY

To obtain information on the international status of access management, inquiries requesting volunteer participation in the survey were sent to transportation professionals and researchers from various nations that have participated in TRB Access Management Conferences, the TRB Access Management Committee, access management sessions at TRB Annual Meetings, as well as other selected international contacts of TRB Access Management Committee members. Individuals from eight countries agreed to complete the survey and document their understanding of access management practices in their country. They are: a) North America: United States, b) Africa: South Africa, c) Asia: China and South Korea, d) Europe: Germany, Greece, Poland and the United Kingdom.

Issues addressed in the survey include historical context, legal framework, processes for addressing roadway function, access control and enforcement schemes, associated land use and development regulation processes, other pertinent policies and standards, transportation modes addressed, and related issues. Perspectives on the future development of roadway access management and control mechanisms, based on the nation's experiences and problems with implementing access control to date, were also requested, as were any special or country-specific access management issues of importance to the development of an access management scheme.

FINDINGS

Below are selected findings and observations on some of these issues,as obtained from the survey responses. Findings are generalized for purposes of comparison of similarities or differences across the nations responding. Table 1 illustrates the general framework for implementing access management in the countries reviewed. These findings are discussed below.

Legal Framework

In the U.S., three decades of continuous efforts to develop the access management concept have led several states to enact a strong legal framework to implement access management in the planning, design and operation of arterial roadways. This has involved codifying access management in state law, policy, standards, regulations, and procedures. An important element of U.S. practice is a clear procedure and written criteria for review of deviations from access standards to provide the flexibility necessary to address site access constraints in different roadside environments, including built-up areas.

China reports the implementation of a strict access regulation scheme based on a socialist land ownership tradition. This regulatory scheme offers little comparison to that of the U.S., given the different societal context and institutional structures. The ability of China to centralize decision making on a national level provides an exceptionally strong legal framework for the integration of a nationally defined concept of access management throughout transportation and land use planning, engineering and regulatory activities. The concept of access management in China is still undergoing development and refinement.

South Africa is proceeding to accomplish an access regulation scheme similar to that of the U.S. and national in scope. The South African Committee of Transport Officials produced a national roadway classification scheme and access management guidelines as documented in the *South African Roadway Classification and Access Management Manual* (Committee of Transport Officials 2012.) The national guidelines are presently advisory and not yet widely applied by the provinces. The provincial government of the Western Cape adopted its own guidelines fifteen years ago that are being applied in that region; these guidelines are presently being reviewed based on experience to date.

In all other countries reviewed, a relatively small-scale access management program is carried out solely through primary rules, acts or bylaws. These legal tools are associated basically with:

- the administration and operation of a functionally classified roadway, which is governed by a national or local competent agency, and/or
- the building rules or bylaws prescribed in the urban or land use development plan.

Table 1.International Comparison of Access Management Frameworks

Countries	**Legal Framework and Tools**			
	Extent of Codifying Access Management	Access Point Engineering Standards	Access Permit Procedures	Criteria and Procedures for Deviation from Standards
China	Full for freeways and arterials. Evolving for overall system.	Yes	Yes	Yes

Germany	Partially for all road categories	Partially in urban settings only	Yes	No
Greece	Partially for freeways and arterials	Partially for specific land uses	In specific cases	No
Poland	Partially for freeways and arterials	No	Partially	No
S. Africa	Partially for freeways and arterials	Yes	Yes	No
S. Korea	Partially for freeways and arterials	Ad hoc, based on U.S. documentation	No	No
U.K.	Partially for all road categories	Yes	No	No
U.S.A.	Full for freeways; varies by state for other road categories	Yes	Yes	Yes

Roadway Function and Access Management

All countries reviewed associate the opportunity to obtain access to a highway from abutting properties with the functional classification of that highway. A common finding for all the countries reviewed is that freeways (aka motorways) are fully access controlled and expressways are partially access controlled. In all countries, intersection spacing, signalization rules, and divided highway criteria apply to these roadways, based on traffic engineering criteria, thereby resulting in an implicit access control scheme. Beyond these roadway categories, access schemes vary widely from country to country. In most countries reviewed, no access control at all is placed on subordinate roadway categories, although traffic volumes, trip generation data and intensity of abutting land uses imply the necessity of access management on these roadways.

In South Africa, three levels of governance define the jurisdictions within which access restrictions apply– the national government, nine provinces and 262 municipalities. Access restrictions vary by roadway category and agency of jurisdiction. Most urban Class 1-3 roadways that are not under the national freeway system fall under provincial authority.

In China, a strict authoritative regulatory scheme defines and guides the public roadway functions and associated access possibilities. Highways are classified into five types: expressway and Class I- IV, based largely on desired travel speed and anticipated traffic volume. Access considerations are addressed in the category definitions. In S. Korea, the basic access patterns for freeways and arterials are as established in pertinent U.S. documentation (e.g., AASHTO "A Policy on Geometric Design of Highways and Streets") that is applied to these roadway classes.

In Germany, general restrictions on access type are set in the urban plan regulations for built-up areas, whereas the corresponding public road law is the guiding basis for access regulation of rural roadways. Germany recently enacted a highway category scheme fulfilling the criteria of "self-explaining roads" (SER) (Fig. 3). The SER concept aims to ensure consistent design features on each class of roadway (i.e., pavement, markings, lighting, signing) so drivers intuitively understand how to behave on that class of roadway.

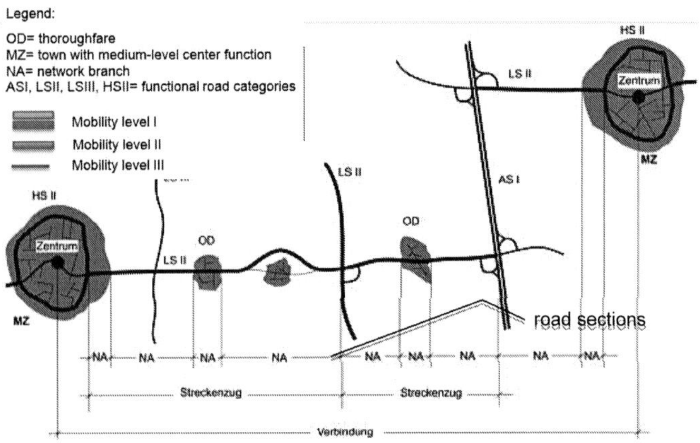

Fig. 3.Functional classification of roadway network in Germany.
(Source: T. Raeder-Grossmann 2013)

In Greece, a limited number of access control rules apply for arterial highways and other streets mainly related to the dimensions of the property frontage. In Poland, other than freeways and arterials, providing public road access to every abutting property is mandatory. Administration of the type and design of access is accomplished on an ad hoc basis.

In the U.K., access control for road categories other than freeways and arterials is accomplished by the proper local authorities. Roadway access control complements the conventional functional category of a road and is augmented by a "complete streets" scheme–an effort aimed at increasing priority to non-auto modes in roadway planning and design that is also gaining considerable traction in the U.S.Both the complete streets scheme and the various roadway context definitions impose explicit or implicit physical measures, as well as various operational measures, that indirectly affect access type and density on U.K. roads.

Land Use Regulation

All countries reviewed have a strong land use regulatory framework. Also, as would be expected, the most concentrated land use regulation is found in urban settings. Zoning and site design criteria, including master plans, can be found practically in all countries. However, distinct differences exist across the various countries reviewed in terms of the nature and extent of access management issues considered.

Beyond U.S. and China, the most comprehensive land use regulatory framework that encompasses access management is found in South Africa ,where land use is stringently controlled through legislation. Every property has a right to access the abutting roadway, but provision of that access is carefully considered. Applications for a change in land use rights or for subdivision of any property for a more intense use must be accompanied by a site plan and traffic impact assessment. Access is evaluated by the road authority with jurisdiction .Retrofitting of poorly designed access is attempted when sites redevelop and access management plans are developed to address substandard property access situations for some higher order routes.

The other countries reviewed have a strong legal framework for land use regulation through which access management issues are presently being addressed only partially or in an ad hoc manner. This existing land use legislation may provide the most effective framework through which to incorporate access management criteria and thereby introduce the concept in countries that lack experience with access management. It is not clear, however, to what extent such changes in national land use legislation would be politically feasible.

Policies, Standards and Enforcement Schemes

An access management manual (Access Management Committee 2003) with extensive description of a variety of access features and design controls was produced in the U.S. in 2003 and is presently being updated. China translated the U.S. manual, but presently does not have one specific to China's context. As noted above, South Africa has a "South African Road Classification and Access Management Manual" (Committee of Transport Officials 2012) that is intended for use as a national guideline, although it is presently not widely applied. One section of Volume 6 of the U.K. Design Manual for Roads and Bridges (DMRB) on Highway Features explicitly refers to access management.

In all other countries, highway design or land development manuals do exist that contain rules for intersection spacing and functional classification of roads, as well as land subdivision instructions. From these countries, Germany and Greece provide explicit procedures for a limited number of driveway designs. In Germany, an access permit procedure is also found. In South Korea, access management is practiced either during review of land development projects or via ad hoc use of the TRB *Access Management Manual*(Access Management Committee 2003) produced by the U.S.

Direct enforcement schemes for access management are provided in the U.S., China, and South Africa. In all other countries, an enforcement scheme whereby adjacent properties must obtain permission for access to roadways is provided indirectly and partially through urban planning criteria or land use regulation in conjunction with the functional classification of the abutting roadways. This urban planning and/or land use legislation forms an appropriate legal tool for further advancement of the access management concept in these nations.

Transportation Modes Addressed

A multimodal approach that is indirectly associated with the concept of access management can be found in Germany and the U.K. Figure 4, for example, demonstrates the redevelopment of Exhibition Road in London, incorporating the concept of shared space. In the U.S., the auto-oriented access management approach that primarily addressed the suburban context is transitioning to a multimodal one that is sensitive to various land use contexts. When completed, it will represent a comprehensive multimodal access management approach. Other countries examined have design policies that include pedestrians, cyclists, and in some cases transit and freight, but with little or no direct association or reference to access management or control.

Fig. 4.Exhibition Road before (left) and after (right) redevelopment as a shared space road in London, U.K.

(Source: IoannisKaparias, City University London, undated)

Perspectives

Perspectives of the countries examined on development or advancement of the access management concept ranged from promising to discouraging. In the U.S., the concept of access management and its beneficial effects on roadway safety and operation are presented in detail in the TRB *Access Management Manual* (Access Management Committee 2003) and numerous national conferences. The U.S. is striving to better include all road users in the corresponding analyses and design solutions and continues to engage in research to advance the practice. Incorporating modal options and land use context more fully into the scope and decision-making process for access management remains a high priority for the U.S. in the immediate future.

South Africa represents another promising example of the systematic advancement of access management in the highway design and land development process. Despite some administrative and practical design hurdles, South Africa continues to engage in advancing the state of the practice on a national and provincial level. In most of the other countries reviewed, the concept of access management may be recognized by some professionals as important to roadway safety, operation, and the environment, but is receiving only sporadic or limited implementation through access regulation and/or the land development process.

Based on U.S. experience, factors and influences that can impede access management and that seem to be relatively universal and therefore important for all countries to consider, include:

- a general lack of familiarity with the contemporary practice of access management and its benefits;
- lack of nationally accepted guidelines, coupled with significant variation of guidelines applied across the country and the many local government jurisdictions;
- limited tools to predict the impacts of access management techniques;
- inadequate number of case studies and examples of successful practices;
- lack of government policy and programmatic support;
- institutional challenges caused by the need for numerous agency functions, jurisdictions, and levels of government to coordinate in planning, design and implementation;
- local stakeholder opposition, particularly by the business community due to perceptions that access restrictions may adversely impact their businesses; and
- lack of resources, along with a continuing need, for outreach and education of agency managers, staff, consultants, public officials, and the public.

In the U.S., these access management impeding factors are being overcome through policy and programmatic measures that have evolved over the last 30 years. A particular emphasis of the U.S. process is the importance of applying a variety of methods in coordination with local governments and allowing "controlled" flexibility in recognition of private property constraints and features. Some success factors in the U.S. include(Access Management Committee 2003,Gluck 2010): a) a strong legal foundation in legislation, b) roadway access classification system for system wide access management, c) integrating access management across agency planning, design and maintenance functions, d) written criteria and procedures for handling deviations from adopted standards, e)staff dedicated to the access management process, f) an access management "champion" within the agency to advance the concept, g)case studies illustrating merits of access management projects, h)intergovernmental and stakeholder coordination, and i) continuous monitoring and self-evaluation to identify issues in implementation and resolve problems.

CONCLUSIONS AND NEXT STEPS

This limited international survey of access management practices has provided information on the status of access management practices in a diverse group of nations. It has revealed that nations are in varying stages of developing a concept of access management for further implementation. Many of the participants in this survey have become aware of access management through professional practice, conferences, and research activities. A broader examination of international practices will likely reveal that a majority of nations are largely unfamiliar with the concept of access management and its benefits. Nonetheless, the building blocks exist in every nation to begin integrating the concept into urban planning and regulation, as well as major roadway planning, policy, and design.

The U.S is currently working toward system wide advancement of access management practices, with an emphasis on improved roadway network planning for all modes and careful control of access in relation to planned roadway function. South Africa and China are also quite familiar with access management and seeking to expand their access management programs and requirements on a national level.

In some of the other countries, such as South Korea, Poland and Greece, the concept has been recognized as important in managing land development and is being introduced partially or occasionally, as opportunities arise. However, various reasons hinder further development of the concept in each of these countries, at least for the present. In Germany and the U.K., access management practices are limited and more strongly focused on serving vulnerable road users.

The findings of this review suggest that a strong platform for initiating the necessary policy and regulatory criteria for access management in those that lack it is the robust land use regulation context that existed in every country reviewed. Building upon this review and the U.S. experience, next steps to be considered by the international community include the following:

1) Prioritize and advance country-specific research on the topic of access management, including both urban planning and transportation engineering considerations.

2) Tailor the programmatic and technical approach to implementing access management to the societal and institutional context of the country.

3) Continue to test and evaluate the impacts of access management projects and actions relative to all modes of transportation and refine practices accordingly.

4) Document case studies and examples of effective practices within the country for further national and international dissemination. Share successful experiences with other nations having similar institutional and political contexts.

5) Regularly convey the results of research and practice to the professional community, government agency staff, and public policy makers through conferences, training, and other means.

6) Prepare a national access management manual to document the national state of the practice and provide a foundation for further advancements.

The Transportation Research Board Access Management Committee has gained an understanding of a successful access management concept and the consequences of inattention to managing roadway access. Through its efforts, the Access Management Committee is committed to working with the international community to define a workable strategy and formulate specific activities and synergies in the upcoming years to advance the concept of access management worldwide.

ACKNOWLEDGEMENTS

The authors gratefully acknowledge the contributions of the following individuals to the compendium survey that formed the basis for this paper(alphabetically by nation):1) China: ZhongyinGuoand Zhen Yang, *Tongji University,* Shanghai, China; Zhongren Wang, California Department of Transportation, Sacramento, California, USA; 2)Poland: Marian Tracz, Stanislaw Gaca, and MariuszKiec, Krakow University of Technology, Krakow, Poland; 3)South Africa: Hein Stander, AECOM, Capetown, South Africa; 4)South Korea: Jaisung Choi, University of Seoul, Seoul, Korea; and 5) United Kingdom: IoannisKaparias, City University London, London, England. Summaries relative to the U.S. (Kristine Williams, University of South Florida, Tampa, Florida, USA) and Germany and Greece (Basil Psarianos, National Technical University of Athens, Athens, Greece) were prepared by the authors.

REFERENCES

Access Management Committee (2003). *Access Management Manual*, Transportation Research Board, Washington, D.C.: National Academy Press.

Committee of Transport Officials (August 2012), *South African Road Classification and Access Management Manual*, Pretoria, South Africa.

Gluck, J. *NCHRP Synthesis 404: State of the Practice in Highway Access Management*, Transportation Research Board, Washington, D.C., 2010.

German Research Association for Road and Traffic Engineering (2006). *Guide for the Design of Urban Streets*, Cologne, Germany.

German Research Association for Road and Traffic Engineering (2011). (Forschungsgesellschaftfuer das Strassen-und Verkehrswesen -FGSV), Jahresbericht, Germany.

Raeder-Grossmann, T.(June 18, 2013) "Colloquium for Rural Road Design," Cologne.

Ross, A., Baguley, C., Hills, B., McDonald, M., and Silcokc, D. (1994) *Toward Safer Roads in Developing Countries: A Guide for Planners and Engineers*, Transport Research Laboratory: Berkshire, England.

A Summary of the Economic Impacts of Raised Medians in Utah

Matt Riffkin, P.E.[1]; Charles Allen, P.E.[2]; Michael Baker[3]; Grant G. Schultz, Ph.D., P.E., PTOE, M.ASCE[4]; and Cameron Kergaye, Ph.D., PMP, P.E.[5]

[1]InterPlan Co., 7719 S. Main, Midvale, UT 84047. E-mail: matt@interplanco.com
[2]InterPlan Co., 7719 S. Main, Midvale, UT 84047. E-mail: charles@interplanco.com
[3]InterPlan Co., 7719 S. Main, Midvale, UT 84047. E-mail: michael@interplanco.com
[4]Dept. of Civil and Environmental Engineering, Brigham Young University, 368 Clyde Building, Provo, UT 84602. E-mail: gschultz@byu.edu
[5]Utah Dept. of Transportation, 4501 S. 2700 West, Salt Lake City, UT 84114. E-mail: ckergaye@utah.gov

Abstract
Raised medians offer a departure from more traditional two-way left turn lanes typically built on collector and arterial streets and can offer advantages in terms of safety, capacity, and aesthetics. At the same time, individual business owners often oppose raised medians due to fear of business loss. This paper summarizes research on the effects of raised medians on retail sales in Utah before and after the installation of raised medians. Sales analysis was also performed for corresponding control corridors in which a similar roadway project was completed that did not include the installation of a raised median. For all of the corridors in which raised medians were constructed, there was an increase in corridor-area retail sales and sales per square foot. Analysis showed that in every case there was no evidence of a negative impact on corridor sales due to the raised median installation. A business impact survey was administrated to qualitatively evaluate the perceived impacts of each road project in front of individual businesses. Business owners on corridors, where the project included the installation of a raised median, typically perceived the actual results of the project more negatively than corridors, where the project did not install a raised median.

INTRODUCTION

The Utah Department of Transportation (UDOT) has implemented a variety of roadway projects involving raised or otherwise non-traversable center medians. Raised medians are a departure from more traditional two-way left turn lanes (TWLTL) typically built on Utah collector and arterial streets and, in certain applications, provide proven advantages in terms of safety, capacity, and aesthetics(Eisele and Frawley 2005; Gluck et al. 1999; Schultz et al. 2007; Schultz et al. 2011). Raised medians are often promoted for their ability to reduce conflict points or offer an improved streetscape appearance. Yet, individual business owners often oppose raised medians due to a fear of business loss (Cunningham et al. 2010; Eisele and Frawley 2000a; Eisele and Frawley 2000b; Maze et al. 1999; Preston et al. 2004; Saito et al. 2005; Vu et al. 2002; Weisbrod and Neuwirth 1998). There was a need to determine the impacts of raised medians in Utah to help inform UDOT in further median implementation.

The purpose of this paper is to summarize the results of the study completed to evaluate the effects of raised medians in Utah on retail sales and specifically examine whether raised medians negatively affect sales (Riffkin et al. 2013).Although the focus was to gather and evaluate sales data, additional traffic data were summarized and a business survey was administered to review perceptions.

STUDY METHODOLOGY

The examination of the business impacts of raised median projects was accomplished primarily through evaluation of taxable sales data and administration of a business owner/manager survey on select corridors in Utah. Three study and control corridor pairs that had recently experienced a road construction project were selected, as summarized in Table 1 and illustrated in Figure 1. Study corridors were roadways in which the UDOT construction project included installation of a raised median. Control corridors were corresponding nearby roadways with a UDOT construction project completed in a similar timeframe that did not include the installation of a raised median.

Table 1. Final Analysis Corridors.

Area	Type	Route	Road	Extents	Length	Year	Notes
Cache County	Study	US-89/ US-91	Main Street	100 W to 900 S	0.64km.	2010	Installed raised median
	Control	US-91	Main Street	1000 N to 1600 N	1.21km.	2009	Resurfacing, portion
Salt Lake County	Study	US-89	State Street	9000 S to 10000 S	2.01 km.	2007/ 2008	Widen 4 lanes + TWLTL to 6 lanes + raised median
	Control	SR-68	Redwood Road	10400 S to 11400 S	2.01 km.	2007	Widen 2 lanes to 4 lanes + TWLTL
Washington County	Study	SR-9	State Street	300 W to 100 E	0.64 km.	2010	Widen 2 lanes + TWLTL to 4 lanes + raised median
	Control	SR-9	State Street	1150 W to 300 W	1.45 km.	2009	Resurfacing. Post-project median added 2010 or 2011

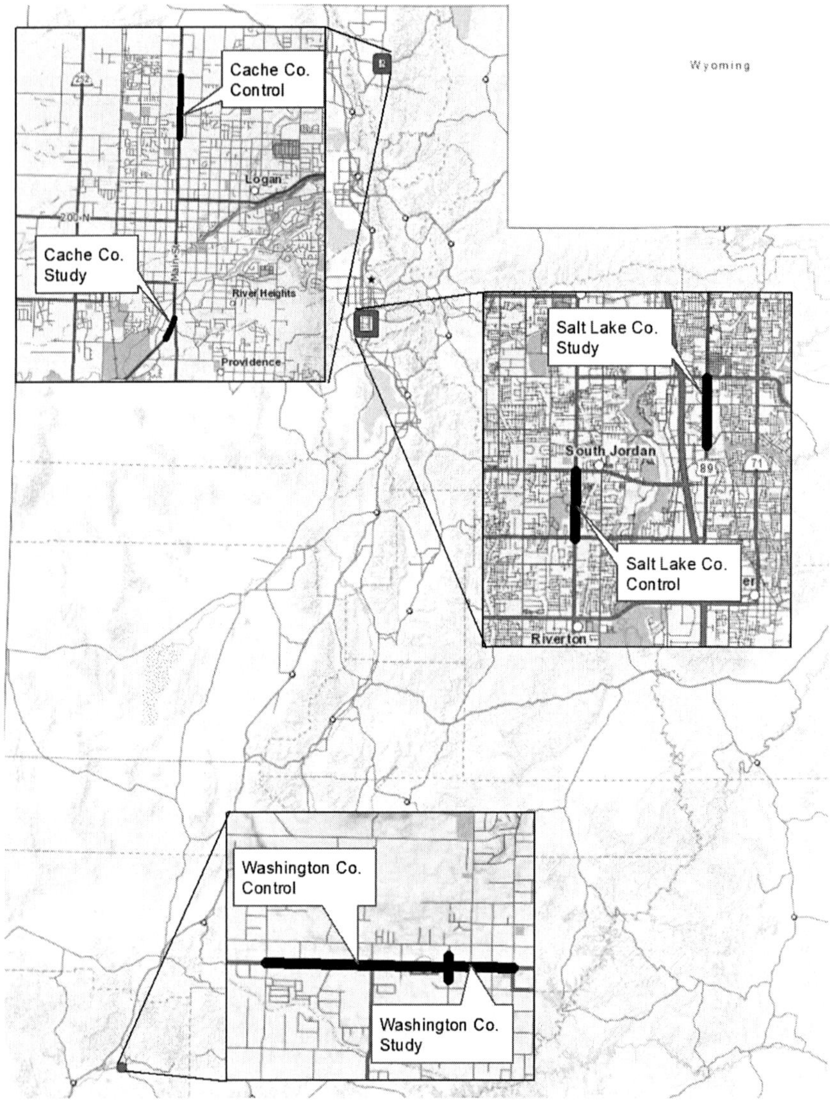

Fig. 1. Analysis Corridors

Analysis was performed using taxable sales data from the Utah State Tax Commission (2013). Sales data were requested for one full calendar year prior to and after the construction of the road project. Data for the selected corridors were aggregated after being located along the street centerline to protect individual business confidentiality. Specific data were then compared against overall sales data at the zip-code level to provide a larger geographic context to the corridor-specific changes in revenues. Gross floor area estimates for businesses along the corridors were used to calculate sales per square-foot.

The business impact survey was administrated to qualitatively evaluate perceived impacts of the paired corridor road construction projects among business owners located along or near the roadway frontage. The survey compared anticipated versus experienced impacts related to the finished form of each project. Survey respondents were asked a series of questions on such topics as business access, customer numbers, and sales. Responses to the survey were primarily collected by canvassing businesses. To be eligible to participate, both the business and survey respondent had to be present prior to and after the road project.

RETAIL SALES ANALYSIS

Analysis of retail sales was completed along the selected corridors to determine if construction of raised medians in the study corridors had a negative effect. The analysis compared aggregate sales and sales by major industrial classification code (i.e., three digit North American Industry Classification System (NAICS)) for the calendar year prior to and after construction. Due to concerns about recessions in the economy, the analysis also compared the before and after retail sales with those from the control corridors. In addition, to measure the extent to which any identified changes in retail sales were the result of economy-wide influences, the study and control retail sales data were compared to those of their own or intersecting zip codes and counties (in some cases the corridors were in different zip codes but within the same county). The zip code data were "unrounded" for comparison purposes though rounded data are publicly available on the Utah State Tax Commission (2013) website. Comparisons were first made between the study and control corridors then against the greater zip codes and counties for each.

The total retail square footage for each of the corridors for the time period analyzed was estimated using aerial photographs and site visits. Retail sales per square foot were calculated to ensure that the comparison was based on sales activity rather than simply changes in retail area.

Sales tax data records were aggregated by the Tax Commission to predefined address ranges that were acquired from street centerline Geographic Information System(GIS) data of streets within the corridor areas. The ranges relate to the local neighborhood area associated with each corridor considered. As such, only street segments for which access could be potentially affected by improvements along the corridor were considered. This process allowed for the examination of sales across sites, zip codes, counties, and time.

The nature of the tax data and pre-defined segmentation of the GIS shape files for the streets presented a challenge in acquiring sales tax data limited to the vicinity of the corridor. In general, sales tax data were acquired for those addresses most directly impacted by the corridor operation.

Sales Analysis Corridor Summary of Results

In all three corridors in which raised medians were constructed there was an increase in corridor area retail sales and sales per square foot as summarized in Table 2 through 4 for Cache County, Salt Lake County, and Washington County, respectively. In all cases the study corridors performed as well or better than the control corridors, study corridor zip codes, and county-wide areas. Although the construction of a big-box retail store within or near each of the corridors complicated the studies, taxable sales data from before and after installation of a median were neutral or positive for each. This should not imply that each business within the corridor did better following installation of the median. In some cases the retail mix changed afterwards.

Table2. Cache County Corridor Sales Tax Results.

	Year	Corridor	Zip Code	Cache County	Corridor as % of Zip Code	Corridor as % of County	Total Taxable Sales/SF
Study Corridor	2009 (Before)	$35,757,453	$682,326,700	$1,409,836,291	5.2%	2.5%	$53
	2011 (After)	$71,707,043	$733,004,133	$1,338,547,784	9.8%	5.4%	$106
	% Change	100.5%	7.4%	-5.1%	86.7%	111.2%	100.5%
Control Corridor	2008 (Before)	$322,028,251	$924,647,511	$1,520,982,619	34.8%	21.2%	$101
	2010 (After)	$283,879,698	$841,646,582	$1,324,009,946	33.7%	21.4%	$87
	% Change	-11.8%	-9.0%	-13.0%	-3.2%	1.3%	-14.2%

Table3. Salt Lake County Corridor Sales Tax Results.

	Year	Corridor	Zip Code	Salt Lake County	Corridor as % of Zip Code	Corridor as % of County	Total Taxable Sales/SF
Study Corridor	2006 (Before)	$124,971,856	$1,462,244,805	$20,328,814,095	8.5%	0.6%	$221
	2009 (After)	$151,607,057	$1,406,888,761	$18,284,173,856	10.8%	0.8%	$265
	% Change	21.3%	-3.8%	-10.1%	26.1%	34.9%	19.9%

Control Corridor	2006 (Before)	$49,300,769	$382,707,591	$20,328,814,095	12.9%	0.2%	$156
	2008 (After)	$60,245,395	$534,268,595	$20,477,875,258	11.3%	0.3%	$186
	% Change	22.2%	39.6%	0.7%	-12.5%	21.3%	18.9%

Table4. Washington County Corridor Sales Tax Results.

	Year	Corridor	Zip Code	Washington County	Corridor as % of Zip Code	Corridor as % of County	Total Taxable Sales/SF
Study Corridor	2009 (Before)	$5,708,465	$125,391,345	$2,240,397,413	4.6%	0.3%	$46
	2011 (After)	$6,056,277	$132,918,645	$2,130,979,356	4.6%	0.3%	$48
	% Change	6.1%	6.0%	-4.9%	0.1%	11.5%	4.5%
Control Corridor	2008 (Before)	$64,185,609	$127,990,834	$2,582,025,982	50.1%	2.5%	$189
	2011 (After)	$49,838,724	$132,918,645	$2,130,979,356	37.5%	2.3%	$141
	% Change	-22.4%	3.9%	-17.5%	-25.2%	-5.9%	-25.5%

Two additional levels of analysis were completed to gauge the impact of the new big-box retail stores. The first compared taxable sales data on a per square foot basis. The second investigated retail sector taxable sales data by comparing the before and after construction taxable sales data for each sector. A close examination of the percent change in taxable sales for all sectors, except retail food and beverage sales in Cache County, provided an indicator for localized impacts of the new big-box retail store. The retail food and beverages sales sectors comprised the majority of the big-box retail store sales that were coded for Cache County. In the other two study areas the big-box retail store was not part of the corridor sales dataset as the stores were not in the study areas. Zip code aggregations in both of these cases proved to be too general and included too many competing businesses to isolate individual effects. Although removal of the retail food and beverage sales category affected the percentage change in the Cache County corridor, the overall change in retail sales remained positive at 17 percent.

BUSINESS IMPACT SURVEY

To supplement the quantitative sales data, canvassing-style surveys of businesses along the paired corridors were performed. The objective was to better understand how individuals anticipated the finished project would impact their business and what impacts were actually experienced. The nature of the questions in the survey required that both the survey respondent and the business were present prior to and after the road project.

The survey methodology was based on similar studies that found improved response rates through an in-person interview-style approach as opposed to exclusive use of a mail-back or online survey (Eisele and Frawley 2000b). If individuals were unable to participate in the in-person survey, they were offered a mail-in or online version. Since the survey was conducted on study and control corridors, the survey form was designed to be applicable to both road projects. The survey contained 15 attitudinal questions regarding anticipated impacts and those experienced afterward the project with respect to traffic congestion, ease of deliveries, number of traffic crashes, business access, sales, and number of customers. Impressions of each impact were collected using a 1-5 scale: 1 being a strongly negative, 3 neutral, and 5 strongly positive. There was also an "unsure/no opinion" option. Since most of the responses were provided in person, respondents could not review actual business data.

Of the total 346 businesses approached, 56 businesses (16 percent) provided a response. Many businesses did not have an owner/manager present throughout the project available to respond. For those businesses that existed before and after the corridor project, many owner/managers had no recollection or opinion about the impacts. This was particularly evident on control corridors, where the project was twice limited to pavement rehabilitation.

The differences in expectation versus experience were gauged by averaging qualitative responses. An average experienced impact score that was lower than the expected indicated that businesses assumed the impact would have been more positive. In contrast, an experienced impact score that was higher than expected indicated impacts were better than anticipated.

Comparison of Results to Quantitative Data

The aggregate survey responses were compared to quantitative data to assess how well business-owner attitudes reflected measurable data. Table 5 compares average impact scores for the number of traffic crashes, business access, sales, and number of customers categories against the change in Average Annual Daily Traffic (AADT), number of crashes, and sales tax data for the study areas before and after project construction. It shows that on study corridors scores for all four response categories dropped, meaning perceived (experienced) impacts were, on average, worse than expected impacts. However, quantitative data often reflects the opposite trend. For example, despite a pessimistic view of the project impact on sales, the average sales tax on study corridors (weighted by corridor length) increased by 32 percent. Likewise, although the perceived experienced impact on number of traffic crashes was lower than the expected impact, UDOT data revealed a 51 percent decrease in crashes. Conversely, on control corridors, anincrease in impact scores for the sales category was accompanied by the opposite trend, a small decrease in actual sales tax data.

Business Impact Corridor Summary of Results

The business-owner survey frames the perception among business owners regarding raised median projects. Overall, control corridors typically had more positive scores than their study corridor counterparts. Other general conclusions include:

- The worst scoring corridor and the best scoring corridor occurred in Salt Lake County.
- On average, overall impressions of study corridor effects were neutral.
- Business owners in the Salt Lake County study corridor expressed frustration at the lack of median breaks, while Washington County study corridor business owners noted both positive and negative opinions of the median landscaping.

Table 5. Survey Impact Score vs. Data Comparison.

		Average Scores			
	Survey Topic	**Number of Customers**	**Business Access**	**Number of Traffic Crashes**	**Sales**
Study Corridor	Expected	2.6	2.2	3.8	2.5
	Experienced	2.4	1.9	3.4	2.4
	Change	-0.2	-0.3	-0.4	-0.1
	Comparison Data	AADT		Number of Traffic Crashes	Sales Tax Data
	Change in Comparison Data	-3.0%		-51%	32%
Control Corridor	Expected	2.9	3.1	3.6	3.0
	Experienced	3.2	3.1	3.9	3.2
	Change	0.3	--	0.3	0.2
	Comparison Data	AADT		Number of Traffic Crashes	Sales Tax Data
	Change in Comparison Data	2.2%		-30%	-0.4%

- Cache County and Washington County control projects consisted primarily of resurfacing work. Responses for these corridors yielded the most neutral scores.

The comparison of survey results to quantitative data showed that perceptions often did not reflect reality. On study corridors, business owners reported neutral to negative perceptions of sales impacts; however, sales tax data showed an overall increase of 32 percent. This discrepancy may be related to changes in businesses between periods. A decrease in sales for some businesses may have been outweighed by an overall increase in sales from new or redeveloped businesses that were not surveyed.

CONCLUSIONS

The purpose of this paper was to summarize the results of a study completed in Utah to evaluate the effects of raised medians on retail sales and specifically examine whether raised medians negatively affected sales. Research was conducted in the area of retail sales and businesses were surveyed to assess perceived impacts.

Retail Sales Conclusions

For all three of the corridors in which raised medians were constructed, there was an increase in corridor-area retail sales and sales per square foot. In addition, in each case the study corridors performed as well or better than the control corridors, study corridor zip codes, and county-wide areas. In every case there was no evidence that installation of a raised median had a negative impact on retail sales though individual businesses were affected differently. In some cases the retail mix changed in response to the economy, new area competition, and other contributing factors.

Abig-box retail store opened within the market area of each study corridor during the study period. While it is difficult to disentangle the influence of each new big-box retail store on study-area sales tax patterns, their effects were investigated. Results showed positive gains in sales per square foot ranging from 5 percent to over 100 percent. In the Cache County study area, the percent change in taxable sales remained positive at 17 percent even without considering retail food and beverage sales .In the other two study areas thebig-box retail store was not part of the corridor sales dataset.

Business Impact Survey Conclusions

Overall, business owners had more negative perceptions of impacts on corridors that included installation of a raised median. They were most likely to report negative experienced impacts regarding the ease of deliveries, business access, sales, and number of customers response categories. Meanwhile, the traffic congestion, number of traffic crashes, and overall impact question categories were more likely to elicit neutral or positive responses. Thus, study corridor business owners had a more optimistic view of the impact on traffic operations than business related factors.

The comparison of sales data to business owner survey results showed that perception often did not reflect reality. This was particularly true in regards to sales. On study corridors, business owners reported neutral to negative perceptions of sales impacts while sales tax data showed an overall increase of 32 percent. This may reflect on the difference between businesses eligible to complete the survey and those that have arisen since the project.

Similarly, the expected impacts of the raised medians on safety were not met according to survey respondents. All study corridors yielded an experienced impact score that was lower than the expected impact for the number of traffic crashes response category (despite a 51 percent reduction in crashes). This pattern may represent a more heightened awareness of crash occurrences after amedian installation.

ACKNOWLEDGMENTS

The authors acknowledge UDOT for funding this research. The authors would also like to acknowledge Christine Richman and Jonathan Dorward for their assistance in the economic analysis. Special thanks to W. Scott Jones, Travis Jensen, Andre Baksh, and Justin Smart who played key roles in helping to guide the research.

REFERENCES

Cunningham, C. M., Miller, M., Findley, D. J., Smith,S., Carter,D., Schroeder,B., Katz,D., and Foyle,R. S. (2010). "Economic Effects of Access Management Techniques." Report No. 2009-12, Institute for Transportation Research and Education, North Carolina State University, North Carolina Department of Transportation, Raleigh, NC.

Eisele, W. L., and Frawley,W. E. (2000a). "A Methodology for Determining Economic Impacts of Raised Medians: Final Project Results." Report 3904-4. Texas Transportation Institute, Texas A&M University, College Station, TX.

Eisele, W. L., and Frawley,W. E. (2000b). "Assessment of Economic Impacts at Selected Raised Median Installation Locations in Texas and Development of Recommended Methodology for Economic Impacts Estimation." Report 3904-5. Texas Transportation Institute, Texas A&M University, College Station, TX.

Eisele, W. L., and Frawley,W. E. (2005). "Estimating the safety and operational impacts of raised medians and driveway density."*Transportation Research Record: Journal of the Transportation Research Board,* No. 1931, Transportation Research Board of the National Academies, Washington DC, 108-116.

Gluck, J., Levinson,H. S., and Stover,V. G. (1999). "Impacts of access management techniques."*National Cooperative Highway Research Program Report 420,* Transportation Research Board of the National Academies, Washington, DC.

Maze, T., Plazak,D., Chao,P., Evans,J. K., Padgett,E., and Witmer,J. (1999). "Access Management Research and Awareness Program: Phase IV Final Report."Management Project 97-1, Center for Transportation Research and Education (CTRE), Iowa Department of Transportation, Ames, IA.

Preston, H., Gute,M., Storm,R., Plazak,D., and Albrecht,C. (2004). "Interstate 394 Business Impact Study: Research Summary and Key Findings." Minnesota Department of Transportation, St. Paul, MN.

Riffkin, M., Allen,C., Baker,M., Richman,C., and Dorwart,J. (2013). "Raised Median Economic Impact Study." UDOT Report No. UT-12.17, Utah Department of Transportation, Research Division, Salt Lake City, UT.

Saito, M., Cox, D. D., and Jin,T. G. (2005). "Evaluation of Recent Traffic and Safety Initiatives, Volume II: Developing a Procedure for Evaluating the Need for Raised Medians." Report UT-04.11, Utah Department of Transportation, Research Division, Salt Lake City, UT.

Schultz, G. G., Lewis, J. S.,and Boschert,T. (2007). "Safety impacts of access management techniques in Utah."*Transportation Research Record: Journal of the*

Transportation Research Board, No. 1994, Transportation Research Board of the National Academies, Washington, D.C., 35-42.

Schultz, G. G., Thurgood, D. J., Olsen, A. N., and Reese,C. S. (2011). "Analyzing raised median safety impacts using Bayesian methods."*Transportation Research Record: Journal of the Transportation Research Board*, No. 2223, Transportation Research Board of the National Academies, Washington, DC, 96-103.

Utah State Tax Commission. (2013).<http://tax.utah.gov/.> (July 10, 2013).

Vu, P., Shankar, V., and Chayanan,S. (2002). *Economic Impacts of Access Management*. Report T103, Task 32. Washington State Transportation Center, Washington State Department of Transportation, Olympia,WA.

Weisbrod, G. E., and Neuwirth,R. (1998). "Economic Effects of Restricting Left Turns."*National Cooperative Highway Research Program: Research Results Digest,* No. 231, Transportation Research Board of the National Academies, Washington, DC.

Impact of Access Management on Traffic Performance in Poland

S. Gaca[1] and M. Kiec[2]

[1]Department of Highway and Traffic Engineering, Cracow University of Technology, Warszawska Street 24, 31-155 Cracow, Poland. E-mail: rbak@pk.edu.pl
[2]Department of Highway and Traffic Engineering, Cracow University of Technology, Warszawska Street 24, 31-155 Cracow, Poland. E-mail: mkiec@pk.edu.pl

Abstract
In this paper, characteristics of access control on Polish national two lane roads, rural and urban, and its evaluation on traffic performance is presented. In order to demonstrate the impact of access control and traffic performance, quantification of factors associated with the road and its surroundings was carried out. The study was based on empirical research and on developed regression models of speed. These models were used to evaluate the expected value of vehicle speeds on sections with different levels of access control. The study indicated a significant effect of the lack of access control on traffic performance. Conclusions based on the regression models were used to emphasize the most important issues caused by the absence of proper access management on existing roads. Moreover, proposals for possible methods to solve indicated problems were presented. As a result, the recommendations were formulated in relation to the following countermeasures, e.g., infrastructure investments, traffic, and access management, whose main goal is to improve traffic performance on Polish roads.

INTRODUCTION

In Poland, despite the applicable legislation, lack of access control on many sections of roads causes significant worsening of traffic performance and increased road safety hazards. Such a problem applies to both sections of rural and urban national roads. The ambiguity of the existed regulations, especially in relation to two lane roads, which set the greater part of the Polish road network, causes a need to introduce appropriate additional regulations and new design solutions. The projects of such regulations have been developed taking into account the possibility of reconstruction of road sections in the existing transportation corridor in order to improve traffic performance and road safety.

The most important problems regarding access control are:

- lack of hierarchical structure of road network, which is one of characteristic features of road network in Poland
- development suburbs along national and regional roads caused by process of urban deglomeration (migration of inhabitants from cities to suburbs). It has resulted in the construction of commercial and residential buildings alongside sections of national roads with direct access to the major road
- locating buildings in the vicinity of major roads, without frontage and local roads network

These problems cause overlapping of commuter and through traffic with local traffic leading to interruptions and worsening road safety and traffic performance.

Such situation concerns mostly two lane roads located in suburban areas and built-up areas with through traffic, where roads with rural characteristics change into stretches of two-lane streets. These sections constitute an important fraction of the national roads network, amounting to 35% of total length of two-lane roads (Gaca and Kiec 2012), which main purpose is to support through traffic. Due to the lack of well-developed network of expressways and freeways, which requires complete access control, two lane roads called GP (major road with fast through traffic) and G (major road) classes serve, as links between regions of Poland, where both high traffic volume (up to 25,000 AADT) and high speeds are observed to an extent not seen in anywhere else. This is strictly related to drivers' expectations who wish to quickly cover large distances.

The purpose of this paper is to describe the problem of management of the road surroundings and road accessibility, as well as the effect of these on speeds determined basing on models developed for rural and two-lane streets (sections of national roads, with rural characteristics, through built-up areas – villages, communities, towns). As results show, the local speeds strongly depend on the road surroundings especially on an urbanized area (Aronsson and Bang 2005; Donald and McGann 1996). Basing on the conducted analyses, suggestions of suitable modifications of the national road network have been made, in order to separate local and through traffic.

CHARACTERISTICS OF ROAD SURROUNDING DEVELOPMENT AND ACCESS CONTROL PROBLEMS

Sections of roads, through small communities and towns surrounded by a rural or suburban environment, usually have the following characteristics:

- long sections with varying density of linear development,
- lack of alternative routes for local traffic,
- high diversification of traffic composition,
- hardly any protection for existing pedestrian traffic, no space provided for concentration of pedestrian traffic,
- high accident rates,
- road surroundings serviced directly from national/regional roads,
- traffic volumes significantly below capacity resulting in high speeds.

Sections of roads which pass through built-up areas surrounded by a typical suburban environment have the following features:

- through sections have an impact on how the development of town evolves; distributor roads or local streets do not provide alternative routes to take away some of the traffic volumes,
- long through sections,
- various functions of the surrounding development including service and commercial functions,
- traffic composition typical for rural roads (noticeable share of heavy vehicles),
- visible pedestrian traffic and in some passages of cycle lanes concentrated in central parts, with pedestrians frequently crossing roads along sections with commercial development,
- traffic performance is frequently unacceptable due to insufficient capacity of sections or intersections,
- high accident rates,
- presence of public transport.

Sections of rural two lane roads usually have no access control, however, most of them are rarely used during the day.

The characteristics specified above are true not only for the Polish conditions but also for most countries in Central and Eastern Europe. These roads differ considerably from collector-distributor roads in the USA, because of location of housing buildings in their close neighborhood and therefore they are more similar to two-lane street (HCM 2010).

Combination of the abovementioned road features with desired high travel speed leads to traffic safety hazards. Solutions, which ensure a more uniform speeds appropriate in a given road surrounding, need to be sought after in order to decrease the number of accidents. Therefore, the influence of the road surroundings on speeds needs to be evaluated.

This is why one of the aims of this paper is to quantify the road surroundings influence and road accessibility on the speeds. Such quantification translates into formulation of speed prediction models, which would include both scalar and qualitative (such as cross-section) variables. It was assumed that these variables should quantitatively describe all possible factors related to the road surroundings, which can influence the driver's speed choice.

Knowledge of the estimated speed will allow appropriate adaptation of planning and design solutions, as well as to examine the factors influencing speed choices.

Rational planning of road surrounding development requires development of models quantifying impacts of features of roads and their surroundings on both accidents and traffic performance. Such models allow to predict the expected effects of application of different means of improving road safety (Gaca and Kiec 2012), including limitations of the accessibility and changes in roadside development.

In Poland the following problems, reported on national rural roads and two-lane street related to speed, can be distinguished:

- numerous intersections, including those with access roads,
- high density of driveways (commercial and residential)

- numerous roads with linear development of different intensity. Share of road sections surrounded by buildings differs depending on region of the country.
- linear form of development and its dispersion make pedestrian safety measures difficult to apply (e.g. installation of pedestrian crossings) due to the lack of concentration of pedestrian traffic,
- high share of through traffic on national roads

In order to describe accessibility of roads and characteristics of road surroundings the authors applied both quantitative and qualitative features. In analyses of access to roads the following was considered:

- residential driveways (to private buildings, except those where business operations are run),
- commercial driveways (to the facilities related to business operations),
- intersections with local roads (non-marked by vertical signs on the major roadway which would inform about the intersection and which are not residential driveways),
- intersections with other roads marked by vertical signs, of various traffic intensities.

Presence of access points and intersections was described by their density. For quantitative describing road surroundings indirectly the impacts of intensity and different types of development located along the road on the local traffic generated and on the behaviour of traffic participants the following indicators were used (Gaca and Kiec 2012).

Development Density indicator DD_{wide} [%] is a quantitative measure defining the level of development concentration along the road within the bands 50m, 100m and 150m *wide* from the road, measured on both sides. Such indicator defines the percentage of built-up area of the total area of the land band analyzed.

Development Type indicator *DTB* [%] is a variable defining the type, number and proportions of buildings located on this area. The development type description is of significant importance, as it affects the amount and nature of traffic generated. Most often in the prediction models of speed the development type description appears as qualitative variables. For scalar descriptions the development type indicator *DTB* [%] has been suggested because it expresses the area percentages of each type of buildings located on this area (*Residential*, *Farm* buildings, *Commercial* included other) of the total built-up area.

Apart from the abovementioned indicators, the following were investigated in respect to their influence on traffic performance: distance of the buildings from the road (*D*), type of cross-section (paved shoulder – *C1*, unpaved shoulder – *C2*, sidewalk – *C3*), volume of pedestrian (*VP*) and share of through-traffic (*T*). Type of cross-section was used as a qualitative variable.

DATA AND METHODS

Models predicting speed quantile (V_{85}) of vehicles were developed using the most popular OLS regression approach (TRB 2011) for both rural roads and two-lane streets. Models allowed to quantify the impact of factors of road surroundings development on traffic speed measured by quantile V_{85}.

Empirical data from 48 sections of rural roads (with speed limit of 90 km/h) and 158 sections of two-lane streets (with speed limit of 50 km/h) was incorporated. Measurement sites were located on straight sections, with low grades of road profile, beyond the impact of large traffic generators (e.g. offices, shops, etc.), intersections, pedestrian crossings and bus stops, without physical measures of speed reduction with general speed limit and without additional speed enforcement, in order to estimate only the impact of road and surrounding accessibility.

The general form of model, consistent with research result was adopted (Baruya 1998):

$$V_{85} = a_0 + a_1 \cdot x_1 + a_2 \cdot x_2 + ... + a_n \cdot x_n \quad (1)$$

where: a_0, a_1, ..., a_n – unknown model parameters, directional coefficients for variables, x_0, x_1, ..., x_n – observed, non-random independent variables.

Models for the dependent variable V_{85}, in which the explanatory variables were scalar variables (i.e. indicators characterizing the development of the road surrounding, traffic data) and qualitative variable of cross-section type were sought.

Only some of the mentioned variables proved to be statistically significant. The constructed models with the highest R^2 values are described by the following models:

- rural roads:

$$V_{85} = 113.95 - 0.89 \cdot AP - 5.05 \cdot API + 3.23 \cdot C1 - 3.23 \cdot C2 \quad (2)$$

$$R^2=0.28$$

- two-lane street:

$$V_{85} = 90.72 - 0.757 \cdot APC - 0.153 \cdot DTR - 0.068 \cdot VP + 0.094 \cdot T + 0.264 \cdot D + \\ + 3.16 \cdot C1 - 1.13 \cdot C2 - 2.03 \cdot C3 \quad (3)$$

$$R^2=0.37$$

where V_{85} – speed quantile V_{85} [km/h], *AP* – density of all access points (excluding intersections) [number/km], *API* – density of intersection access points [number/km], *APC* – density of commercial access points [number/km], *DTR* – development type indicator for residential [%], *VP* – volume of pedestrian [ped/h] *T* – share of through traffic [%], *D* – distance between edge of roadway and buildings [m], *C1, C2, C3* – cross-section symbol, respectively two-lane roads with paved shoulder, unpaved shoulder, sidewalks. A qualitative variable; equal to *1* when a given cross-section is present, and *0* if it is not.

Table 1 shows independent variables present in the models along with regression coefficients and statistics describing their statistical significance (*Wald* statistics, *p-value*). The grey cells contain variable *C2* which was found not to be statistically significant (with the *p-value* > 0.05), however, the general impact of cross section variable is statistically significant.

Table 1. Data and Parameters of Regression Models for Speed Prediction.

(a) two-lane rural roads (speed limit – 90 km/h) – data and parameters

Variable	Value range in the dataset		Average in the dataset	Standard deviation
V_{85}[km/h]	84.09	143.15	105.54	10.92
AP [No/km]	0	23.3	5.4	5.20
API [No/km]	0	1.8	0.5	0.46

(a) two-lane rural roads (speed limit – 90 km/h) – regression model (2)

Variable	Coefficient in the equation	Standard error *SE*	Wald statistics	*p-value*
intercept	113.95	2.543	2006.53	0.000
AP	-0.89	0.256	12.03	0.000
API	-5.05	3.046	2.75	0.047
C1	3.23	1.469	4.82	0.028
C2	-3.23	1.511	4.58	0.032

(b) two-lane streets (speed limit – 50 km/h) – data and parameters

Variable	Value range in the dataset		Average in the dataset	Standard deviation
V_{85} [km/h]	59	114	87.01	11.07
APC [No/km]	0	15	3.4	2.48
DTR [%]	22	75	48.8	10.91
VP [ped/h]	0	352	25.99	37.13
T [%]	7.1	88	47.4	15.17
D [m]	4	50	14.4	7.90

(b) two-lane streets (speed limit – 50 km/h) – regression model (3)

Variable	Coefficient in the equation	Standard error *SE*	Wald statistics	*p-value*
intercept	90.72	5.281	295.06	0.000
APC	-0.757	0.317	5.69	0.017
DTR	-0.153	0.070	4.71	0.030
VP	-0.068	0.021	10.18	0.001
T	0.094	0.053	3.06	0.050
D	0.264	0.103	6.59	0.010
C1	3.16	1.202	6.92	0.008
C2	-1.13	1.026	1.21	0.269
C3	-2.03	0.811	3.25	0.045

The analyses of the impact of road surroundings development on speed of vehicles were carried out with the reference to free flow speed. Thus only the influence of road surroundings on behavior of drivers was taken into consideration, without the influence of traffic volume.

IMPACT OF ACCESS MANAGEMENT ON SPEED

In analysis of traffic performance on sections of roads passing through built-up areas (two-lane streets) and rural roads only operating speed factor (V_{85}) was taken into account.

In the case of developed models for speed quantile (V_{85}) estimation in free flow traffic, coefficient of determination R^2 is 0.28 for rural roads and 0.37 for roads in built-up areas. Low value of R^2 resulted from the significant impact of human factors (e.g., drivers' behavior) on vehicle speed. Therefore there is need for verifying the proper selection of measurement segments. Despite the low degree of explanation of the random variable it can be stated that the surrounding development is an important determinant of speed selection by the drivers. The nature of the impact of the independent scalar variables, under the study, on V_{85} was consistent with expectations arising from the physical interpretation of the role of these variables. Developed models indicate a need for further studies using even more uniform road sections. Thanks to such approach the impact of road surrounding on speed would be better considered and therefore the reliability of the models may be higher.

For rural roads, variables related to the density of access points have an influence on degree of explanation of variable V_{85} in Eq. 2. A qualitative variable, namely cross-section, has also appeared to be important. The density of road intersections also has a significant influence on speeds: a single road intersection can cause speed reduction by approximately 5 km/h.

As expected, more factors related to the road surrounding proved to be statistically significant for estimating the speed V_{85} on roads located in built-up areas (3). Among these variables it can be identified, those which are associated with the accessibility to the road (*APC*) and related to buildings along the road (*DTR*), its distance from the edge of the roadway (*D*) and pedestrian traffic (*VP*). An important factor causing increase in speed is variable describing the share of through traffic. The speed increase with the increase of the share of transit traffic can be explained by a desire to travel at higher speeds especially during long journey.

The variables associated with development of road surrounding cause the speed reduction, which seems to be lower than expected.

The study results confirm that management of the road accessibility can also affect the speed. At the same time it should be noted that with the drivers' tendency to drive at high speeds, it is necessary to control the accessibility, which may reduce traffic conflicts and to improve road safety

CONCLUSION

Challenges in planning, improper regulations and lack of the real supervision of road surroundings development result in uncontrolled development. Building of housing estates and commercial development directly by major roads, without development of networks of local or service roads is a frequent phenomenon in

Poland. This is one of important causes of increasing accident hazard (Gaca and Kiec 2012) and poor traffic performance. The situation demands a radical solution i.e. for reconstruction of the chaotic network structure into a hierarchical structure. Since it is a long-term task, there is an immediate need to find answers to the following questions:

- to what extent is it possible to combine different functions (traffic, distributor or access) of roads running through built-up areas with preserving both good traffic performance and road safety?
- how should new multifunctional roads be designed, and how existing ones reconstructed?
- in which cases separation of different road functions is necessary, e.g. by building road bypasses?

The analyses conducted by the authors confirmed the essential influence of lack of accessibility management and its impact on traffic performance. The developed regression models enable quantitative assessment of the influence of various factors on the speed.

The influence of the accessibility of a road and kind of road surroundings development has a complex character. On the one hand development is generating additional traffic and increases the number of possible conflicts in a stream, but at the same time it affects speed reduction considerably.

No doubt, the combination of different functions of roads is not right, but acceptable at low traffic volumes of local traffic. At larger traffic volumes several local by-passes were constructed. Also reconstruction of existing road network in order to adapt existing roads to functions they really perform or should perform is needed.

The presence of high speed of vehicles, both on sections of rural roads and in built-up was proved and therefore the need for more strict control of access to higher classes of roads was also confirmed. Such control is conducive to reducing the number of potential conflicts, and thus results in improved traffic safety.

Introduction of traffic calming and additional left turn lanes is a short-term measure to reduce the number of traffic conflicts on major roads. This form of road design is becoming more popular in Poland. However, such design decreases traffic capacity and considerable deteriorates the effectiveness of road transport by extending travel times. Nevertheless, substantial decrease of accidents and number of casualties was noticed on roads with additional median lane to left turn.

REFERENCES

Aronsson K. and Bang K. (2005). „Factors influencing speed profiles on urban streets.“ *Proc., 3rd International Symposium on Highway Geometric Design,* TRB, Chicago.

Baruya A. (1998) „Speed-accident relationship on European roads. Proceedings of the Conference Traffic Safety in Europe.“ *VTI konferens 10A, Part 10.* Swedish National Road and Transport Research Institute, Sweden.

Donald D. and McGann A. (1996). „Relationship between initial speed and final speed at the approaches to rural cities and towns.“ *18th ARRB Transport Research Conference – Roads 96*, New Zealand.

Gaca S., and Kiec M. (2012) „Quantification of impacts of road accessibility and selected factors of road surroundings development on road traffic safety.“ *Transportation Research Board 91st Annual Meeting 2012*, Washington.

„Highway Capacity Manual 2010“, (2010) Transportation Research Board, National Research Council Washington, D.C.

„Modeling Operating Speed, Synthesis Report“ (2011) *Transportation Research Circular E-C151*. Transportation Research Board of the National Academies, Washington D.C.

Rapid Diagnostic Model of Road Lifeline Seismic Damage Based on Seismic Intensity

Donglan Su[1]; Benmin Liu[2]; Zhongyin Guo[3]; and Zhizhong Li[4]

[1]The Key Laboratory of Road and Traffic Engineering of the Ministry of Education, School of Transportation Engineering, Tongji University, No. 4800 Cao'an Road, Shanghai 201804, P. R. China. E-mail: 2011sudonglan@tongji.edu.cn
[2]The Key Laboratory of Road and Traffic Engineering of the Ministry of Education, School of Transportation Engineering, Tongji University, No. 4800 Cao'an Road, Shanghai 201804, P. R. China. E-mail: liubenming@mail.tongji.edu.cn
[3]The Key Laboratory of Road and Traffic Engineering of the Ministry of Education, School of Transportation Engineering, Tongji University, No. 4800 Cao'an Road, Shanghai 201804, P. R. China. E-mail: zhongyin@tongji.edu.cn
[4]Transportation Research Institute of Yunnan Province, Kunming, Yunnan 650011, P. R. China. E-mail: 852217086@qq.com

Abstract

Earthquake is one of the serious geological disasters in China, it would cause varying degrees of damage to road lifeline. Due to the difficulty of quickly getting the damage information of road lifeline after earthquake, it has a great impact on the rescue work. The paper provided the seismic damage grades of subgrade, bridge and tunnel according to seismic damage type, and the influence on road traffic function. Then, the relationship between seismic intensity and road damage was analyzed. The rapid diagnostic model of road lifeline seismic damage based on seismic intensity was constructed from the following aspects: division of assessment unit, construction of seismic damage index set, determination of weight set, and model validation. The model can improve the emergency rescue efficiency through rapidly forecast the highway damage risk.

1. INTRODUCTION

Road traffic lifeline means evacuation and rescue channel which can directly access to disaster center and maintain afflicted people's safety, it is the most important channel to carry out rescue work after earthquake. However, due to the destruction of communications, road seismic damage can't be judged rapidly.

Currently, road seismic damage discrimination mainly depends on remote sensing, more and more scholars start to use remote sensing technology to identify seismic damage and extract road information. Zhao (2010), Wang(2012) proposed an object-oriented information extraction method of road damage; Ma(2013) put forward the rapid extraction method of seismic damage information based on road edge; Liu (2013) proposed the rapid extraction method of road damage from high resolution remote sensing images.

However, there exist some restricting factors in using remote sensing technology to identify seismic damage after earthquake: it requires available satellite or aircraft to obtain remote sensing information in good weather condition and takes long times, and its deciphering technique is complex. In order to fill the lack of rapidly forecast technology on road seismic damage, according to inherent relationship and rule between seismic intensity and road traffic lifeline damage, using seismic intensity as the main index, the paper proposed risk index and grading standards of seismic damage, which could provide the basis for rapid assessment of road traffic lifeline damage after earthquake.

2. TYPES AND GRADES OF ROAD LIFELINE SEISMIC DAMAGE

2.1 Subgrade seismic damage and its grades

According to subgrade seismic damage data in Wenchuan earthquake, combined with the major influences of subgrade damage on road traffic functions(Su 2013), it can be divided into four grades, seen in table 1.

Table 1. Grades of Subgrade Seismic Damage

Grades	Characteristics of Subgrade Seismic Damage			Traffic function loss
	Subgrade Damage	Slope Damage	Support Structure Damage	
Grade I - Slight	There are minor cracks and slight folding and depression in the road	The damaged area of slope protection structures is less than 5%	Seismic damage area(crack length)< retaining wall area(length)10 %.	No effect on vehicles passing.
		No landslide and slump in unprotected slope.		

Grade II - Medium	There are slab staggering and folding in the road, with crack width < 10cm, small-scale collapse at embankment edge, rock fall in cutting slope.	The damaged area of slope protection structures is 5%~20%. There is landslide and slump in unprotected slope, support structure is slightly damaged.	Seismic damage area(crack length)accounts for 10%~30%,displacement of retaining wall is significant.	Opening to traffic by simple treatment
Grade III-Serious	There are significant slab staggering and folding in the road, with crack width > 10cm, embankment edge collapsed, landfall and rock fall in slope, landslide buries the road, road space is narrow and is hard to pass by.	The damaged area of slope protection structures accounts for 20%~50%, partial function failed. There are serious landslide and slump in unprotected slope, support structure and subgrade are damaged and overall buried.	Seismic damage area(crack length)accounts for 30%~60%,displacement of retaining wall is remarkable.	Vehicles can't pass by, needs certain time to clean up before opening to traffic
Grade IV-Destroyed	Subgrade completely failed with slab staggering and folding, most parts of subgrade collapsed, landfall and rock fall in slope and block the road, landslide buries the road.	The damaged area of slope protection structures accounts for more than 50%, protection function completely failed. There are large-scale landslide and slump in unprotected slope, support structure and subgrade are seriously damaged and overall buried, river is blocked and barrier lake is formed.	retaining wall is completely destroyed.	Vehicles can't pass by, needs a long time to clean up before opening to traffic

2.2 Bridge seismic damage and its grades

The seismic damage survey of bridges can be divided into two phases, the first phase is urgent repair stage, in this phase, the equipment is not used, only the external manifestation of the seismic damage phenomenon. The second phase is ensuring access stage, in this phase, the equipment is used in comprehensive bridge detection, and systematic bridge inspection report is formed. For the urgent repair stage, grades of bridges seismic damage seen in Table 2.

Table 2. Grades of Bridges Seismic Damage

Grades	Characteristics of bridges seismic damage		Loss of bearing ability
	Main beam	Bridge pier	
Grade I - Slight	There is not significant residual displacement and damage in main beam	There is no significant crack in pier body,bent cap,tie beam and nodes(crack width≤ 0.1mm).	There is no damage in bearing member, no loss in bearing ability

Grade II - Medium	There is certain residual displacement in main beam, the bridge still have reliable bracing system and beam is in little risk of falling	There are a few significant cracks in pier body,bent cap,tie beam and nodes.	Although the main bearing member is damaged, there is no significant loss in bearing ability,and the bridge is also available for emergency access without reinforcement.
GradeIII- Serious	There is large residual displacement, main beam slides away from centerline of bearing, bearings completely loss its function, main beam has no reliable bracing system.	There are throughout cracks in pier body,bent cap,tie beam and nodes, plastic hinge appears in pier body and bent cap,bridge pier inclined.	The main bearing member is damaged seriously, there is serious loss in bearing ability,and the bridge is available for emergency access after reinforcement.
Grade IV- Destroyed	Beam falling	There is rupture or shear failure in bridge pier.	Bearing ability is almost lost, the bridge is unable to provide traffic passing capacity.

2.3 Tunnel seismic damage and its grades

According to the position of damage, tunnel seismic damage can be divided into two types: lining structure damage and bottom structure damage.

(1)Lining structure damage: lining crack(clear crack with certain trend),lining crack(flaky or reticular crack without certain trend), concrete spalling, lining dislocation, secondary lining collapse, tunnel collapse, construction seam crack, lining seepage.

(2)Bottom structure damage: pavement cracks(clear crack with certain trend), pavement cracks(flaky or reticular crack without certain trend), inverted arch dislocation, inverted arch uplift, pavement seepage.

Due to the materials of lining and its damage degree, grades of tunnel seismic damage seen in Table 3.

Table 3. Grades of Tunnel Seismic Damage

Grades	Reinforced concrete lining	Plain concrete lining	Loss of road traffic function
Grade I - slight	Construction seam crack; small amount of discrete cracks in secondary lining.	Construction seam crack; small amount of discrete cracks in secondary lining.	No effect on vehicles passing
Grade II - medium	Some cracks deep inside secondary lining with large width, parts of them are penetrating cracks.	Cracks deep inside secondary lining with large width, parts of them are penetrating cracks.	Traffic safety may be endangered.
GradeIII- serious	Continuous longitudinal concrete spalling in	Some reticular cracks in secondary lining, concrete	Traffic safety may be seriously

	secondary lining with steel showing up.	spalling in secondary lining, parts of lining collapsed.	endangered, countermeasures should be taken.
Grade Ⅳ - destroyed	The whole tunnel collapsed.	A large area of lining collapsed; the whole tunnel collapsed.	Vehicles cannot pass by

3. RELATIONSHIP BETWEEN SEISMIC INTENSITY AND ROAD DAMAGE

2008 Wenchuan earthquake is the most devastating natural disaster after the founding of New China, series of research projects about damage survey on all the highways, national and provincial trunk, and some county roads were carried out in the hardest-hit area, results had formed series of "Report on Highways' Damage in The Wenchuan Earthquake". To find inherent law in rapid diagnostic technology of road damage after earthquake, further statistical analysis was carried out in this paper on the basis of this investigation. Beside the seismic intensity, the distance from road to fault zone also effect damage significantly, this factor would also consider in the rapid diagnostic model.

3.1 Relationship between seismic intensity and subgrade seismic damage

Due to its structure type and location, subgrade seismic damage can be divided into three kinds: subgrade damage, slope damage and support structure damage. According to Wenchuan earthquake subgrade seismic damage investigation(Chen,2012a), the regulation was shown in table 4.

Table 4. Situation of Subgrade Seismic Damage in Different Seismic Intensity Regions

Seismic intensity	Ratio of subgrade seismic damage (%)		
	Subgrade damage	Support structure damage	Slope damage
IX~XI	86	81.8	78.7
VIII	4	3.5	11.2
VII	8	12.8	10.1
VI	2	2.7	

Survey results on 579 places of subgrade damage during Wenchuan earthquake have the following characteristics:

(1) Subgrade damage, slope damage and support structure damage mainly happened in areas where seismic intensity is between IX and XI, nearly accounts for 80% of seismic damage;

(2) Subgrade and support structure on soil foundation have weak seismic

performance; rock slope is more prone to collapse.

3.2 Relationship between seismic intensity and bridge seismic damage

According to Wenchuan earthquake bridge seismic damage investigation(Chen 2012b), the regulation was shown in table 5.

Table 5. Bridge Damage Statistics in Different Seismic Areas during Wenchuan Earthquake

Grades / seismic intensity	Grade Ⅰ		GradeⅡ- Medium	GradeⅢ- Serious	GradeⅤ- Completely destroyed	Total
	No damage	Minor damage				
Ⅶ	304	432	42	0	0	778
Ⅷ	115	115	50	7	0	287
IX	33	35	73	21	3	165
X,XI	7	24	47	33	9	120
Total	459	606	212	61	12	1350

Statistical results show that with the increase of seismic intensity, ratio of GradeIII damage and GradeⅤ damage gradually increase, the relationship is clear. In the seismic area, the extent of bridge damage is also related with bridge's type, size, and alignment.

3.3 Relationship between seismic intensity and tunnel damage

Table 6. Relationship Between Seismic Intensity and Tunnel Damage

<table>
<tr><th rowspan="2">Seismic intensity</th><th rowspan="2">Tunnel fortification grade</th><th rowspan="2">The whole tunnel damage</th><th colspan="3">Tunnel portal damage</th></tr>
<tr><th>Slope around tunnel portal</th><th>Tunnel Portal structures</th><th>Tunnel portal lining</th></tr>
<tr><td>VI</td><td>VI、VII</td><td>No damage</td><td rowspan="3">No damage</td><td rowspan="3">Capstone of tunnel portal is smashed by falling rocks</td><td rowspan="3">Minor damage (few tunnels have minor cracks), secondary lining and tunnel have not collapsed</td></tr>
<tr><td>VII</td><td>VI、VII</td><td rowspan="2">Almost no damage(few tunnels have minor cracks)</td></tr>
<tr><td>VIII</td><td>VI、VII</td></tr>
<tr><td>IX</td><td>VII</td><td rowspan="2">Secondary lining has severe cracks or collapsed</td><td rowspan="3">Landslide or slump</td><td rowspan="2">There are cracks in capstone of tunnel portal</td><td rowspan="2">Secondary lining collapsed, while tunnel hasn't collapsed</td></tr>
<tr><td>X</td><td>VII</td></tr>
<tr><td>XI</td><td>VII</td><td>Tunnel collapsed</td><td>capstone of tunnel portal collapsed</td><td>Tunnel collapsed</td></tr>
</table>

According to survey results, tunnel damage has the following general features(Chen 2012c):

(1) In VI seismic areas, no damage occurred to tunnels; in VI~ X seismic areas, no tunnel is subject to grade Ⅳ damage; in XI seismic areas, all tunnels are subject to different degrees of damage.

(2) Tunnels with fortification grade VI almost have no damage when seismic intensity is lower than VIII.

(3) Tunnels with fortification grade VII almost have no damage when seismic intensity is between VI and VIII; in IX~X seismic areas, these tunnels are subject to grade Ⅲ damage; and in XI seismic areas.

4. RAPID DIAGNOSTIC MODEL OF ROAD LIFELINE SEISMIC DAMAGE BASED ON SEISMIC INTENSITY

Aiming at severity and suddenness of the earthquake disaster, researches on rapid report of seismic intensity have been carried out at home and abroad. With the development of earthquake discipline, rapid report of seismic intensity in a relatively short time after the earthquake has been achieved, there are some typical methods: Empirical Green's Function method, the finite fault source model method, and American Shake-Map method. The development and maturation of this technology have provided a foundation for putting forward the road damage assessment model based on seismic intensity.

To estimate road seismic damage more accurately, road seismic damage assessment unit was divided according to the structural characteristics of road network. Bridge and tunnel are used to be independent assessment unit respectively.

4.1 Preparatory work for the rapid diagnostic model

In order to improve the rapidity and effectiveness of rapid diagnostic, the following preparatory work should be done well.

(1) Establishing road network basic database

This part includes network distribution, road technical standards, design parameters, the current status.

(2) Analyzing the key sections' anti-seismic capacity

This part includes carrying out evaluation work of road infrastructure before earthquake, analyzing the anti-seismic capacity of bridges, tunnels and important intersections, especially those age-old bridges and tunnels.

(3) Make sure road damage information acquisition and transmission modes

This part is to make sure road damage information acquisition and transmission modes which could improve the veracity of diagnostic results.

4.2 The index set of road seismic damage

Through summary and analysis of survey data on road seismic damage after Wenchuan earthquake, the indicator structure of seismic damage estimation of different structures was put forward on the basis of the division of road seismic damage assessment unit, shown as follows:

(1) The index set of subgrade seismic damage

The selection of the index set should obey the following principles such as concise, comprehensive, high relative testability, independence of each index. According to the survey results of subgrade seismic damage after Wenchuan earthquake, the index set of subgrade seismic damage was established as follows: seismic intensity and distance to fault zone U^S_1 (seismic intensity U^S_{11}, distance from road to fault zone U^S_{12}), road technical level and subgrade types U^S_2 (road technical level U^S_{21}, filling and digging form U^S_{22}, subgrade material U^S_{23}), support structure types U^S_3 (retaining wall material U^S_{31},retaining wall height U^S_{32}),slope types U^S_4 (slope protection form U^S_{41}, slope height U^S_{42}, slope angle U^S_{43}).

Based on the relative mathematical statistics analysis of subgrade seismic damage, single index classification structure which could influence road lifeline subgrade seismic damage was established, seen in Table 7.

Table 7. Index Structure of Subgrade Seismic Damage

Damage grade	Influence indicators									
	U^S_{11}	U^S_{12} (km)	U^S_{21}	U^S_{22}	U^S_{23}	U^S_{31}	U^S_{32} (m)	U^S_{41}	U^S_{42} (m)	U^S_{43} (°)
Completely destroyed	XII	<1	< III	semi-filling and semi-excavating at hillside	Soilsupport structure	Dry-laid cutting	> 8	No protective structure	30~40	45~65
Serious destruction	IX~XI	1~5	III	semi-filling and semi-excavating at the toe of slope	Soil subgrade	Dry-laid embankment	6~8	Entity facing wall	20~30	40~45
Partial destruction	VII~VIII	5~10	II	cutting	Rock slope	mortar flag stone (block stone)	4~6	suspended mesh gunite	10~20	35~40
Basically well-re	< VII	> 10	> II	embankment	Rock under soil	flag (block stone)	< 4	Other protective	0~10	0~35

main						concrete		structu re		

(2) The index set of bridge seismic damage

The index structure of bridge seismic damage: intensity and distance to fault zone U^{B}_{1} (seismic intensity U^{B}_{11}, distance from bridge to fault zone U^{B}_{12}, bridge fortification intensity U^{B}_{13}), bridge type and size U^{B}_{2} (bridge type U^{B}_{21}, bridge length U^{B}_{22}), bridges alignment U^{B}_{3} (curve radius U^{B}_{31}, main beam angle U^{B}_{32}).

Based on the relative research results(Zhang 2012; Zhang 2011) and the relative mathematical statistics analysis of bridge seismic damage, single index classification structure was established, seen in Table 8.

Table 8. Index Structure of Bridge Seismic Damage

Damage grade	**Influence indicators**						
	U^{B}_{11}	U^{B}_{12} (km)	U^{B}_{13}	U^{B}_{21}	U^{B}_{22}(m)	U^{B}_{31}(m)	U^{B}_{32}(°)
Completely destruction	≥XI	<1	VI	Arch bridge	>500	<300	<30
Serious destruction	IX~X	1~5	VII	Simple supported girder bridge	100~500	300~500	30~60
Partially destruction	VIII	5~10	VIII	Continuous beam bridge	30~100	500~1000	60~75
Basically well-remain	≤VII	>10	≥IX	Other types	<30	>1000	75~90

(3) The index set of tunnel seismic damage

The index structure of tunnel seismic damage: intensity and distance to fault zone U^{T}_{1} (seismic intensity U^{T}_{11}, distance from tunnel to fault zone U^{T}_{12}, tunnel fortification intensity U^{T}_{13}), surrounding rock and lining U^{T}_{2} (surrounding rock grade U^{T}_{21}, lining types U^{T}_{22}), tunnel portal protection and mountain condition U^{T}_{3} (tunnel portal protection forms U^{T}_{31}, mountain vegetation and rock condition U^{T}_{32}).

Based on the mathematical statistics analysis of tunnel seismic damage, single index classification structure was established, seen in Table 9.

Table 9. Index Structure of Tunnel Seismic Damage

Damage grade	**Influence indicators**						
	U^{T}_{11}	U^{T}_{12} (km)	U^{T}_{13}	U^{T}_{21}	U^{T}_{22}	U^{T}_{31}	U^{T}_{32}
Completely destroyed	≥XI	<1	VII	I~II	No secondary lining	No protection	Loose, bare weathered rock / no vegetation
Serious	IX~X	1~5	VII	III	No	Spray plain	Soft rock /

destruction					reinforced concrete lining	concrete / gunite	small amount of vegetation
Partially destruction	VII~VIII	5~10	VIII	IV	Reinforced concrete lining	Anchor pile suspended mesh gunite / concrete	Hard rock / large amount of vegetation
Basically well-remain	≤VI	>10	IX	V~VI	Composite lining	Reinforced concrete	Hard rock / dense vegetation

4.3 Comment set of road seismic damage

Through fuzzy judgment of the above single index, road damage danger risk level could be obtained, and fuzzy comment set could be determined. Fuzzy comment set was composed of four grades: completely destroyed, serious destruction, partially destruction, and basically well-remain, It can be expressed by formula $T=\{100,75,50,25\}$.

4.4 Determination of weight set

Each above index set structure contains multiple indexes, AHP is an effective method of determining a plurality of weight vectors. On the basis of seismic data analysis of Wenchuan earthquake, combined with expert advice, judgment matrix is constructed and the weight is calculated, the results are as follows:

(1) Index weight of subgrade seismic damage: index weight of $U^S=\{U^S_1,U^S_2,U^S_3,U^S_4\}$ is $w^S=(0.5120,0.1045,0.1588)^T$; index weight of $U^S_1=\{U^S_{11},U^S_{12}\}$ is $w^S_1=(0.6667,0.3333)^T$; index weight of $U^S_2=\{U^S_{21},U^S_{22},U^S_{23}\}$ is $w^S_2=(0.5396,0.1634,0.2970)^T$; index weight of $U^S_3=\{U^S_{31},U^S_{32}\}$ is $w^S_3=(0.3333,0.6667)^T$; index weight of $U^S_4=\{U^S_{41},U^S_{42},U^S_{43}\}$ is $w^S_4=(0.5396,0.1634,0.2790)^T$. Consistency ratio is less than 0.1 by inspection, up to the requirement.

(2) Index weight of bridge seismic damage: index weight of $U^B=\{U^B_1,U^B_2,U^B_3\}$ is $w^B=(0.6337,0.1919,0.1744)^T$; index weight of $U^B_1=\{U^B_{11},U^B_{12},U^B_{13}\}$ is $w^B_1=(0.5936,0.2493,0.1571)^T$; index weight of $U^B_2=\{U^B_{21},U^B_{22}\}$ is $w^B_2=(0.7500,0.2500)^T$; index weight of $U^B_3=\{U^B_{31},U^B_{32}\}$ is $w^B_3=(0.2500,0.7500)^T$. Consistency ratio is less than 0.1 by inspection, up to the requirement.

(3) Index weight of tunnel seismic damage: index weight of $U^T=\{U^T_1,U^T_2,U^T_3\}$ is $w^T=(0.6144,0.2684,0.1172)^T$; index weight of $U^T_1=\{U^T_{11},U^T_{12},U^T_{13}\}$ is $w^T_1=(0.5936,0.2493,0.1571)^T$; index weight of $U^T_2=\{U^T_{21},U^T_{22}\}$ is $w^T_2=(0.3333,0.6667)^T$; index weight of $U^T_3=\{U^T_{31},U^T_{32}\}$ is $w^T_3=(0.7500, 0.2500)^T$. Consistency ratio is less than 0.1 by inspection, up to the requirement.

4.5 Fuzzy information relations between influence indexes and road seismic damage risk level

The roadbed and bridge seismic damage index includes a few quantitative indexes. It would hide the regulation with the exact boundaries. Fuzzy set theory adopts the membership function and can solve this contradiction. Through the further analysis about quantitative seismic damage indexes for subgrade and bridge, it can be assume that the indexes were linear distribution (subordinate function) in the corresponding comment set evaluation grades probability. The road seismic damage risk level and index membership function can be expressed:

$$r_{i1} = \begin{cases} 1, a_{i1} \le x < a_{i2} \\ (a_{i3} - x)/(a_{i3} - a_{i2}), a_{i2} \le x \le a_{i3} \\ 0, x > a_{i3} \end{cases} \qquad r_{i2} = \begin{cases} (x - a_{i1})/(a_{i2} - a_{i1}), a_{i1} \le x \le a_{i2} \\ 1, a_{i2} \le x < a_{i3} \\ (a_{i4} - x)/(a_{i4} - a_{i3}), a_{i3} \le x \le a_{i4} \\ 0, else \end{cases}$$

$$r_{i3} = \begin{cases} (x - a_{i2})/(a_{i3} - a_{i2}), a_{i2} \le x \le a_{i3} \\ 1, a_{i3} \le x < a_{i4} \\ (a_{i5} - x)/(a_{i5} - a_{i4}), a_{i4} \le x \le a_{i5} \\ 0, else \end{cases} \qquad r_{i4} = \begin{cases} (x - a_{i4})/(a_{i5} - a_{i4}), a_{i4} \le x \le a_{i5} \\ 1, a_{i5} \le x \\ 0, else \end{cases}$$

Within the formulae: $r_{i1}(x), \cdots, r_{i4}(x)$ are risk level of road lifeline subgrade and bridge seismic damage : FS-I (basically well-remain), FS-II(partial destruction), FS-III(serious destruction), FS-IV(completely destroyed); $a_{i1}, a_{i2}, \cdots, a_{i5}$ are the index threshold value of corresponding risk level; x is the parameter values of the evaluated object.

For the classification indexes, the score can be obtained according to classification stander. Combined with the risk level standard: the FS-I(75~100),FS-II(50~75),FS-III(25~50),FS-IV(<25), we can judge the damage risk level, which were corresponding to the four level: basically well-remain, partial destruction, serious destruction, completely destroyed. Thereby, the user can rapidly judge the seismic damage of road traffic lifeline.

CONCLUSIONS

(1) There are so many influence factors to road traffic lifeline seismic damage risk. In order to assess the road damage rapidly, the road damage types and classification

were proposed in this paper.

(2) On the basis of Wenchuan earthquake road damage survey data, by using the mathematical method of fuzzy evaluation to analyze various evaluation index synthetically, the grading standard of the road traffic lifeline seismic damage risk characteristic index was proposed.

(3) The influence factors of road seismic damage risk were extremely complex, currently, the fault zone trend and the distance from fault zone to road haven't been considered. The index, grading and weight of rapid assessment model of road traffic lifeline seismic damage need to be researched deeply.

ACKNOWLEDGMENTS

This research was supported by the Social Development Science Programs of the Department of Science and Technology of Yunnan Province (No.:2012CA004), Yunnan Province China and Science-Technology Programs of Transportation Commission of Yunnan Province (No.:2011(B)06-b).

REFERENCES

Chen, L.S. (2012a). *Report on Highways' Damage in The Wenchuan Earthquake-Subgrade*. Beijing: China Communications Press. (in Chinese)

Chen, L.S. (2012b). *Report on Highways' Damage in The Wenchuan Earthquake-Bridge*. Beijing: China Communications Press. (in Chinese)

Chen, L.S. (2012c). *Report on Highways' Damage in The Wenchuan Earthquake-Tunnel*. Beijing: China Communications Press. (in Chinese)

Liu, M.Z., Ghang J.F.,Li C.L.,Liu G.L. (2013). "Rapid Extraction of Road Damages from High Resolution Remote Sensing Images". *Earthquake*, 02:79-86. (in Chinese)

Liu, Y., Zhang, L., Zhang, Y.H., Zhang, X.J.(2012). "Difference Solution for Shear Lag Effect of Twin-cell Box Girder". *Journal of Lanzhou Jiaotong Univcrsity*, 03:27-30+34. (in Chinese)

Ma, H.J., Lu, N., Li X.X. (2013). "Rapid Extraction of Earthquake Road Damages Based on Sidelines from Remote Sensing Images". *Earthquake*, 02:71-78. (in Chinese)

Su, D.L., Guo Z.Y., Liu B.M. (2013). "Traffic Function Loss of Highway Lifeline under Major Natural Disasters". *Proceedings of the ICTE2013*. 2598-2604

Wang, Y.P., Jiang J.Y., Lin L.L. (2012). "Road Damage Recognition Method in High Resolution Remote Sensing Image Information". *Computer Engineering and Applications*, 03:173-175+227. (in Chinese)

Zhao, F.J. (2010). "Seismic Disaster Information Extraction from Remotely Sensed Imagery". Harbin:Institute of Engineering Mechanics, China Earthquake Administration. (in Chinese)

Zhang, Y., Zhang H., Ma, J.Z.(2011). "Calculations and Analysis of Bevel Angle Influence on Internal Force of Skewed Bending Bridge". *Journal of Shijiazhuang Tiedao University (Natural Science)*,02:6-9+45. (in Chinese)

Roadway Access Management and Its Importance to the Transportation System of Developed and Developing Nations

K. M. Williams, AICP[1]; and V. G. Stover, Ph.D., P.E.[2]

[1]Program Director, Planning and Corridor Management, Center for Urban Transportation Research, University of South Florida, 4202 E. Fowler Avenue, CUT100, Tampa, FL. E-mail: kwilliams@cutr.usf.edu
[2]Affiliated Faculty, Center for Urban Transportation Research, University of South Florida, 4202 E. Fowler Avenue, CUT100, Tampa, FL; and Emeritus member, TRB Access Management Committee. E-mail: vgstover@gmail.com

Abstract
Access management can be generally defined as the coordinated planning, regulation, and design of access between roadways and adjacent land. The practice of access management has steadily evolved over the past century as an important element of the U.S. transportation policy. Since the 1980s, dramatic advances in access management have taken place along arterial roads, particularly in suburban settings. Colorado established the first statewide access management program in the U.S. (1980), followed by statewide codes or increased efforts in several other states. The American Association of State Highway Transportation Officials *Policy on Geometric Design of Highways and Streets* expanded its treatment of access management in 2001 and 2003, the first U.S. *Access Management Manual* was published by the Transportation Research Board. The documented safety and operational benefits of access management techniques have also led to policies in several other nations. These advancements continue to raise awareness of the need to manage traffic conflicts through functionally designed roadway networks with varying levels of access and mobility. This presentation explores the U.S. concept of access management, including recent developments, and addresses the importance of access management to the quality and safety of every nation's transportation system.

OVERVIEW OF ACCESS MANAGEMENT

Access management can be generally defined as the coordinated planning, regulation and design of access between roadways and adjacent land. It encompasses a range of methods that maintain the efficient and safe movement of people and goods by reducing conflicts on the roadway system and at its interface with other modes of travel. These methods include improvements to benefit transit, pedestrians, and bicyclists, as well as different treatments for urban, suburban, and rural settings.

The purpose of access management is to provide vehicular access to land development in a manner that preserves the safety and efficiency of the transportation system. The safety benefits of access management have been clearly documented in more than four decades of research (Gluck, Levinson and Stover, 1999). These safety benefits are attributable to three key issues: 1) improved access design, 2) fewer traffic conflict locations, and 3) higher driver response time to potential conflicts (TRB Committee on Access Management, 2003).

An important principle of access management is to provide a specialized roadway system that is planned, designed and regulated based on the level of mobility versus access that is appropriate for each roadway type. Roadways are classified for access control based upon their importance to local and regional mobility. The greatest access control is applied to roadways intended to serve through movement of traffic over relatively long distances and higher speeds. The least access control is applied to local streets, where through movement is curtailed and speeds are reduced for pedestrian and bicycle safety and amenity for residential neighborhoods. These streets are intended to provide direct access to property and allow traffic to circulate at lower speeds. This concept is not new – it is the foundation of transportation planning.

IMPORTANCE TO THE TRANSPORTATION SYSTEM

Access management presents a clear opportunity for effective and economical management of any nation's transportation infrastructure. It is a relatively low cost set of strategies that preserve the quality of the transportation system by:

- Improving safety in roadway design and operation, through traffic conflict management;
- Promoting economic vitality by preserving market area for business and maintaining major roadways for higher speed, higher volume traffic and freight movement;
- Increasing connectivity between transportation modes through improved network planning and increased coordination of land use and transportation planning;
- Promoting efficient site access and circulation design along major roadways;
- Reducing congestion and delay on major roadways, while enhancing mobility within and between urban areas.

Nations throughout the world can achieve goals related to improved transportation management, and coordinated land and transportation planning, through policies, plans and designs that address the following basic principles of access management.

1. Provide a specialized road system in which different roads serve different purposes. A balanced roadway network serves a range of functions from higher speed, long distance movement, where access must be controlled (e.g. freeways, expressways), to local streets (e.g. local or minor collector streets), where extensive access is provided and speeds and traffic volumes are curtailed.

2. Limit direct access to major (arterial) roads. Direct property access should be limited along roads intended to serve higher volumes of traffic over longer distances at higher speeds; it should be denied whenever reasonable alternative access can be provided.

3. Locate traffic signals to favor through traffic movement. Signalized access points should fit into an overall traffic signal coordination plan.

4. Locate access connections away from road intersections. Driveways and street connections should be located outside the functional area of road intersections or interchanges to preserve intersection safety and operations.

5. Remove turning traffic from through traffic lanes. This is accomplished through the design of major roadways with auxiliary lanes. Warrants for right and left turn lanes are also methods of requiring auxiliary lanes in the permit and development review process.

6. Provide a supporting street and circulation system. An interconnected network of collector and local streets improves local mobility, removes local trips from arterial roads, and reduces the need for direct property access to arterials.

The importance of access management has been a topic in the literature for many years. An early U.S. National Cooperative Highway Research Program report (Marks, 1971), for example, stated, "The lack of access control along arterial highways has been the largest single factor contributing to the obsolescence of ….an entire generation of new arterial facilities built only a short while ago." U.S. highways were rendered functionally obsolete by a proliferation of driveways, street connections, and traffic signals. The adverse effects of not managing roadway access are identified in Table 1.

Table 1. Adverse Effects of Failure to Manage Roadway Access.

• Increase in vehicular crashes and collisions involving pedestrians and cyclists
• Accelerated reduction in roadway efficiency and increased delay and travel times for private and public transportation
• A more frequent need for roadway reconstruction and right of way acquisition
• Reduced aesthetics from frequent driveways, cluttered signage, and inadequate area for landscaping
• Increased fuel consumption and vehicular emissions as numerous driveways and traffic signals intensify congestion and delay

The relationship of unmanaged access to roadway safety has also been clearly documented. *NCHRP Report 420* is the most comprehensive study to date of the impacts of access management techniques. It includes composite crash rate indices derived from analysis of 37,500 crashes, and compared with a synthesis of the literature. The indices represent average crash rates by access density using the crash rates for 10 access points per mile as a base. These indices are shown in Figure 1.

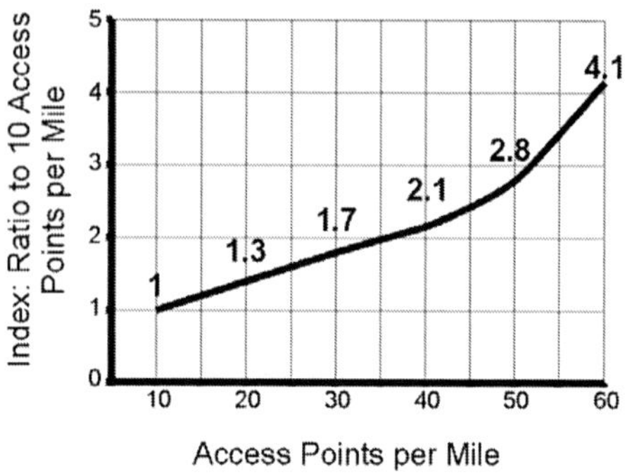

Fig. 1. Composite crash rate indices.
(Source: Gluck, Levinson and Stover, 1999)

The specific relationship varies due to differences in road geometry (lane width, presence or absence of turn lanes and medians), operating speeds, and driveway and intersection traffic volumes. Nonetheless, these indices clearly show a correlation between access density on the margin of the roadway and crash rates. They suggest, for example, that an increase from 10 driveways to 30 driveways per mile would increase crash rates by roughly 70 percent. The report also includes a more refined procedure for estimating the relative change in crash rates by access density.

IMPORTANCE TO THE BUILT ENVIRONMENT

Roadside strip development with a sparse or disconnected local street network is a key problem contributing to poor access design along the roadway system. Strip development with uncontrolled access not only reduces roadway safety, as noted above, it also has a clear adverse effect on community aesthetics and reduces the potential for walking, bicycling and transit use. Strip development and poor connectivity between land uses are defining characteristics of urban sprawl - a situation in which development occurs without consideration of urban form or transportation system needs. Access management includes land use planning policies and strategies that discourage incremental strip development and promote planning of land uses into activity centers or towns with supporting networks and unified circulation. These strategies prevent the poorly planned conversion of rural land to urban uses and encourage a built environment that supports economic growth while promoting bicycle, pedestrian, and transit mobility.

Figure 2 is a schematic of two scenarios that exemplify the issues discussed above. The top half of Figure 2 (left to right) demonstrates how separating land uses into stand alone developments with disconnected local networks increases local traffic circulation on the arterial system and can result in more arterial conflict (access) points. The bottom half of Figure 2 shows how local traffic and conflicts on the arterial are reduced when land uses are organized into activity centers with a unified street network. The bottom example is also more conducive to walking, bicycling and transit use.

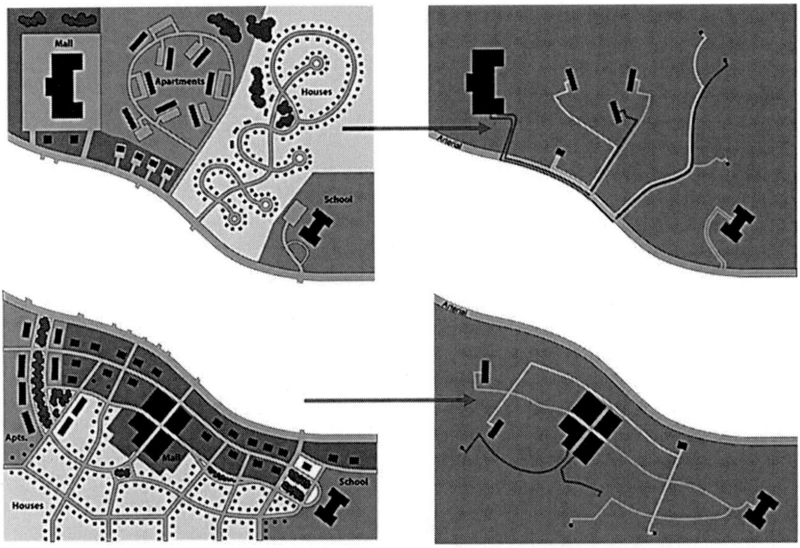

Fig. 2. Network connectivity, access and arterial traffic.
(Source: Duany Plater-Zyberk and Company as supplemented by AECOM)

To reduce strip development along the roadside, local and provincial urban planning agencies can seek to plan and organize land use activity centers that are highly accessible both regionally and locally via a variety of transportation modes and multiple alternative paths. These paths could include (Williams and Levinson, 2011):

- Freeways, expressways, and other access-controlled major arterial highways, along with regional transit service (e.g. commuter rail, rail rapid transit, bus rapid transit on dedicated lanes) to support regional mobility between major activity centers;
- Regularly spaced arterial and major collector roadways, complemented by local transit service (e.g. bus circulators, street cars, light rail) to support mobility within and across urbanized areas; and
- A dense, connected network of minor collector and local streets, multi-use paths, sidewalks, and user facilities (e.g. bicycle racks, benches, water fountains, etc.); to support neighborhood mobility within and between local activity centers and surrounding residential areas.

Through attention to these issues, urban planners in coordination with transportation agencies can help protect the flow of vehicular traffic along major transportation routes and support the transportation needs of commercial development, while providing improved mobility for travelers. Access permitting and traffic impact assessment procedures and requirements provide a mechanism for managing the impacts of development on the transportation system as sites develop and redevelop. These mechanisms can be used to require changes to site access that improve overall roadway safety and operations.

CONTEMPORARY ACCESS MANAGEMENT PROGRAMS

Since the 1980s, dramatic advances in access management have taken place along arterial roads in the U.S., particularly in suburban settings, as well as in national policy and guidance documents. Colorado established the first statewide access management program in the U.S. (1980), followed by statewide codes or increased efforts in several other states. The American Association of State Highway Transportation Officials *Policy on Geometric Design of Highways and Streets* expanded its treatment of access management in 2001. In 2003, the first U.S. *Access Management Manual* was published by the Transportation Research Board, establishing the state of the practice in the U.S. and serving as a resource for many other nations interested in advancing the concept.

NCHRP Project 15-43 was recently completed and produced a 2nd edition of the TRB *Access Management Manual*. This edition includes increased attention to network planning strategies for access management, interchange area access control, and the application of access management techniques in different land use contexts (e.g. urban, suburban, rural and main street environments). It also emphasizes the importance of access management to all transportation modes and increases guidance to transportation agencies on the effective administration of roadway access policies and regulations in a multimodal context.

Contemporary access management programs in the U.S. are characterized by the following key elements: (1) classifying roadways into a logical hierarchy by function, (2) defining allowable access for each class of roadway, (including standards for spacing of signalized and un-signalized access points), (3) applying appropriate geometric design and traffic engineering criteria to each access point, and (4) establishing policies, regulations, and permit procedures to carry out and enforce the program.

The access classification system provides the foundation for contemporary access management programs. It defines when, where, and how access can be provided between highways, cross streets and driveways and relates the allowable access to each roadway's purpose, importance, and functional characteristics. The functional classification system is the starting point in assigning access categories to highways.

Several basic access categories or "levels" can be applied to any roadway system. They range from full control of access (freeways), to little or no access control on local streets. Modifying factors in assigning these categories include existing development, driveway density, and geometric design features such as the presence or absence of a physical median. Access spacing, location and design standards for interchanges, signalized intersections, unsignalized intersections, and median openings are keyed to the access categories. The standards apply to new development and when a significant change is made in the size and nature of an existing development.

Existing substandard access design or spacing is upgraded to the extent feasible when a site is redeveloped. In addition, changes to median design and site access may be made during the roadway improvement process. Cities with short blocks and frequent local street connections may be addressed through subcategories in the access classification system.

Access is provided to parcels that do not conform to spacing criteria when there is no alternative reasonable access; however, the basis for such deviations must be clearly documented to avoid setting undesirable precedents. Conditions may also be included in the access permit for removal of the access where alternative access becomes available.

Signalized intersection spacing criteria along roadways apply to both intersecting streets and driveways. The goal is to limit signals to locations where the progressive movement of traffic will not be significantly impeded and the "window" for progression at desired travel speeds is maintained (Williams and Levinson 2010). Excessively long cycle lengths (usually over two minutes) indicate a need for corrective actions such as interchanges, grade separations, rerouting left turns, adding lanes, or improving the secondary street system to reduce arterial left-turn volumes.

Unsignalized driveway spacing is based on safe stopping sight distance, operating speed, overlapping right-turns, and more recently, decision sight distance. Spacing and design standards reflect roadway level of importance (access categories), roadway speeds, and the size of traffic generators. The design (length) of left-turn and right turn bays also influences spacing.

Medians reduce safety hazards posed by frequent access to major high volume roads by limiting the left-turn movements to locations expressly designed for them. Unsignalized directional openings between signalized intersections provide convenient access to abutting properties and reduce U-turns and conflicting left-turns at signalized intersections. Replacing unsignalized full median openings with directional openings can substantially reduce crash rates.

The typical access application process includes consideration of the access classification of involved roadways, and the ability of the proposed property access to meet spacing requirements. Access review may also involve traffic impact analysis, and circulation and safety assessments. Key issues in administering an access management program include setting fees for applications and permits, handling deviations from standards, dealing with small lots, and upgrading access to land uses that redevelop.

Examples of specific access management techniques include:

- locate traffic signals to support signal coordination and permit efficient traffic progression over a wide range of traffic volumes and speeds;
- use a nontraversable median to limit the exposure of through traffic, pedestrians, and bicyclists to left-turning vehicles and to provide a refuge for mid-block crossings by pedestrians and bicyclists;
- design access points to minimize conflicts at the entrance of a site and support smooth entry and exit at speeds appropriate to the connecting roadway – consider separating left turn ingress and egress at sites with large traffic volumes for improved operations and safety;
- provide right- and left-turn deceleration and storage lanes so drivers can wait safely to complete a turn and so that turning vehicles do not cause serious conflict with through traffic movement;

- limit and separate driveways and other access points to major roadways to simplify the driving task and reduce the potential for driver confusion and crashes;
- restrict driveways in the vicinity of signalized intersections to reduce intersection conflicts and crashes;
- provide an adequate network of local and collector roadways and promote connections between adjacent land developments to reduce the need for driveway access on major roadways and allow vehicles to circulate within neighborhoods and centers rather than on the arterial system; and
- provide bicycle and pedestrian connections to maintain continuity of non-vehicular pathways and provide direct connections to transit or midblock crossing locations.

CONCLUSION

Access management has gained increased attention in recent years due to continued efforts in the U.S. to advance the concept both nationally and internationally. In addition, the adverse impacts of poorly managed access on mobility, safety and the built environment are becoming increasingly apparent to both developed and developing nations as automobile usage increases. An appealing aspect of access management is that it is flexible, and involves relatively low cost planning, regulatory and design strategies that can be carried out incrementally or systematically. Application of these strategies has the potential to extend the life of existing highways and thereby reduce the fiscal pressures on transportation agencies. It also provides demonstrated transportation system benefits at low cost for both system users and government agencies.

Positive economic effects on communities include improved design of the built environment and the ability to maintain a safe, efficient and sustainable transportation system for all modes and users. Consequences of not managing access include accelerated reduction in roadway efficiency, increased rate and frequency of collisions, and increased need for costly highway reconstruction. These consequences are not sustainable.

Although the TRB *Access Management Manual* and manuals under development in other nations offer guidance on approaches, these concepts will require adaptation to the societal and institutional context in which they are applied. No single set of international standards can address the needs of every nation, region, or community. Nonetheless, the principles and techniques suggested in these resources can be adapted and applied under a variety of circumstances and would be of great benefit to any nation's transportation system. Transportation and urban planning professionals and researchers must continue to engage in research to evaluate specific applications to determine how access management principles and techniques can best be systematically integrated into their nation's planning, design and regulatory processes. An ultimate goal of this effort is a national manual that conveys the results of the work and forms a basis for ongoing advancement and implementation of access management.

REFERENCES

Committee on Access Management (2003). *Access Management Manual*, Transportation Research Board of the National Academies, Washington, DC.

Gluck, J.S., Levinson, H.S., and Stover, V.G. (1999). *NCHRP Report 420: Impacts of Access Management Techniques*, Transportation Research Board, Washington, D.C.: National Academy Press.

Marks, H. (1971). *NCHRP Report 121 Protection of Highway Utility*, Transportation Research Board, Washington, D.C.: National Academy Press

Williams, K.M. and Levinson, H.S. (2010). "Access Management – An Overview," ITE Journal, pp. 24-28.

Williams, K.M. and Levinson, H.S. (2011). "The Role of Access Management in Sustainable Development," *Proceedings of the First T%DI Congress,* ASCE, Chicago, Illinois.

The Changing Focus of Access Management

G. Sokolow[1]; K. Williams [2]; and H. Levinson[3]

[1] Gary Sokolow, Florida Department of Transportation, 605 Suwannee Street, Tallahassee, FL 32399. E-mail: gary.sokolow@dot.state.fl.us
[2] Kristine Williams, AICP, Center for Urban Transportation Research, University of South Florida, 4202 E. Fowler Avenue, CUT100, Tampa, FL 33620. E-mail: kwilliams@cutr.usf.edu
[3] Herbert Levinson, Suite 5305 Ashlar Village, Wallingford, CT 06492. E-mail: hslevinson@aol.com

Abstract
This paper seeks to demonstrate the importance of expanding the orientation of access management of the major roadway system beyond the goals of maintaining high-speed through movement. Speeds can be managed without increasing congestion. The paper also addresses how access management and design can better relate to pedestrians, bicyclists, and public transportation. The profession must be more aware of the land use and community context, where access management techniques are being applied. Context changes over time and it also changes by location. Therefore, the same approach will not be appropriate on all segments of a roadway or in all countries or regions. A change in the focus of access management with regard to context is consistent with recent trends in transportation and urban planning.

INTRODUCTION

In the U.S., the 1950's were a time of building the Interstate Highway System - a system of multi-lane freeways that allow the free flow of traffic from and to every major city (and some not so "major') in the country. The increased speed of intercity travel excited the transportation profession for decades. This came with the belief that getting people and goods to their destination faster is the goal of all roadway projects.

The practice of access management mirrored this American philosophy with its emphasis on maximizing speed and through movement through access control. Early writings in access management focused on the increased "side-friction" caused by driveways with the implicit assumption that fewer driveways equal better traffic flow. Of course, that was not the entire story. The American Association of State Highway and Transportation Officials (AASHTO 2004) devotes the entire first chapter of its design policy (aka the "Green Book" - a primary resource for highway design principles in the USA) to the issue of roadway "hierarchy". The principle of roadway hierarchy states that roadways should be designed for the functions they are intended to serve.

The principle of roadway hierarchy is based on the idea that systems of roadways serving various functions will meet the needs of society. Freeways are at the top of the hierarchy where speed and long distance travel is the top priority. These freeways should connect to at-grade arterial roads, which distribute traffic to other lessor roadways. These lessor roadways serve traffic that is entering or exiting residential areas and business districts and are generally slower and with easier access to homes and businesses. The primary difference in the function of these different roadway types is defined largely in terms of speed and access.

This principle has led to numerous diagrams explaining the practice of functional classification and access management. Figure 1 is an example of how the access management concept has been illustrated and conveyed in terms of access and mobility–with mobility defined as high-speed through movement.

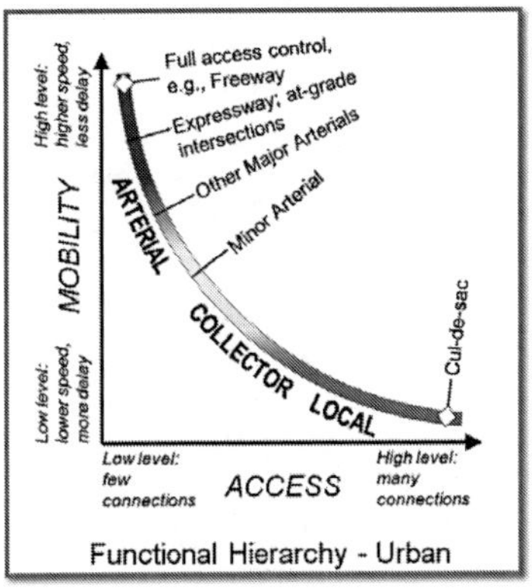

Fig. 1. Relationship Between Access and Mobility
(Source: J. Gattis (2013), University of Arkansas, unpublished notes)

WHAT LED THE U.S. TO THE EMPHASIS ON SPEED?

With the construction of the Interstate Highway System in the U.S., many were impressed with the speed of travel across regions and the safety (measured in crashes per vehicle mile traveled). The lesson gained from this experience was that similar benefits could be achieved if major arterial roadways were access controlled like freeways. Some of the design principles that came about as a result were:

- Pedestrian service as an impediment to vehicle travel; and
- Minimize access points as much as possible, but when they are provided (as with freeway interchanges) these accesses should be channelized to provide conflict points only at flat angles.

In the 1980's, an influential report called the *Intersection Channelization Design Guide* (Neuman 1985) was published by the U.S. Transportation Research Board. Some of the features the report encouraged for use in roadway design were:

- Channelization of intersections to allow cars to merge without stopping. This ignores the fact that pedestrians may be entering the merge area not in the driver's field of vision; and
- The use of only wide radii on driveways to allow cars to quickly enter and exit the roadway, leaving the pedestrian to deal with vehicles turning into their sidewalk space at higher speeds.

The *Guide* (Neuman 1985) also includes plenty of good advice on the use of channelization where slower speeds and harmony with pedestrians is important. Yet some of the advice was taken out of context and overused in the urban situation. Figure 2 is an example of one of the principles of good channelization.

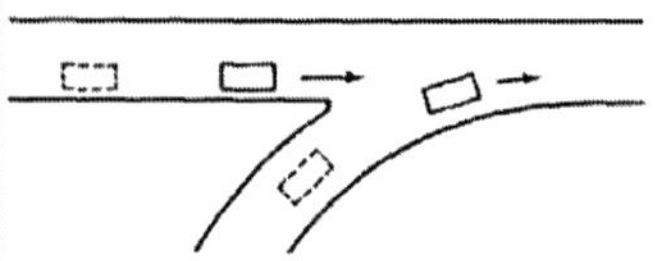

Roadways that merge at flat angles facilitate the actual merging maneuver. In addition, flat angle merges lessen impact energy, thereby resulting in less severe accidents and conflicts.

Fig. 2. Intersection Channelization Design Guidance
(Source: Neuman 1985, p. 30)

In the quest to reduce the difference in speed between vehicles travelling through and those that are turning, highway professionals also encouraged wide driveways with large turning radii. Turn lanes were added to reduce conflicts between turning and through vehicles, and long and uniform signal spacing was adopted to ensure efficient through movement. The end result was high speed traffic on wide roadways with few safe or convenient options for pedestrians to cross.

THE IMPACT ON URBAN FORM

The impact of the Interstate Highway System on land use patterns in the U.S. has been widely documented. The ease of driving long distances and growth in private automobile ownership led to homes, offices and businesses being built wherever land was cheaply available. Little or no consideration was given to the distance between land uses. Zoning, a regulatory technique for the separation of incompatible land uses, required the segregation of places where people live from where they work and shop.

Large landholdings were gradually divided into lots and parcels of varying sizes without consideration of overall network and infrastructure needs or costs. Roadway strip development grew along major routes as metropolitan areas expanded. And developments were designed to accommodate automobile access and required to provide abundant on-site parking.

The result has been a dramatic increase in land consumption for urbanization, growth in transportation costs, and adverse effects of unplanned access on both urban form and highway operation. The separation of land use and transportation authority in the U.S. contributed to this problem. It is clear that our approach to transportation influences land use and that the two issues are interdependent. As we seek more sustainable approaches to urban development and transportation, the American view of how to accomplish access management is also changing.

A CHANGE IN PHILOSOPHY

Changes in the philosophy of access management have been evolving for many years. These changes have paralleled changes in roadway design philosophy, urban design, and transportation and land use planning practice over the past decade. The shift is toward a less regulatory and more design-oriented approach to access management that accommodates compact, high-density urban environments, and not just the suburban context. Below are a few publications that exemplify this trend:

- *Main Street…when a highway runs through it: A Handbook for Oregon Communities* (Oregon Department of Transportation 1999)
- *Flexibility in Highway Design* (U.S. Federal Highway Administration 1997)
- *Designing Walkable Urban Thoroughfares: A Context Sensitive Approach*, (Institute of Transportation Engineers 2010)
- Abu Dhabi *Urban Street Design Manual* (Abu Dhabi 2012)

The change in philosophy relates to how we view roadway hierarchy and function for classification purposes. The concept of hierarchy goes to the heart of roadway design. To properly design a roadway, we need to answer this question; "How do we want this roadway to function?" This question is important, but we need to broaden the perspective. Perhaps the transportation profession has favored speed for cars and trucks too much. We should also be asking ourselves, how do we want this roadway to function in this particular context?

That said, the concept of roadway hierarchy is just as valid in today's multi-modal and pedestrian-sensitive world, as it was during the growth of the automobile age. Some major roadways will still need to accommodate the automobile and express bus service over long distances and at high speeds through urban areas, and some major roadways will need to serve slower speeds with more emphasis on pedestrians, cyclists, and accessibility to other modes, such as local buses and other forms of public transportation. The diagram in Figure 1 presents a broad theory that although reasonable, does not fully explain the character and needs of major roadways in the urban context.

In the past, those practicing arterial access management were accused of trying to make all at-grade arterial roadways into freeways. Looking back, this was correct to a certain extent. The primary issue was the overriding focus on speed and through movement, without equal regard to the context in which those roadways were to function. Many of the design features necessary to accommodate high-speed through movement of automobiles, though acceptable on some roads and in some contexts, create untenable hazards and obstacles for the pedestrian, bicyclists and transit riders.

In the 1980s and 1990s, the U.S. Federal Highway Administration was being routinely questioned about the flexibility of their roadway design standards. This came from questions about bridges in rural areas which, when upgraded, would impact a larger surrounding area and possibly change the character of the surrounding community. To provide guidance on the issue, the Federal Highway Administration published *Flexibility in Highway Design* (FHWA 1997). In this publication, FHWA analyzed the minimum and maximum design specifications and showed how the roadway designer can work within acceptable standards to produce roadways for safety, mobility, and sensitivity to the surrounding environment, both natural and social.

The publication includes case studies where previous lane width minimums were relaxed to allow the installation of roadway medians. One of the case studies was Lincoln Beach Parkway, a section of the Oregon Coast Highway also known as U.S. Route 101. This highway corridor is both rural and urban and is used extensively by tourists and residents. Medians were constructed and some of the median openings were specially designed to accommodate tourist buses. One of the lessons learned in the case study was the importance of preparing more detailed access management plans (FHWA 1997).

In *Designing Walkable Urban Thoroughfares* (ITE 2010), the Institute of Transportation Engineers teamed with architects and urban designers to produce guidance on what is now widely known as "context sensitive solutions (CSS)" for roadway corridors. The CSS framework matches elements of context "zones" with features of thoroughfare types that yield compatibility between roadway design, traffic operations, and context. Figures 3 and 4 illustrate efforts to address contextual issues associated with access management.

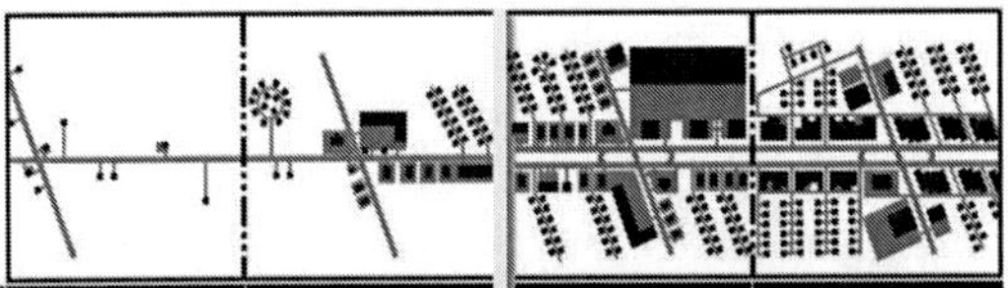

	Rural	Suburban Fringe / Exurban	Suburban	Compact Urban
Land Use	Residential/agriculture, commercial nodes	Residential subdivisions, commercial strips	Residential/commercial, Redevelop w/ infill TOD	Residential/urban commercial mixed-use
Street	2 lanes, center turn lane select locations	2 or 3 to 5 lanes, center turn lane	5 or 6 to 8 lane blvd, some on-street pkg.	6 or 8 lane blvd with on-street parking
Access	1-1/2 Mile signals, 455' drive spacing w/ service drives	455' drive spacing, retro-fit to eliminate some access near signals	Infill access, retro-fit and new cross-access, esp. signals & poor offsets	Retro-fit to reduce number drives, 1-sided signals

Fig. 3. Context based summary of access management strategies.
(Source: Strader 2012)

Fig. 4. A Context Based View of Access Management
(Source: United Kingdom Department of Transport 2007)

In 2013, the iconic *Highway Functional Classification: Concepts, Criteria and Procedures* (FHWA 1989) was rewritten to be more sensitive about the importance of context in the planning and design of roadway systems. This document offers suggestions to integrate context into highway classification. Instead of attempting national guidance, it offers examples of how other states have attempted to achieve this integration with regard to regional at-grade arterials.

Access management practice provides for variation in access management standards as one moves from rural to more densely populated areas. Nonetheless, pedestrians and bicyclists benefit from restriction on access in areas where we want to encourage high pedestrian movement. Figure 5 conceptually illustrates the additional crash potential frequent driveways and uncontrolled left turns for pedestrians.

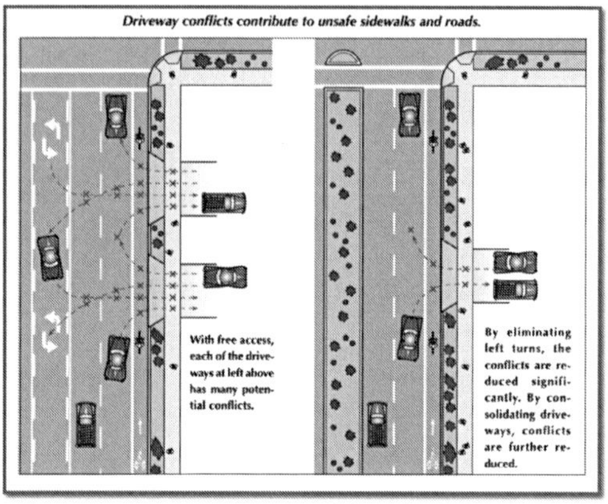

Fig. 5. Effect of access management on bicycle and pedestrian exposure to crashes.
(Source: Oregon Department of Transportation 1999)

Alternatively, curb turning radii and driveway design should be adjusted in areas with pedestrian traffic to help communicate the superior position of the pedestrian where vehicle paths cross sidewalks. This is accomplished with smaller turning radius and sidewalks at a higher elevation, making the driver slow down to access their desired destination. In areas with heavy pedestrian traffic, driveways could also be limited to side streets for parking only and buildings could be moved closer to the sidewalk allowing direct pedestrian access and reducing pedestrian/conflicts.

Previously, in publications and training, large turning radii in driveways were encouraged to help the vehicle move in and out of traffic as quickly as possible. Now this practice is being changed to give more thought to the pedestrian (see Figure 6). Some of the problems associated with overly large radii in pedestrian areas are (Figure 7):

- Poor visibility for the pedestrian to judge when to cross,
- Increased distance for the pedestrian to cross the driveway, increasing the exposure to crashes,
- Vehicles speeding up to arrive at the driveway before the pedestrian.

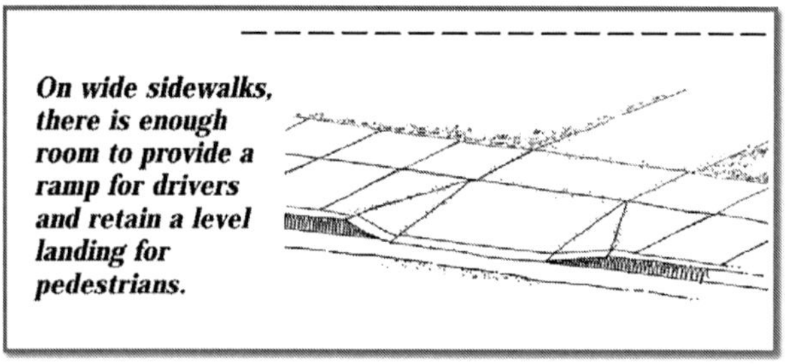

Fig. 6. Driveway design favoring pedestrians
(Source: Oregon Department of Transportation 1999)

Fig. 7 Problems with large radii in pedestrian areas.
(Source: National Center for Safe Routes to School)

Some communities are correcting the designs to make them more pedestrian friendly. The *Abu Dhabi Street Design Manual* (2012) is one of the newest roadway design guides to incorporate numerous land use and transportation contexts. This manual provides design and access management guidance for major arterials and other facilities where the pedestrian is the predominant mode of travel. A goal of this and other recent guidance is to achieve an appropriate balance between auto and pedestrian access. Figure 8 is the corner radius advice. The guidance takes into account residential as well as commercial areas.

The *Intersection Channelization Design Guide* (Neuman 1985) was heavily used, and possibly led to, over-channelizing roadways. Similar to the issue of radius, channelization theoretically helps the driver enter and exit the through movement lanes easier and at flatter angles. However, in practice, problems are associated with this over channelization. The at-grade arterial is not a freeway and the vision of the driver is actually reduced as shown in the Figure 9.

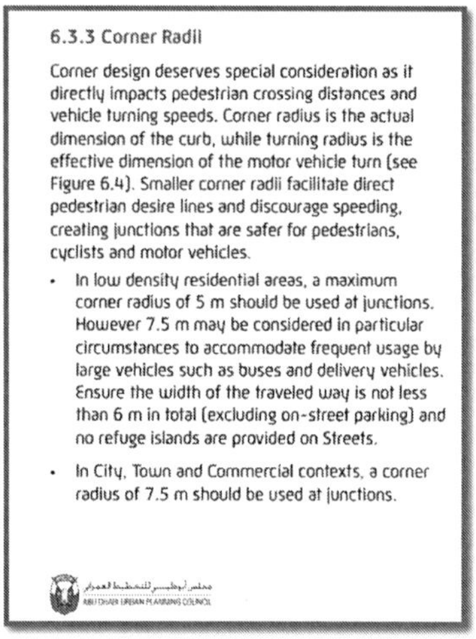

6.3.3 Corner Radii

Corner design deserves special consideration as it directly impacts pedestrian crossing distances and vehicle turning speeds. Corner radius is the actual dimension of the curb, while turning radius is the effective dimension of the motor vehicle turn (see Figure 6.4). Smaller corner radii facilitate direct pedestrian desire lines and discourage speeding, creating junctions that are safer for pedestrians, cyclists and motor vehicles.

- In low density residential areas, a maximum corner radius of 5 m should be used at junctions. However 7.5 m may be considered in particular circumstances to accommodate frequent usage by large vehicles such as buses and delivery vehicles. Ensure the width of the traveled way is not less than 6 m in total (excluding on-street parking) and no refuge islands are provided on Streets.
- In City, Town and Commercial contexts, a corner radius of 7.5 m should be used at junctions.

Fig. 8. Restricting curb radius to blend auto and pedestrian traffic.
(Source: Abu Dhabi Street Design Manual 2012)

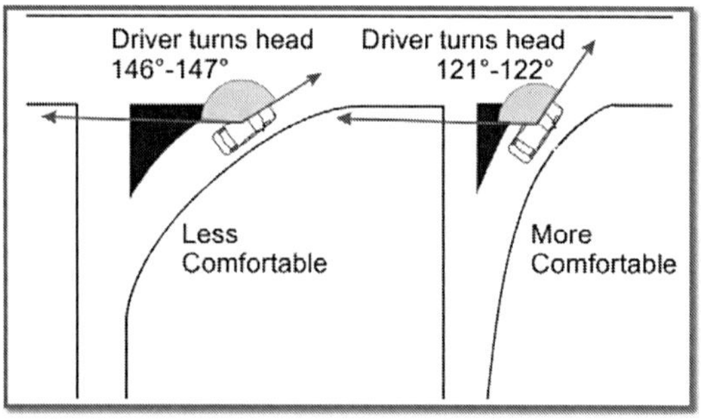

Fig. 9. Recommended channelization guidance.
(Source: Michael Wallwork, unpublished notes)

City blocks have always reflected the transportation mode of the times. Old European cities had blocks built to horse and pedestrian scale. With the advent of mechanized vehicles there has been a tendency for larger block sizes. On the surface, the larger city blocks, called "Super Blocks", are sensible. The outer boundary would be reserved for heavy and fast vehicles. This leaves a large area that could be used for living, recreation, walking, and gardens.

Over use of super blocks has created certain problems for the pedestrian. This concept was used extensively in the "Garden Cities" and "New Cities" of Brasilia, Brazil and Greenbelt, Maryland, USA. These practices constrained the ability of vehicular traffic to use the outer perimeter roadways focused more traffic onto the perimeter roads and resulted in excessively wide streets. These streets pose a problem for the traffic engineer due to the time it takes for vehicles to clear an intersection, and the pedestrian is also faced with multiple lanes of traffic. More connectivity on the highway system can lead to benefits for motorists as well as pedestrians as seen in Figure 10. The lack of a supporting street structure has also been shown to be a detriment to access management (Access Management Committee 2003).

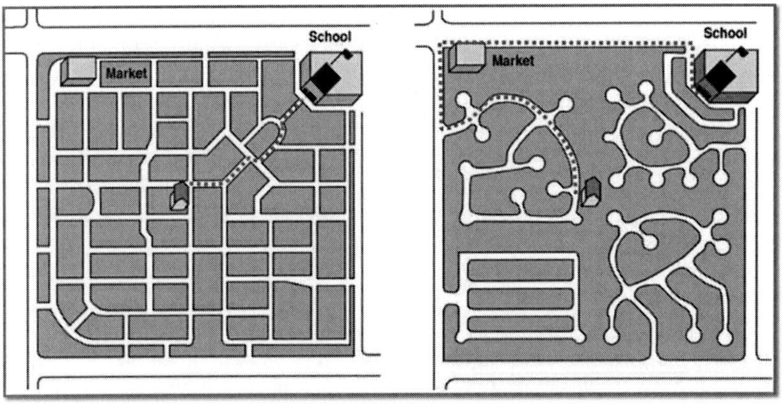

Fig. 10. Comparison of a walking trip from home to school in two distinct networks.
(Source: Kentucky Transportation Cabinet 2009)

National, Cultural, and Institutional Context Matters

Just as the context for access management changes over time, it also changes from location to location. As previously discussed, what is appropriate for a large city may not be appropriate for a small town. In addition, what is appropriate in one nation may not be appropriate in another. A number of considerations must be taken into account, including urbanization of the region, community or nation; available technical resources, availability of professionals to plan and engineer solutions, availability of people and money to help maintain the systems set up to manage access.

One resource for international access management activities is "*Sustainable Interurban Roads for Tomorrow*" produced by the PIARC Technical Committee (PIARC 2009). This publication details the access management activities practiced in India, Iran, South Africa, and China. The report also encourages ongoing training and technical exchanges between nations. One of the lessons contained in the report is to identify local experts within each nation or persons with education in access management. They may have engaged in access management practices with few resources.

CONCLUSION

This paper addresses the need to be sensitive to land use context in highway design and access management. It supports the concept of access management and functional classification as essential aspects of sustainable transportation planning. It further purports that access management is principles are consistent with urban planning goals of achieving urban places that support pedestrian, bicycle and transit activity. The pertinent issue is that roadways serve many functions and functional classification has always been more art than science. Context sensitive solutions are being proposed that offer insight into how to better accommodate context in roadway classification and with that have come ideas on how to manage access in a given context.

Finally, in the effort to serve pedestrians and cyclists, we must not forget that urban places also require major arteries and connectors that handle vehicular traffic efficiently (including autos, buses and trucks). If we do not carefully manage access to these thoroughfares, it will not be possible to accommodate the densities needed to counter sprawl development and to support public transportation. In doing so, the profession will need to keep the principles of access management in mind as documented in the literature (Committee on Access Management 2003) and seek effective ways of advancing these principles in each given context.

ACKNOWLEDGMENTS

The authors wish to acknowledge Vergil Stover for defining the state of the practice in access management and his continuing leadership in guiding how it can be effectively applied in various roadway and land use contexts.

REFERENCES

Abu Dhabi Urban Planning Council (2012), *Urban Street Design Manual*, Abu Dhabi, United Arab Emirates.

American Association of State Highway and Transportation Officials (various ed.s). *A Policy on the Geometric Design of Highways and Streets*. Washington D.C.

Federal Highway Administration. (1997) *Flexibility in Highway Design*. Washington, D.C.

Federal Highway Administration. (1989) *Highway Functional Classification: Concepts, Criteria and Procedures*, Washington, D.C.

Federal Highway Administration (2013). *The Highway Functional Classification: Concepts, Criteria and Procedures*, Washington, D.C.

Institute of Transportation Engineers (2010). *Designing Urban Thoroughfares for Walkable Communities*, Washington, D.C.

Kentucky Transportation Cabinet. (2009) *Street Connectivity, Zoning and Subdivision Model Ordinance*. Frankfort, Kentucky.

Neuman, T. R. (1985). *NCHRP Report 279: Intersection Channelization Design Guide,* Transportation Research Board, Washington, D.C.: National Academy Press

Oregon Department of Transportation. (1999) *Main Street…when a highway runs through it: A Handbook for Oregon Communities*, Salem, Oregon.

PIARC (2009). "*Sustainable Interurban Roads for Tomorrow*" Technical Committee 2.2 Interurban Roads and Integrated Interurban Transport World Road Association. Strader, Brad (2012). "Context based access management across the transect: complete streets," *Proceedings of the Texas Planning and TRB Access Management Conference*, Dallas, Texas.

The Development of Access Management Guidelines for the Western Cape Government, South Africa

M. L. Watters, Pr.Eng[1]; and G. R. Tyndall, Pr.Eng[2]

[1]Department of Transport & Public Works, Western Cape Government, P.O. Box 2603, Cape Town 8000, RSA. E-mail: malcolm.watters@westerncape.gov.za
[2]Kantey & Templer Consulting Engineers, P.O. Box 3132, Cape Town 8000, RSA. E-mail: royt@ct.kanteys.co.za

Abstract

During the mid-nineteen nineties, the need to properly define a safe and consistent approach for dealing with property access applications was identified. To this end, a unique approach was adopted in an attempt to find a balance between the demand for access to encourage development and the need to protect the rights of the wider community for sustainable transportation, more particularly road infrastructure development, while at the same time ensuring adequate mobility in support of accessibility to economic opportunities. This resulted in the production of a document "Road Access Guidelines". This document is still in use. During 2012, it was decided to produce a new guideline on access management, taking into account the development of the science of access management around the world and the changed environment within the region during the ensuing years. This new document serves to set out guidance for the overall management of access for people and goods from the road network to adjacent land uses and to determine the type and overall spacing of such accesses so as to minimize the disruptive effect of access conflicts on the operating performance of all users of the relevant elements of the road network.

INTRODUCTION

Purpose:

The document under discussion has been prepared as the outcome of an extensive review of the original Road Access Guidelines Second Edition (September 2002). The document builds on the original document and it expands its scope so as to move beyond simply providing guidance relating to the determination of the type, location and control processes to new and expanding land uses to the road network.

The document provides guidance to practitioners planning access for particular developments to officials considering the acceptability and approval of proposed access as well as to those planning, reviewing and approving the extension of the road network to serve newly developing areas beyond the boundaries of current development.

During the period that the present Road Access Guidelines have been in use it has played a significant role in promoting and optimizing the ongoing development of housing and economic activity in the metropolitan, urban and rural areas of the province. The rational and consistent methodology adopted by the Road Access Guidelines has contributed to development while minimizing the disruptive effects of access conflicts on all road users.

It has become important to consider road access from a broader perspective. More specifically it includes the operational requirements and impacts of Non-Motorised Traffic (NMT), Public Transport (PT) and Bus Rapid Transit (BRT).

Greater consideration of the road system in a network context has also become critical to the optimization of access to development.

The need was thus identified to undertake a comprehensive review of the Road Access Guidelines in view of many years of active experience, the changed environment and the need for a broader approach. The result of this review has led to the creation of this document, "The Access Management Guidelines", with its more encompassing content and purpose.

Vision, Mission, Goals and Objectives

The **vision** of the Department of Transport and Public Works of the Western Cape Government is -

> To create an open opportunity society for all in the Western Cape so that people can live the lives they value.

The **mission** related to the abovementioned vision is –

> To achieve a road based transport hierarchal network which promotes social and economic development by providing appropriate safe access to adjacent land uses in a way that optimizes the needs of all their users as well as the passing traffic, irrespective of mode.

The **goals** that have been identified are -

- To encourage and facilitate viable land use developments which promote social and economic development.
- To ensure that appropriate access is provided to adjacent developments from the road network that is safe and efficient for all road users.
- To ensure that the needs of all road users, irrespective of mode, are optimized in the determination of the nature, type, location and control mechanisms of accesses from the road network.
- To encourage the use of consistent and sound technical policies by all road authorities in their consideration and approval of access to all elements of the road network.

Leading from the above it is the **objective** of the Western Cape Government to –

- Provide a consistent and technically sound methodology for the determination of accesses from the road network.
- Ensure that all road and planning authorities within the Western Cape Province apply the principles and recommendations contained in the Access Management Guidelines.

DEFINITION AND BENEFITS OF ACCESS MANAGEMENT

The definition of access management used in the Access Management Guidelines is– *"the systematic control of the location, spacing, design, and operation of driveways, median openings, street connections and interchanges to a roadway. It also involves roadway design applications such as median treatments and auxiliary lanes and the appropriate spacing of traffic signals and roundabouts"*.

Access management is particularly important along principal, major and minor arterials and other primary roads that are expected to provide safe and efficient movement of traffic as well as some access to property. Access management's prime requirement on lower order roads is to address safety considerations.

Roads are an important public resource. They are expensive to construct, upgrade and maintain. In a socio-economic environment where there are many competing demands on public funding it is imperative to ensure there is effective and efficient management of the transportation system. It is irresponsible to allow major arterials which are critical for efficient economic activity to become heavily congested and fail to play their intended role effectively.

By managing road access road authorities can prolong the lives of roads, improve public safety, reduce traffic congestion and improve the appearance and quality of the built environment. Not only does proper access management preserve the transportation functions of the road network, it helps preserve the long-term property values and economic viability of abutting developments. From an environmental perspective improved traffic flow translates into better fuel efficiency and lower vehicular emissions.

Ensuring optimal effectiveness and efficiency in the transportation system benefits all stakeholders and users of the system and therefore proper access management provides society as a whole.

ROAD DESIGN LAND USE PRINCIPLES

Introduction

"… as global society swings into action to reduce carbon emissions, the data ever more clearly points to the need to reduce dependence on vehicular mobility and to remake the urban environment as transit- and pedestrian-friendly places of dense economic and social interaction"

\- *Elizabeth Plater-Zyberk (one of the founders of the New Urbanism)*

"in SA such an approach will also be more conducive to promoting economic growth and employment creation as well as urban integration and social cohesion, qualities that are (desperately) needed in South Africa over the next decade."

- *Simon Nicks*

Demands to improve socio-economic development, upgrade the quality of the urban environment and reduce energy consumption have been increasing and are likely to result in significant changes during the next decade, especially in developing nations.

The key relationship between transport and social economic development, the upgrading of the quality of the urban environment and the reduction in energy consumption are likely to result in significant changes during the next decade in the developing world.

The key relationship between transport and social economic development is the land use activity pattern abutting the transport network. The following has to be considered during the development of policies –

- The promotion of economic development;
- Opportunities for economic development that are affordable and enjoy low barriers to entry;
- Employment creation needs;
- The road network itself has a responsibility towards directly facilitating social and economic development.

LEGISLATIVE FRAMEWORK

Roads and associated infrastructure located within the road reserve and used by the public are generally owned and managed by one of the spheres of government, namely local, provincial and national. The responsibilities for the ownership and administration of roads by these entities are set out in the Constitution of the Republic of South Africa.

The National Land Transport Act, 9 of 2009 (NLTA) provides the legislative framework for statutory planning of all transport systems, including transportation policy, strategic plans and projects. The expectation of the NLTA is that before any sphere of government projects are approved for expenditure on transport projects these must be dealt with in a planning process. These projects are listed in approved plans such as the National Land Transport Strategic Framework, Provincial Land Transport Frameworks and Municipal Integrated Transport Plans.

ROAD FUNCTIONAL CLASSIFICATION SYSTEM

In a report titled "Road Infrastructure Strategic Framework for South Africa" (RISFSA), published in 2006, the National Department of Transport set out details of a functional hierarchical classification system for roads in South Africa. This system was later taken further by the Committee of Transport Officials (COTO) and was incorporated and amplified in its publication entitled "TRH 26 South African Road Classification and Access Management Manual", August 2012.

This system has been accepted and applied in the Access Management Guidelines as the definitive functional hierarchical classification system to be applied in assessing the requirements of an access management system for roads in the Western Cape Province.

The functional classification system comprises six classes of road with a differentiation between urban and rural areas. These are shown in Table 1.

Table 1. TRH 26 Road Classification System

Road Class	Function	Description
Class 1	Mobility	Principal Arterial
Class 2		Major Arterial
Class 3		Minor Arterial
Class 4	Access / Activity	Collector Street
Class 5		Local Street
Class 6		Walkway

The functional classification of roads is a process which is carried out by the examination of the entire road network and assessing the function of each element of the network. The function is determined by consideration of the actual road and traffic characteristics of an element and comparing them with the various physical and operational criteria described in TRH 26. It is not to be done on an ad hoc basis or by considering only the elements which fall under the control of a particular road authority.

The outcome of the classification process should be documented and made public in order for all that are involved with the road network and adjacent land development are aware of the classification in place.

ROADSIDE DEVELOPMENT ENVIRONMENT

Background

The Roadside Development Environment (RDE) concept was evolved during the compilation of the Road Access Guidelines (2002).

Five different roadside development environments were identified to recognize that road access guidelines need to vary according to the nature of the urban or rural environment through which a particular road passes.

The prime factor used to distinguish between the different types of roadside development environment was development density. This was seen as an appropriate proxy to reflect the intensity of a number of other associated conditions, for example retail, commercial and residential activities. Table 2 sets out the five Roadside Development Environments and their main distinguishing features. These are based on a bulk factor (floor area ratio). This has been extended to indicate a bulk square metre per hectare.

Table 2. Roadside Development Environments: Density Characteristics

Roadside Development Environment	Floor Area Ratio (FAR)	Bulk (m^2)
CBD	>1	10 000 m^2 / ha
Intermediate	0.3 - 1	3 000 m^2 – 10 000 m^2 / ha
Suburban	0.1 – 0.3	1 000 m^2 – 3 000 m^2 / ha
	Urban Edge	
Semi-rural	<0.1	
Rural	<1 bldg / 10 ha	

Roadside Development Environments as a vehicle to assist transport policy

Transport planning policy and proposals have a fundamental impact on the ability of urban and rural environments to realize the economic and social demands of society and the environment. This responsibility should be borne in mind at all times and policy makers should be mindful of the danger of promoting transport goals as ends in themselves rather than in terms of their larger societal and environmental responsibility.

The use of Roadside Development Environments as overlay zones in Integrated Transport Plans and Spatial Development Frameworks can be a useful tool in this regard as they can provide the same common platform for both transport and land use guidelines, thus helping to achieve their alignment.

The Integrated Development Plan process in which local authorities evaluate and monitor all projects occurring within their boundaries has highlighted contradictions between various sector plans, including the ITPs and SDFs. These plans and frameworks have tended to be prepared in silos with, at best, only superficial cross references to other sectors' plans and, at worst, no reference whatsoever.

Road Development Environments (RDEs) can provide a vehicle for aligning ITP's and SDF's because characteristics of the road the Access Management Guidelines have a real impact on the type of development that can occur on abutting land –

- One of the main challenges facing urban growth management is the coordination of land use and transport planning;
- A transversal interface is required to facilitate the relationship between land use intensity levels, the functional road hierarchy and access management. This interface can be provided by the Roadside Development Environment concept.

Roadside Development Environment principles have the potential to provide the basis for a rigorous, systematic and consistent policy approach to managing the relationship between access management and land use controls.

The application of Roadside Developments Environments can be achieved by using them as overlay zones in Spatial Development Frameworks and precinct plans.

INTERSECTION SPACING

Every intersection or access on the network introduces conflicting movements into the traffic stream. With these conflicts come some level of safety risk as well as potential delay and congestion. Access management seeks to space out such conflict and congestion areas so as to reduce their negative impacts while still allowing adequate access to adjacent properties and other elements of the network whilst retaining an appropriate level of mobility in keeping with the classification of the road.

Intersection controls can generally be one of stop control, traffic signals or roundabouts.

This paper does not discuss any of the technical details pertaining to the development of intersection spacing but typical examples of the results of the studies undertaken are shown in Tables 4 and 5.

Alternatives to signalization

The fact that the installation of traffic signals may be warranted does not mean that signalization is the best or optimum solution to a specific problem.

Alternative solutions may well exist that are viable and feasible and which, if implemented, may obviate the need for traffic signals.

The provision of a traffic circle (roundabout) may not only increase the capacity of a junction but also significantly improve traffic safety.

Table 3. Planning-level Thresholds for Single Lane and Two Lane Roundabouts

Number of Circulatory Lanes	ADT* (Design Year)	% Traffic on Major Road ** (opening & design year)
Single Lane	< 25 000	< 90
Two-lane	< 45 000	< 90

* Based on traffic entering the circulatory roadway for a 4-leg roundabout.
** The volume of traffic entering the roundabout from the major road divided by the total traffic volumes entering the roundabout as a percentage.

Table 4. Minimum Spacing of Signalized Intersections

Development Environment	Road category			
	Class 1	Class 2	Class 3	Class 4
Urban	540 m	370 m	270 m	210 m
Intermediate	800 m	540 m	370 m	270 m
Suburban	1 200 m	800 m	540 m	370 m
Semirural	1 600 m	1 200 m	800 m	540 m
Rural	Values are not included as operational speeds exceed 80 km/h which is an inappropriate speed for signalized control[1]			

[1] Where signals are to be installed in semi-rural or rural environments it is normal practice to post a speed limit not exceeding 80 km/h

Table 5. Spacing Distances for Class 3 Roads

Spacing		Roadside Development Environment Distance (m)		
From	To	CBD	Intermediate	Suburban
Signalised Intersection	Full Unsignalised Intersection	180 m	225 m	260 m
Signalised Intersection	High Volume Driveway	60 m	80 m	Not Permitted
Full Unsignalised Intersection	Full Unsignalised Intersection	180 m	225 m	260 m
Full Unsignalised Intersection	High Volume Driveway	60 m	80 m	Not Permitted
Full Unsignalised Intersection	Signalised Intersection	180 m	225 m	260 m
High Volume Driveway	Full Unsignalised Intersection	180 m	225 m	Not Permitted
High Volume Driveway	High Volume Driveway	60 m	80 m	Not Permitted
High Volume Driveway	Signalised Intersection	180 m	225 m	Not Permitted

PUBLIC TRANSPORT

A public transport orientated road, designed to accommodate and balance the needs of all modes of transport can contribute to the attractiveness and efficiency of public transport as well as add contribute to the livability of the environment and the social well-being of the surrounding community.

Public transport operations and infrastructure impacts on roads in various ways –

- When a general traffic lane is dedicated to public transport it can reduce the overall capacity available to general traffic;
- When public transport services require priority at intersections along an arterial the available green band along the route is reduced for general traffic or signal plans are so interrupted that coordination is not possible;
- Right-turners (for motorists driving on the left of the road) are affected through reduced turning opportunities. The following are options to deal with this challenge –
    - Full signalised intersections can be rationalized;
    - Full unsignalised intersections can be closed or converted to marginal intersections (left-in left-out);
    - Partial intersections can be introduced;

- The following alternative movements can be considered –
  - 3 left turns plus a through movement;
  - Through movement plus a downstream u-turn.

Various studies have been undertaken to identify the elements of the functional area of intersections. The functional area at signalised intersections with particular forms of public transport priority also impacts the functional boundaries of intersections and access management. This is specifically relevant where Bus Rapid Transit is being used as a form of public transport.

NON-MOTORISED TRANSPORT (NMT)

Traditionally non-motorised transport has not been included in transport planning. However, the National Land Transport Act (NLTA) and resulting Regulations has made it statutory to include NMT in the development of Integrated Transport Plans (ITP's) and Public Transport Plans.

It is the objective of transport authorities to enable pedestrians and cyclists to form an important feeder for public transport. The ease and convenience with which NMT users can access and use public transport systems promotes and enhances the usage of both modes of transport simultaneously.

These initiatives have resulted in a significant increase in the provision of NMT facilities such as bicycle lane / paths, pedestrian crossings, sidewalks and facilities to make the built environment more accessible to people with special needs. The NMT-friendly legislative framework and the resulting infrastructure provision are proactive steps in addressing the vulnerability of NMT users on roads.

Various opportunities for conflict exist at intersections, including conflict between NMT users and other modes of transport. NMT-vehicular conflict at intersections can be reduced through the introduction of certain constraints on turning movements.

Accordingly, the Access Management Guidelines takes cognizance of NMT user needs on the road network.

CONCLUSIONS

Due to a constant increase in traffic with resultant congestion road access management has become a necessary tool in the provision of a safe and consistent approach to dealing with property access and the extension of the road network. The increasing usage of other modes of transport within the same environment adds to the challenge. Within the developing world pressing socio-economic considerations need to be taken into account whilst retaining an appropriate level of mobility, the primary goal of many roads.

ACKNOWLEDGMENTS

The authors wish to thank the other members of the task team responsible for the compilation of the Access Management Guidelines. Their input and guidance has been invaluable.

REFERENCES

Committee of Transport Officials (2012). "*TRH 26 South African Road Classification and Access Management Manual*." Version 1.0.

Department of Transport (2006). *"Road Infrastructure Strategic Framework for South Africa"*.

Provincial Administration Western Cape, Department of Economic Affairs, Agriculture & Tourism, Transport Branch (2002). "*Road Access Guidelines*", Second Edition.

Republic of South Africa, (2009) "*National Land Transport Act"*.

Watters, M.L., Abrahamson A.D., Rose C.A. (2012) "*The Evolution of Access Management at the Southern Tip of Africa"*. The Transportation Research Board Annual Meeting, Washington DC.

A Sensitivity Analysis of Crash Modification Factors of Access Management Techniques in Highway Safety Manual

Michael Williamson[1]; Mohammad Jalayer, S.M.ASCE[2]; Huaguo Zhou, Ph.D., P.E.[3]; and Mahdi Pour Rouholamin[4]

[1]Instructor, Department of Civil Engineering, Indiana State University, Terre Haute, IN 47809-1902. E-mail: michael.williamson@indstate.edu
[2]Ph.D. Graduate Student, Department of Civil Engineering, Auburn University, Auburn, AL 36849-5337. E-mail: jalayer@auburn.edu
[3]Associate Professor, Department of Civil Engineering, Auburn University, Auburn, AL 36849-5337. E-mail: hhz0001@auburn.edu
[4]Ph.D. Graduate Student, Department of Civil Engineering, Auburn University, Auburn, AL 36849-5337. E-mail: mahdipn@auburn.edu

Abstract
Access management involves control, planning, and coordination of access along roadways to provide safe and efficient operation of transportation networks. According to existing literature, reducing conflict points and improving traffic safety are the main goals of access management techniques. Increase in traffic volumes and rapid growth of roadway networks have made these efficient techniques as a crucial part of roadway planning and design process. Some of the access management techniques employed in the recently published Highway Safety Manual (HSM) include: median treatments, traffic signal spacing, auxiliary lanes, corner clearances, driveways, U-turns as alternatives to direct left-turns, and frontage roads. Moreover, the Crash Modification Factors (CMFs) of these techniques, as estimators of the actual safety effect of countermeasures, are utilized in the HSM predictive methods to approximate the potential changes in crash frequency and severity after treatment implementation. The results of this study provide transportation agencies, practitioners, and decision-makers with a better understanding of how the HSM methods quantify the safety effects of access management techniques on different roadways. This is accomplished by providing output in the form of tabulated ranking of the impact of access management parameters for three roadway facilities defined in the HSM using the graphical method of sensitivity analysis.

INTRODUCTION

Access management (AM) involves balancing the two competing functions of roadways, mobility and accessibility. The mobility provides the ability to readily move people from one place to another place and the accessibility eases to reach to the destination. The primary positive effects of proper implementation of AM techniques includes: increasing roadway capacity, decreasing travel time and congestion, improving access to properties, reducing work zones due to lack of need for new roadways, and widening projects (FHWA 2012; Jones et al. 2014). These techniques have been utilized for many years in urban, suburban, and rural areas to improve roadway safety and reduce congestion. The first edition of the Highway Safety Manual (HSM) includes AM techniques for three facility types consisting of rural two-lane, two-way roadways, rural multi-lane highways, and urban/suburban arterials (AASHTO 2010). Each of these techniques for separate facilities are discussed in this paper. The HSM estimates the average crash frequency of a roadway segment or an intersection through a system of equations known as Safety Performance Functions (SPFs). The underlying concept is to calculate crash frequency using SPFs, assuming the facility is built in accordance with the predefined base conditions; however, Crash Modification Factors (CMFs) can then be employed to account for any deviations from the base conditions defined for the SPFs.

For implementing HSM, it is beneficial to understand how HSM methods quantify the safety effects of the AM techniques to select the most appropriate one for study locations. To better comprehend these AM techniques and assist project-level decision making, state or municipal departments of transportations (DOTs) rely on sensitivity analysis. Although numerous studies evaluated the safety and operational effects of AM techniques, few have solely focused on those that have been employed in the HSM and their effect on road-safety performance measures. A comprehensive literature review is utilized to review AM techniques and their CMFs in the HSM and cmfclearinghouse website for the purpose of this paper. Following this, the reliability of these CMFs are analyzed and discussed, as is an analysis of how the sensitivity of these techniques are related to crash frequency predictions among three different facilities.

LITERATURE REVIEW

AM is the control or restriction of where vehicles can enter a roadway, with common techniques including: intersection spacing, driveway density, median openings, interchanges, and street connections. The benefits from these techniques expand to preserving the integrity of a roadway system, improving safety and capacity, extending the functional life of the roadway, preserving public investment in infrastructure, and protecting private investments. Current AM methods consist of corridor planning legislation, permitting, medians, auxiliary lanes, traffic signal spacing, driveway locations, corner clearances, cross and joint access, and frontage road use. Many municipalities have adopted various forms of AM suggested by the Federal Highway Administration (FHWA) to improve safety and to reduce congestion along roadways (Harwood et al. 2000).

HIGHWAY SAFETY MANUAL

Currently, there are seven AM techniques utilized with the HSM methodology (e.g., median treatments, traffic signal spacing, auxiliary lanes, corner clearances, driveways, U-turns as alternatives to direct left-turns, and frontage roads) (FHWA 2014). CMFs for these techniques were developed through different studies using various methods. It is important to understand the different impacts of AM regarding crash occurrence and severity. The following literature review gives insights into the development of CMFs for various AM techniques.

ACCESS MANAGEMENT TECHNIQUES

Before-after with comparison groups, empirical Bayes before-after, full Bayes, cross-sectional, case-control, and Cohort studies are a few of the most frequently used methods in CMFs development (Gross et al. 2010). Several of the less reliable alternative methods for developing CMFs that may be used are meta-analysis, expert panel, and surrogate measure studies. These listed methods are the approved study designs per the FHWA (Gross, Persuad, & Lyon, 2010) in the development of CMFs and are commonly used in the HSM. When data is available, the before-after method with comparison groups becomes applicable as a single treatment is implemented at specific locations. Untreated sites are also needed to account for crash trends that may alter the results. The simplicity of this method makes it manageable to consider time periods and any changes in traffic volumes that may affect crash patterns. This method's fault resonates in the difficulty to account for regression-to-the-mean bias. The empirical Bayes before-after method is suggested for sites that use similar treatments in all the study locations and when quality before-after data are available for both the treatment sites and the reference groups. The empirical Bayes study is dominant when an SPF is employed to account for regression-to-the-mean bias, when there are changes in traffic volume, and when the non-treatment related time trends. The weaknesses of the empirical Bayes method include the study's complexity, the inability to utilize prior studies, the lack of consideration for spatial correlation, and the inability to specify the complexity of the model forms. The full Bayes method is third method that can be used with before-after or cross-section data when a complex model is required and when there is a need to consider the spatial correlation between sites. CMFs are first developed after the full Bayes method is used. A strength to the full Bayes method is that it only requires a small sample size to obtain reliable results. Knowledge from other studies demonstrates the inclusion of spatial correlation and complex models used in this process. The complexity of this model requires a high degree of training, creating a weakness to the full Bayes method. Cross-sectional methods are used when limited before-after data are available. This method requires a large number of reference sites, but only a few treatment sites are needed for reliable results. This method is prevalent in that CMF can be developed by an estimation even when treatment sites are rare, providing a useful approach to predict crashes; however, a weakness can be an inaccurate portrayal of CMFs due to inappropriate functional forms, variable bias, or correlation among variables. An additional method is the case-control studies, which can become established when a particular treatment has disproportionally been distributed between sites. This method is used when a particular crash type or location is targeted for improvement and the likelihood of the treatments' effectiveness is established by using an odds ratio. Two benefits to this method are that it can be used for rare events and it can be used to investigate multiple treatments per site. Some of the weaknesses are that only one outcome per site can be established, crash numbers cannot be differentiated, and causality is not demonstrated. The final method for the purpose of this study is Cohort, which estimates the relative risk, an indicator of the outcome after a given treatment. Strengths in Cohort studies are that they can be applied to rare treatments and its ability to demonstrate causality. Two weaknesses for this model include only analysis of the first crash's time and the requirement for large sample sizes.

One of the most common ways to develop reliable CMFs is with meta-analysis, a method combining previously developed CMFs and taking into consideration the quality of each study. Strengths of meta-analysis include the development of CMFs without data and the combination of studies from various jurisdictions to create a weighted average. The weaknesses are that previous studies have to be identified, a statistical process must be followed, and previous studies must have similar methodology. The accuracy of the meta-analysis is dependent on the past studies used.

CRASH MODIFICATION FACTOR STUDY DESIGN AND RANKING

The crash modification factor clearinghouse offers 96 CMFs for AM (FHWA 2014). The CMFs can be grouped into the following categories: driveway density, median type, cross road spacing, lane type, turning movements, and grade separation. Most CMFs used in the HSM do not show the specific applicability of the research targeted or the reliability of the study methods. Many previous research projects utilize various methodologies and estimations of safety improvements to obtain CMFs. The reliability of the factors are based on study design, sample size, standard error, potential bias, data source, and then they are rated by a scale of 1 to 5. The standard error appears to have the greatest effect on the rating process for the CMFs in the HSM (FHWA 2014). While several CMFs are published on the cmfclearinghouse website, consideration should be given to the reliability of the CMFs. Table 1 provides an example of the different types of CMF available for AM.

Table 1. An Example of CMFs (FHWA 2014).

AM CMFs				
Crash Severity	**CMF**	**Area**	**Star Rating**	**Countermeasure**
All	0.69	Urban	3	Provide Raised Median
All	0.77	Urban	3	Replace TWLTL with Raised Median
All	0.80	Urban	4	Replace Direct Left-turn With Right-turn U-turn
All	0.93	Rural	1	Closure or Complete Relocation of Driveways in Functional Area of Intersections

The simple before-after method receives a ranking between 1 and 3. In the before-after approach, the crash data is recorded for two to three years, both before and after the implementation of a treatment is utilized. In before-after studies, the weather conditions, traffic volumes, and many other factors are not considered, making this study the least reliable of the methods. Comparison group studies are ranked between 2 and 4 as they compare the treated sites before and after the implementation using group control to account for yearly variations in traffic, crashes, weather, and other similar factors. The geography and traffic volume of the control sites must be comparable to the treatment sites, because the reliability of the comparison group method depends on the degree of excellence with the relationship between treated and control sites. A rank between 2 and 4 is also given to the cross-sectional method, which investigates the effects of different roadway characteristics on accident rates using multivariable regression models. The problem with the cross-sectional method is the categorization of the impacts each variable has on the analysis. The interpretation of the results is the significant factor in ranking and will affect the reliability.

The most reliable method is the Empirical Bayes (EB) which statistically predicts the amount of crashes at a specific location for the after period before implementation of a countermeasure; in this method, both historical crash data and SPF are used to predict the amount of crashes. Based on the methodology, which research teams use to conduct their study, a rank between 3 and 5 can be assigned to the EB method (Monsere et al. 2005). Selection of a CMF with a low-star rating expresses the quality of the study developing the CMF, which will most likely not provide the anticipated results when performing a safety analysis of a roadway. In some studies, the standard error can be as high as 0.56 for a CMF of 1.94. CMFs with high-star ratings tend to have small standard errors. For instance, a standard error of 0.02 for a CMF of 1.09 would result in a high-star rating. Higher ratings provide an accurate prediction of the crash frequency after implementation of a countermeasure (FHWA 2014). The CMF Clearinghouse offers many CMFs for parameters related to AM(i.e., with a 10 percent increase in distance between downstream U-turns and driveways, the amount of all crash types can be reduced by 3.3 percent.)

Installing a raised median can reduce all types of crashes of different severity by 39 percent while reducing 3.1 percent of the crashes involving bicycles. Replacing a two-way left-turn lane (TWLTL) with a raised median can reduce various types of crashes such as angle, rear-end, sideswipe, and head-on. The presence of parking lot entrances will increase crashes between vehicles and bicycles by 0.5 percent. Reducing driveway density from 10-24 to less than 10 per mile, from 26-48 to 10-24 per mile and from 48 to 26-48 per mile will reduce the amount of crashes with injuries by 25, 31, and 29 percent, respectively (FHWA 2014). Different CMFs are needed for each roadway type (e.g., if the roadway is changed from rural to urban, then two different CMFs will be needed. Table 2 provides a list of driveway density CMFs that have been approved by the U.S. Department of Transportation for use with the HSM. Driveway density is one of the AM inputs in all of the HSM models regardless of roadway type. As shown, crashes can be reduced from 31 to 25 percent in urban areas by reducing the number of access points. The reduction is not as significant for rural areas due to higher speeds; however a reduction of 21 to 22 percent is expected by reducing the number of access points in rural areas.

Many CMFs are not in the HSM but are listed on the CMF Clearinghouse website. Many of the CMFs listed on the website have a ranking of 3 or less. Table 3 gives a brief example of the reliability and rank of some studied CMFs.

Table 2. Driveway Density CMFs (FHWA 2014).

Driveway Density CMFs				
Crash Severity	**CMF**	**Area**	**Star Rating**	**Countermeasure**
All	0.75	Urban	5	Reduce Driveway from10-24 to Less than 10 per Mile
All	0.69	Urban	5	Reduce Driveway from 26-48 to Less than 10-24 per Mile
All	0.71	Urban	5	Reduce Driveway from 48 to Less than 26-48 per Mile
All	0.79	Rural	3	Reduce Driveways by 10 for 2-lane Segments
All	0.78	Rural	3	Reduce Driveways by 5 for 4-lane Segments

Table 3. Example CMFs and Ranking Related to AM (FHWA 2014).

CMF	Data Range (Years)	Star Rating	Methodology	Sample Size	Countermeasures
e^0.0152 (Y-X)	5	3	Regression, Cross Section	520 Sites	Change Driveway Density from X to Y
e^0.0096 (Y-X)	5	3	Regression, Cross Section	Unknown	Change Driveway Density from X to Y
0.97	7	3	Regression, Cross Section	5607 Sites	Presence of Median

METHODOLOGY

Sensitivity analysis methods

North Carolina State University identified sensitivity analysis methods that could be used on prediction models (Frey et al. 2002). Sensitivity analysis is said to develop a comfort level of a model in question. Testing the model will determine if the models' response to a particular change of input is reasonable. Sensitivity analysis is utilized to establish whether a model's behavior to input changes reacts in an acceptable way. North Carolina State University identified three possible ways of determining the sensitivity of a model mathematically, statistically, and graphically. The mathematical method establishes the model's sensitivity to inputs and it does not consider variance in the output due to variance of the input. Statistical methods involve multiple simulations based on probability distributions, but these simulations are not common in sensitivity analysis. Graphical methods provide a visual representation of sensitivity and are displayed in graphical form (Gattis et al. 2007; Willimaosn et al. 2012; Jalayer et al. 2013; Zhou et al. 2013). A study completed for the FHWA on the prediction of expected safety performance of rural two-lane highways was completed in 2000 (Harwood et al. 2000). In that study, a sensitive analysis was conducted testing the significance of each parameter used in the prediction model. Each parameter was altered from base-condition to establish the model's reaction, while other parameters were held constant.

In this study, using the graphical method, sensitivity analysis was conducted on AM techniques within the HSM to see which parameters were the most sensitive in terms of crash frequency. The analysis was performed on various parameters related to AM in the HSM, including driveway density, median types and width, and turn lane presence. To accurately compare each parameter, a scatter plot, one of the most common forms of graphical methods, was created, was shown to represent the inputs' impact had on crash frequency. A scatter plot is employed for various purposes due to its simplicity, visualization, and reasonable accuracy. With the gradient of the line, which passes through the points, it is easy to see how sensitive the output (average crash frequency) is to change the input parameters (e.g. lane width, shoulder width). Notably, the values for input variables were selected in consistence with base-condition, which comes from the HSM. Additionally, in the HSM, two input units were commonly used: the length system (feet) and the numeric system (number of turn lanes present). In doing so, the normalization method was utilized to compare various parameters in terms of crash severity and frequency without any dimensions. The method of normalization utilizes a division of scores in each category by some function of the case value, such as maximum value or sum value. In this study, the former method was considered.

ANALYSIS RESULTS

As outlined in the previous sections, with the HSM there are three predictive methods covering rural two-lane, two-way roadways, rural multi-lane roadways, and urban and suburban arterials. The study purpose was to investigate the impact of AM techniques on safety in the HSM. The techniques include driveway density and turn lane presence for rural two-lane, two-way roadways; median width and turn lane presence for rural multi-lane roadway; driveway type, driveway density, and median width for urban and suburban arterials. The results of the statistical analysis were obtained for frequency of crashes, total fatal crashes and property damage crashes to identify the correlations between safety and AM techniques. The CMFs, which have been approved by the CMF Clearinghouse, do not all have a high-star rating. An investigation of the rating of all CMFs related to AM found that only 7 percent of the CMFs had a five-star rating, and only 3 percent had a four-star rating. It was determined that most of the studies producing CMFs had a three-star rating, which accounted for 76 percent of the CMF on AM. Lower star ratings of 1 and 2 only accounted for 1 and 13 percent, respectively. The lowest star ratings are from studies on driveways and bike crashes, providing medians in rural areas, and converting frontage roads from two-way to one way traffic. The distribution of the star ranking can be seen in Table 4.

Table 4. Star Rating for All AM CMFs (FHWA 2014).

Star Rating				
*	**	***	****	*****

1%	13%	76%	3%	7%

RURAL TWO-LANE, TWO-WAY ROADWAYS

There are only three AM techniques in the HSM for rural two-lane, two-way roadways: driveway density and the presence of left and right-turn lanes. The safety effect of these techniques can be quantified with the CMFs in the HSM. When the sensitivity of driveway density was tested for segments, it was determined that as driveway density increases one unit, the crash frequency increases by 0.119 units. The small increase in crash frequency indicates that driveway density accounts for only a small portion of the overall safety of a roadway. A graph depicting the results can be seen in Fig. 1.

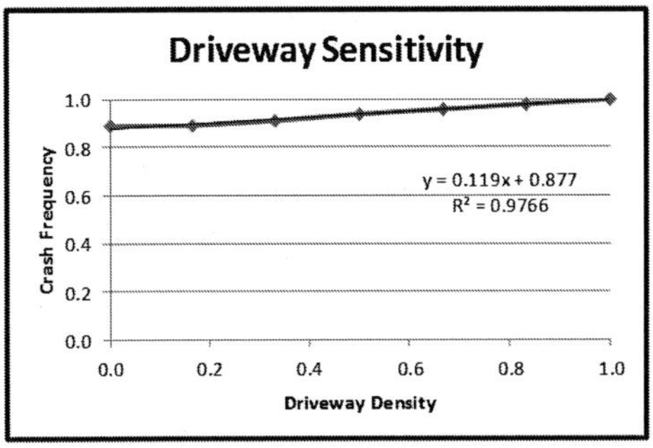

Fig. 1. Driveway Sensitivity for Rural Two-lane Roadways.

When rural intersections were tested, left-turn lane presence was determined to have the largest impact on safety. As the number of left-turn lanes increase, the crash frequency decreases by 0.974 for up to two left-turn lanes; after two left-turn lanes are present, the crash frequency remains constant with the addition of three or four left-turn lanes. Right-turn lanes have less of an effect on safety than that of left-turn lanes. The decrease in crash frequency was determined to be 0.513 with the addition of right-turn lanes from zero to two, with the same pattern holding as left-turn lanes with the presence of more than two. Table 5 displays the ranking in sensitivity for rural two-lane, two-way roadway AM techniques, where one offers the most safety benefit and three offers the least.

Table 5. Sensitivity Ranking for Rural Two-lane Two-way Roadways.

Facility Type	Parameter	Rank
Two-lane two-way Roadway	Left Turn Lane	1
	Right Turn Lane	2
	Driveway Density	3

RURAL MULTI-LANE HIGHWAYS

The HSM offers three AM techniques for rural multi-lane roadways: median width and the presence of right and left-turn lanes. For this type of roadway, the median width varies from 0 to 100 feet. The results demonstrated that the sensitivity of the median width is linear with a decrease in crash frequency per unit increase in width of 0.109. A graph displaying the sensitivity of median width can be seen in Fig. 2.

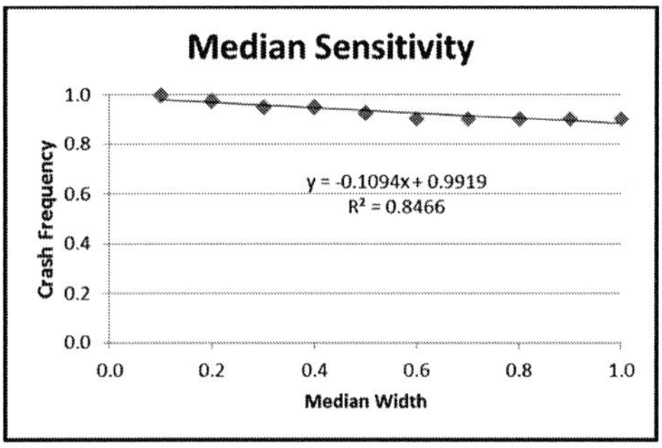

Fig. 2. Median Width Sensitivity for Rural Multi-lane Highways.

Furthermore, there are three configurations for rural multi-lane intersections: three- and four-way stop controlled and four-way signalized. For three-way stop controlled intersections, left turn lane sensitivity proved to be the most sensitive with a 0.920 reduction in the crash frequency with the presence of one left-turn lane; however, the presence of more than one left-turn lane did not further reduce the crash frequency. Right-turn lanes on three-way intersections also provided a reduction in the expected number of crashes. The same pattern was followed in that only one right-turn lane provides a benefit and reduces crashes by 0.360 per unit increase in right-turn lanes.

Four-way stop controlled intersections were also tested to show the presence of left-turn lanes offering an increasing benefit by reducing crashes with the addition of each lane. From zero to two left-turn lanes, the benefit was determined to be a 0.490 reduction in the crash frequency for each unit increase in lane presence. For right-turn lanes, a benefit was only seen for up to two turn lanes. The addition of more than two right-turn lanes did not decrease the number of predicted crashes, and the benefit was determined to be a 0.520 decrease in crash frequency per unit increase in right lane presence. The last intersection tested for rural multi-lane roadways was a four-way signalized. When left and right-turn lanes were tested, no benefit was seen with their presence, using the current CMFs. Table 6 displays the results for rural multi-lane roadway intersections' sensitivity. Similar outcomes are seen in rural two-lane roadways in that turn lane presence has the greatest effect on roadway safety.

Table 6. Sensitivity Ranking for Rural Multi-lane Highway Intersections.

Facility Type	Parameter	Rank
Three-way Stop Controlled	Left Turn Lane	1
	Right Turn Lane	2
	Median Width	3
Four-way Stop Controlled	Right Turn Lane	1

	Left Turn Lane	2
	Median Width	3
Four-way Signalized	Left Turn Lane	N/A
	Right Turn Lane	N/A
	Median Width	N/A

URBAN AND SUBURBAN ARTERIALS

For urban and suburban arterials the HSM addresses three AM techniques: median width, number of driveways, and driveway type. There are five configurations for urban and suburban roadways. For the three lane arterials with TWLTL, major industrial driveways proved to be the most sensitive with a 0.083 reduction in the crash frequency per unit decrease in the industrial driveways. Major commercial driveways were found to have a 0.080 reduction in crash frequency per unit decrease.

For five lane arterials with a TWLTL, major industrial driveways proved to be the most sensitive with a 0.099 reduction in the crash frequency per unit decrease in driveways. Major commercial driveways were second in the sensitivity ranking with a 0.094 reduction in crash frequency with a decrease in one unit of major commercial driveways. Two-lane, undivided arterials were also tested, culminating the results demonstrating that major industrial driveways are the most sensitive parameter with a 0.120 reduction in crash frequency per unit decrease. Four-lane, undivided arterials produced similar results, with a one unit decrease in industrial driveway causing a reduction in crash frequency of 0.120. The last test for urban and suburban arterials was on divided four lanes. Similar to the previous findings, major industrial driveways proved to be the most sensitive with a 0.046 reduction in the crash frequency per unit decrease. An additional technique for urban and suburban roadways was identified to be median width ranging from 0 to 100 feet. The sensitivity of the median width was linear with a decrease in crash frequency of 0.099 as width increased. A graph displaying the sensitivity of median width can be seen in Fig. 3. When urban and suburban intersections were tested, the left-turn lane presence was revealed to have the largest impact on safety. As the number of left-turn lanes increased, the crash frequency decreased for both signalized and un-signalized intersections. The four-leg, un-signalized intersection had the most sensitivity to the presence of left-turn lanes. Table 7 displays the results of the urban and suburban arterial sensitivity rankings. As shown, industrial driveways have the greatest effect on the safety of urban and suburban arterials.

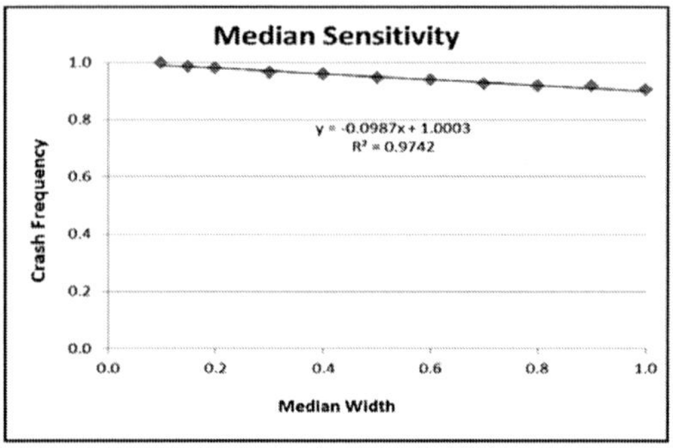

Fig. 3. Median Width Sensitivity for Urban and Suburban Roadways.

Table 7. Sensitivity Ranking for Urban and Suburban Roadways.

Facility Type	Parameter	Rank
Three-lane with TWLTL	Major Industrial Driveway	1
	Major Commercial Driveway	2
	Major Residential Driveway	3
Five-Lane with TWLTL	Major Industrial Driveway	1
	Major Commercial Driveway	2
	Major Residential Driveway	3
Two-lane Undivided	Major Industrial Driveway	1
	Major Commercial Driveway	2
	Major Residential Driveway	3
Four-lane Undivided	Major Industrial Driveway	1
	Major Commercial Driveway	2
	Major Residential Driveway	3
Four-lane Divided	Major Industrial Driveway	1
	Major Commercial Driveway	2
	Major Residential Driveway	3

DISCUSSION

It should be noted that treatments at a particular location were selected by practitioners depending on their assessment of problems. Although turn lanes may be effective for addressing a particular crash problem and closing driveways may be useful in different circumstances, this study attempts to give DOTs and local agencies a better understanding of what factors may have the largest impact on safety; therefore, additional data such as AM can be collected if not available. Most importantly the findings provide a tool to convey AM impacts to the public (e.g., non-engineers).

CONCLUSIONS

While the HSM addresses several AM techniques, there are many that are not included in the current edition for the multitude of facility types such as reducing the number of median crossing and intersections, closing or relocating access points in intersection functional area, and modifying signalized intersection spacing. CMFs for other AM techniques need to be further developed for the next edition of HSM. This study discussed the different AM techniques that are included in the HSM's crash predictive models. A sensitivity analysis using the graphical method was conducted due to its simplicity, multi-purpose, and visualization to identify which parameters are most sensitive in terms of crash frequency. For the purpose of providing a guide to practitioners, this information can be used to quantify the monetary choices regarding AM techniques for safety improvements, which can be easily conveyed to the public. The current AM techniques available for rural two-lanes, rural multi-lanes and urban and suburban arterials were all investigated, with each ranked in order of greatest effect. The sensitivity results demonstrated that for rural two-lane, two-way roadways, left-turn lane, right-turn lane, and driveway density showed more sensitivity, respectively. For rural multi-lane roadways, left-turn lane, right-turn lane, and median width were the most sensitive parameters. For urban and suburban arterials, major industrial driveways, major commercial driveways, and major residential driveways were the most sensitive. It is recommended that the ranking established in this study be used when addressing transportation safety regarding AM to obtain the maximum project benefit of each transportation safety project.

REFERENCES

American Association of State Highway and Transportation Officials (AASHTO) (2010). *Highway Safety Manual*, Washington D.C.

Federal Highway Administration (FHWA) (2012). "Access Management," Transportation Research Board, Washington, D.C. < www.accessmanagement.info> (Jan. 5, 2014).

Federal Highway Administration (FHWA) (2014). "Crash Modification Factors Clearinghouse," < http://www.cmfclearinghouse.org> (Jan. 5, 2014).

Frey H. C., and Patil, S. R. (2002) "Identification and Review of Sensitivity Analysis Methods", North Carolina Department of Transportaion (NCDOT), NC.

Gattis, J. L., Tooley, M. S., Minh, L., Brewer, A. S., and Duncan, L. K. (2007). "Cross-Section Design, Operational, Access, and Safety Relationships on Urban Mulitlane Roadways", *86th Annual Meeting of the Transportation Research Board,* Washington D.C.

Gross, F., Persuad, B., and Lyon, C. (2010). "A Guide to Developing Quality Crash Modification Factors," Federal Highway Administration, FHWA-SA-10-032, Vienna, VA.

Harwood, D.W., Council, F. M., Hauer, E., Hughes W. E., and Vogt, A. (2000). "Prediction of the Expected Safety Performance of Rural Two-Lane Highways," Federal Highway Administration, FHWA-RD-99-207, McLean, VA.

Jalayer, M., and Zhou, H. (2013). "A Sensitivity Analysis of Crash Prediction Models Input in the Highway Safety Manual", *The 2013 ITE Midwest District Conference*, Milwaukee, WI.

Jones, S. L., Meads, J., Skipper, D., Stephenson, J., and Long, C. (2014). "Access Management Manual ," Alabama Department of Transportation (ALDOT), AL.

Monsere, C. M., Bertini, R. L., Houssaye, D., and Bosta, P. (2005). "Improving the Highway Safety Process: An Update and Enhancement to the Oregon DOT's Crash Reduction Factors List", Oregon Department of Transportation (ODOT), OR.

Williamson, M., Jalayer, M., and Zhou, H. (2012). "A Study of Crash Modification Factors of Access Management Techniques in Highway Safety Manual", *The 2012 ITE Midwest District Conference and TRB 4th urban Street Symposium*, Chicago, IL.

Zhou, H., Jalayer, M., Gong, J., Hu, S., and Grinter, M. (2013). "Investigation of Methods and Approaches for Collecting and Recording Highway Inventory Data", Illinois Department of Transportation (IDOT), Springfield, IL.

Accident Analysis Based on Traffic System Stability in Third-Tier City of Developing Country

Lingyu Li[1]; Yuren Chen[2]; and Yaowang Deng[3]

[1]Research Assistant, School of Transportation Engineering, Tongji University, Traffic Office Building A604, 4800 Cao'an Highway, Shanghai 201804, China.
E-mail: liuxinglly@gmail.com
[2]Professor and Ph.D., School of Transportation Engineering, Tongji University, Traffic Office Building A604, 4800 Cao'an Highway, Shanghai 201804, China .
E-mail: chenyr@tongji.edu.cn
[3]Research Assistant, School of Transportation Engineering, Tongji University, Traffic Office Building A604, 4800 Cao'an Highway, Shanghai 201804, China. E-mail: dyw2683908@126.com

Abstract

Traffic safety is one of the principal factors considered by whom accessing the traffic system, such as drivers and pedestrian. Intending to reduce the probability of traffic accidents in the process of access, this study elucidates a new prospective on an accident analysis from a view of considering various accident-related factors as a system, including three major parts as road environment, traffic facilities, and vehicles. The instability of traffic system indicates the accident potential in certain system. By investigating the correlation between accident and relative factors, describing the system instability and comparing different systems, it is accessible to find out the cause of traffic accidents in urban road network of third-tier cities in China. After analyzing accident statistics from 2006 to 2010 in Urumqi, China, six types of urban roadways accidents are selected as a dependent variable, meanwhile, independent variables include road environment, traffic facilities, and vehicle types. By analyzing the impact degree of each factor on single crash type based on binary logistic model, the weight of each factor in the system is decided. Then, a multimodal logistic model is built to describe the accident probability distribution in different

systems with different combinations of factors. Based on the result of binary and multinomial logistic regression, it is approachable to reach an understanding of system instability and analysis the cause of traffic accidents in urban traffic systems in third-tier cities of developing country.
Keywords: Urban roadways; Accident analysis; Traffic system instability; Logistic regression model.

INTRODUCTION

As we all know, one of the most serious problem in the management of traffic access is accidents. In a flawed designing traffic system in which there are improper traffic marks or lines, unreasonably traffic control devices or poor road conditions, traffic efficiency will be largely reduced and people's lives and property are threatened. Under such circumstance, it is necessary to find out the main causes of traffic accidents and decrease accident rate, especially in developing area in China. Actually, the solution of the issue not lies on a single factor but attributes to the whole traffic system, in composition of environment, facilities, and vehicles. The influence of traffic safety cannot be independently taken into account, instead, it should be considered systematically, as a new point of view, which will help us to raise more effective suggestions on improvement of traffic safety in developing country.

Considering instability of system is characterized by accident rate, actually the operation of traffic system is just like the thermodynamic system. Factors within environment, facilities, vehicle and behavior are system elements and instability of system is like the entropy. While the movements of elements become more chaotic, the entropy of system increases, indicating the instable system with high accident rate. In contrast, when most of the factors are reasonably operate, the system becomes consistent and accidents seldom happen.

The study is based on the traffic accident statistic data of Urumqi in China from 2006 to 2010, consisting of three steps. I: Select the factors in the four parts and investigate the impact degree different factors exert upon stability of traffic system, by means of binary logistic model .II: Describe the accident probability distribution based on multinomial logistic model and calculate the indicator of system instability. III: Compared different traffic systems and the cause of instability in urban road of third-tier city.

LITERATURE REVIEW

In recent years, many scholars propose views and methods from different perspectives to analyze correlation between traffic accidents and influential factors. In the first place, some scholars focus on investigating significant accident-relevant factors by clustering method, such as The rough set Theory (Yao et al. 2005), outstanding factor method (Pei et al. 2005), fuzzy diagnosis method (Liu 2009), fuzzy clustering method (Xiao et al. 2002) and grey correlation analysis(Pan et al. 2008), dis-aggregation-aggregation method (Aron et al. 2013), algorithm k-means clustering (Bocarjo et al. 2011) are used to analysis the influential factors and the main cause of traffic accidents. They try to unveil the characteristics of accidents and find out the main reason by illustrating the blackspot and built indicators of cluster center.

As the study got further and diverse, researchers find that the safety relative factors are quite diverse in traffic system. Some complex and efficient regression models are built to mine the statistic data. Such linear models as geographically weight regression (Zhang et al. 2013), ordered and probit models (Jiang et al. 2013) and simultaneous equation model (Wang et al. 2012) and random parameters tobit regression (Anastasopoulos et al. 2012) are applied for diverse studies on traffic accident, including the analysis of non-motorist crashes, crash injury propensity and severity, the relation between speed limit and accidents. At the same time, the popular Bayesian network method is comes into usage (Zhao et al. 2011), accessing the parameter learning to study influential accident factors.

In recent years, a group of scholars propose that a generalized nonlinear model is better and more accurately used to illustrate the correlation between causal factors and accident (Lao et al. 2014). Conditionally, logistic models become popular due to its applicability on the characteristics of accident data. Many scholars apply different types of logistic models in their research and improve the model to a better fitness and accuracy.

Logistic regression is well fit to mining accident data which is nonlinear and sometimes binomial. Time, location, road conditions and environment, vehicle types and drivers are serious factors to cause accidents (Tay et al. 2008; Martensen et al. 2013; Thompson et al. 2013). Additionally, more and more scholars develop the application of various logistic models on accident analysis, such as multinomial logistic model (MNL), partial proportional odds (PPO) logistic model and mixed logistic model (ML). In many research it is approved that logistic models is appliable to analysis the diverse situations of traffic accident, for instance, accidents on non-urban roads, crashes about large trucks, single-vehicle crash and the sever-injury crashes (Nowakowska 2010; Qin et al. 2013; Harb et al. 2008; Ye et al. 2011; Kaplan et al. 2012; Romo et al. 2013; Anastasopoulos et al. 2011). In addition, the conditional logistic model built under Bayesian framework is innovated to investigate effects of microscopic traffic, weathers and roadway geometry information (Yu et al. 2010). This method rapidly becomes popular in accident analysis while researches try to make better adjustment of the models and parameters.

However, reviewing all the literatures above, although scholars are making the models to analyze accidents better fit and effective, most of the studies emphasize on finding out the major factors to traffic safety and their impact degree. Few researchers are capable to analyze the accident from a view of traffic system, which is essential to propose an understanding between traffic system instability and traffic safety.

According to the statistical data and traffic condition in Urumqi of China, apparently, logistic regression model is suitable for the binary data analysis. With an advanced angle of view in this article, the author not only pick out the influential factors in traffic system but also synthetically investigate the stability of system, impacted by different parts as environment, facilities and vehicles, which is applicable to achieve a favorable understanding on urban traffic condition in developing country.

METHODOLOGIES

Binary Logistic Regression model

Each crash type was separately set as the dependent variable Y (value of Y is 0 or 1), which obeys binomial distribution. Y=0 represents no occurrence, and Y=1 represents occurrence. Set (x_1, x_2, …, x_n) as the independent variables, which represent the influential factors, and then logistic regression model could be described as:

$$P(Y=1|X1,X2,,Xn)=\frac{1}{1+e^{\alpha+\sum_{i=1}^{n}\beta_i X_i}} \tag{1}$$

Where α is the intercept, β is the regression coefficient, P is the occurrence probability of this event. $\alpha+\sum_{i=1}^{n}\beta_i X_i$ is defined as the linear function of occurrence probability, and $\frac{P}{1-P}$ is called the Odds Ratio of this event, which means the ratio of probability of occurrence to no occurrence. Calculate the natural logarithm of $\frac{P}{1-P}$, and we will get a linear function as:

$$In\frac{P}{1-P}=\alpha+\sum_{i=1}^{n}\beta_i X_i \tag{2}$$

The Odds $\hat{O}$ means the ratio of probability of expectation occurrence to expectation no occurrence, which could be described calculated as:

$$\hat{O}=\frac{\hat{p}}{1-\hat{p}} \tag{3}$$

Where $\hat{p}$ is the expectation of the event's occurrence probability.

According to equation (3), we know

$$\hat{p}=\frac{\hat{O}}{1+\hat{O}} \tag{4}$$

According to equation (1) and (3), we can calculate the Odds Ratio as:

$$\text{In}\left(\frac{\hat{p}}{1-\hat{p}}\right)=\text{logit}(p)=\text{In}(\widehat{O})=b_0+b_1X_1+b_2X_2+\ldots+b_mX_m \quad (5)$$

Where (b0, b1, b2…bn) is the regression coefficient, which means the variation of logit ($\hat{p}$), when the independent variable Xi changes by 1.

Multi-logistic model

In Multi-logistic model, the dependent variable Y includes several classes. In this paper, there are 9 classes of crash types, set the occurrence probability of each class as (π1, π2, π3,…π9), and they obey $\sum \pi_j = 1$. Set (x_1, x_2, …, x_p) as the independent variables, and Multi-logistic model could be described as:

$$\text{In}\left(\frac{\pi_j}{1-\pi_j}\right) = \alpha_j + \beta_{j1}X_1 + \cdots + \beta_{jk}X_k + \cdots + \beta_{jp}X_p, \quad j=1, \ldots, J-1 \quad (6)$$

Where αj is the constant term and βj is the regression coefficient of this class of dependent variable.

The calculation model based on the predicted data could be described as:

$$\text{In}\left(\frac{\hat{p}_j}{1-\hat{p}_j}\right) = a_j + b_{j1}X_1 + \cdots + b_{jk}X_k + \cdots + b_{jp}X_p, \quad j=1, \ldots, J-1 \quad (7)$$

Where $\hat{p}_j$ is the expectation of occurrence probability of each class.

The predictionprobability of dependent variable can be calculated as:

$$\hat{p}_{ij} = \frac{\exp(a_j+b_{j1}X_1+\cdots+b_{jk}X_k+\cdots+b_{jp}X_p)}{\sum_{h=1}^{J}\exp(a_h+b_{h1}X_{i1}+\cdots+b_{hk}X_{ik}+\cdots+b_{hp}X_{ip})}, \quad i=1, 2, \ldots, m, j=1,2,\ldots,J-1 \quad (8)$$

According to equation (8), we can know$\sum \hat{p}_{ij} = 1$.

DATACOLLECTION

The data used in this study is based on the traffic accident statistic data of Urumqi from 2006 to 2010. According the latest statistics, the total area of Urumqi is 14,216 square kilometers, and the total population is 3,112,559. The per-capita annual income of Urumqi citizens is 21,294 Yuan, and there are 545,000 motor vehicles in Urumqi (Baidu Baike 2013). The accident data includes accident time, location, injuries, crash type, accident cause and some other information. We removed a little scale of accidents that were incompletely recorded, and took the rest 1374 accidents as this study's data basis.

Considering driving behaviors are all in an ordinary level, three parts of factors are selected to compose the traffic system in urban area. Weather, lighting condition and road surface condition attribute to road environment ,road type, traffic signals, marks and guide lines, central and side isolated strips attribute to traffic facilities, motor bus, medium bus, car, large truck, Medium truck, Light van, Motor cycle, Motor tricycle attribute to vehicle type. There are six different crash types are recorded of in Urumqi such as head on crash, sideswipe crash, rolling, rear-end crash, fixture crash and same-direction scratch. These six types of accidents are selected into logistic analysis with observable frequency. TABLE 1 described accident types, variable and their frequency, percentage and code in models.

Table 1.1 Description of Selected Data

Varible	Description	Code	Frequency	Per. (%)
Road Environment				
Weather	Clear/Cloudy(0)	0 0 0	1300	94.61%
	Rainy(1)	1 0 0	47	3.42%
	Snowy(2)	0 1 0	20	1.46%
	Foggy(3)	0 0 1	7	0.51%
Lighting Condition	Day(0)	0 0	684	49.78%
	Night with lighting(1)	1 0	629	45.78%
	N without L(2)	0 1	61	4.44%
	Wet(3)	0 0 1	116	8.44%
Traffic Facility				
Location	Intersection(0)	1 0	26	1.89%
	Elevate(1)	0 1	5	0.36%
	Normal road(2)	0 0	1343	97.74%
Traffic Control	Traffic signal	1	91	6.62%
	No signals	0	128.	93.38%
	Traffic M/L	1	1092	79.48%
	No traffic M/L	0	282	20.52%
Isolated Strip	Central IS	1	322	23.44%
	No central IS	0	1052	76.56%
	Side IS	1	46	3.35%
	No side IS	0	1328	96.65%
Vehicle Type				
	Motor bus(1)	1 0 0 0 0 0 0	120	8.73%
	Medium bus(2)	0 1 0 0 0 0 0	58	4.22%
	Car(3)	0 0 1 0 0 0 0	851	61.94%
	Large truck(4)	0 0 0 1 0 0 0	95	6.91%
	M-truck(5)	0 0 0 0 1 0 0	41	2.98%
	Light van(6)	0 0 0 0 0 1 0	99	7.21%
	Motor tricycle(7)	0 0 0 0 0 0 1	52	3.78%
	Motorcycle(0)	0 0 0 0 0 0 0	58	4.22%

Dependent variable	Code	Frequency	Per. (%)
Crash Type			
Head-on Crash	1	978	71.18%
Sideswipe Crash	2	213	15.50%
Rolling Crash	3	24	1.75%
Rear-end Crash	4	64	4.66%
Fixture Crash	5	12	0.87%
Same-direction Scratch	6	17	1.24%
Others	7	66	4.80%

Results Analysis & Discussion

Results Analysis of Binary Logistic Models

In this section, in order to analyze the impact of each selected factor in the traffic system, it is necessary to achieve a correlation between factors and accident rate base on logistic regression. Thus, seven factors in three aspects in including Environment, Facilities and Vehicle are investigated in six crash models. Increasingly, to make the analysis reasonable and precise, accidentsin Urumqi are divided into six crash types, representing disorder of traffic system. Every crash type model applies a two-step logistic regression in order to make a better goodness of fit and more precise regression result. In the first step, all the factors listed in TABLE 1 are selected as variables in each crash mode. In the second step, factors irrelevant to certain crash model (value of Wals. Test is much greater than 0.1) are eliminated and each logistic model is rebuilt. The final logistic regression results of each crash type are listed in TABLE 2~TABLE 7.

Table 2. Coefficients Estimates of Head-on Crash Model

Variable	Coef.B	Std.Err	Wals.Sig.	Odd Ratio	Tests o Model			
Road Surface Condition					Ominbus Tests of Coefficients			
Dry(0)			0.973			Chi-square	df	Sig.
Snow and ice(1)	-0.096	0.358	0.788	0.908	step	75.383	15	0
Muddy(2)	-21.128	40192.97	1	0	Block	75.383	15	0
Wet(3)	0.085	0.225	0.705	1.089	Model	75.383	15	0
Locations					Hosmer and Lemeshow tests			
Normal Road(0)			0.091			Chi-square	df	Sig.
Intersection(1)	0.528	0.512	0.302	1.696		11.176	6	0.083
Elevate(2)	-2.192	1.14	0.054	0.112				

Traffic				
Facility				
Traffic Signal	0.448	0.266	0.092	1.565
Traffic M/L	0.457	0.157	0.004	1.579
Central IS	-0.428	0.15	0.004	0.652
Vehicle Type				
Motorcycle(0)			0	
Motor bus(1)	-0.519	0.377	0.169	0.595
Medium bus(2)	-0.616	0.426	0.148	0.54
Car(3)	-0.149	0.328	0.649	0.861
Large truck(4)	-1.387	0.382	0	0.25
M-truck(5)	-0.621	0.467	0.184	0.538
Light van(6)	-0.407	0.387	0.293	0.666
Motor tricycle(7)	-0.507	0.439	0.249	0.603
Constant	1.059	0.414	0.011	2.882

Table 3. Coefficients Estimates of Side wipe Crash Model

Variable	Coef.B	Std.Err	Wals.Sig.	Odd Ratio	Tests of Model			
Weather					Ominbus Tests of Coefficients			
Clear/Cloudy(0)			0.23			Chi-square	df	Sig.
Rainy(1)	-1.504	0.731	0.04	0.222	step	24.755	13	0.025
Snowy(2)	-0.2	0.638	0.754	0.819	Block	24.755	13	0.025
Foggy(3)	-0.022	1.087	0.984	0.978	Model	24.755	13	0.025
Lighting Condition					Hosmer and Lemeshow tests			
Day(0)			0.237			Chi-square	df	Sig.
Night with L(1)	0.048	0.156	0.759	1.049		5.011	7	0.659
Night no L(2)	0.781	0.486	0.108	1.458				
Traffic Facility								
Central IS	0.359	0.168	0.033	1.433				
Vehicle Type								
Motorcycle(0)			0.158					
Motor bus(1)	0.175	0.461	0.705	1.191				
Medium bus(2)	0.195	0.519	0.707	1.215				
Car(3)	0.05	0.396	0.9	1.051				
Large truck(4)	0.729	0.454	0.108	2.073				
M-truck(5)	0.355	0.551	0.52	1.426				
Light van(6)	-0.471	0.521	0.366	0.625				
Motor tricycle(7)	0.36	0.522	0.49	1.433				
Constant	-2.667	1.218	0.029	0.069				

Table 4. Coefficients Estimates of Rolling Crash Model

Variable	Coef.B	Std.Err	Wals.Sig.	Odd Ratio	Tests of Model			
Traffic					Ominbus Tests of Coefficients			
Facility								
T-Marks/lines	-1.041	0.432	0.016	0.353		Chi-square	df	Sig.
Vehicle Type					step	21.035	8	0.007
Motorcycle(0)			0.042		Block	21.035	8	0.007
Motor bus(1)	0.446	0.837	0.594	1.563	Model	21.035	8	0.007
Medium bus(2)	-0.566	1.245	0.65	0.568	Hosmer and Lemeshow tests			
Car(3)	-1.351	0.806	0.094	0.259		Chi-square	df	Sig.
Large truck(4)	-0.681	1.021	0.505	0.506		2.603	5	0.761
M-truck(5)	0.617	0.946	0.514	1.853				
Light van(6)	-0.61	1.018	0.549	0.544				
Motor tricycle(7)	-17.782	5.510	0.997	0				
Constant	-2.58	0.772	0.001	0.076				

Table 5. Coefficients Estimates of Rear-end Crash Model

Variable	Coef.B	Std.Err	Wals.Sig.	Odd Ratio	Tests of Model			
Traffic					Ominbus Tests of Coefficients			
Facility								
T-Marks/lines	-0.731	0.299	0.014	0.482		Chi-square	df	Sig.
Central IS	0.47	0.295	0.11	1.601	step	38.557	9	0
Vehicle Type					Block	38.557	9	0
Motorcycle(0)			0.001		Model	38.557	9	0
Motor bus(1)	0.452	1.168	0.699	1.571	Hosmer and Lemeshow tests			
Medium bus(2)	1.991	1.101	0.071	7.324		Chi-square	df	Sig.
Car(3)	0.661	1.028	0.521	1.936		3.079	5	0.688
Large truck(4)	2.064	1.056	0.051	7.877				
M-truck(5)	0.895	1.248	0.473	2.447				
Light van(6)	1.713	1.071	0.11	5.545				
Motor tricycle(7)	0.194	1.43	0.892	1.214				
Constant	-3.633	1.042	0	0.026				

Table 6. Coefficients Estimates of Fixture Crash Model

Variable	Coef.B	Std.Err	Wals.Sig.	Odd Ratio	Tests of Model			
Weather					Ominbus Tests of Coefficients			
Clear/Cloudy(0)			0			Chi-square	df	Sig.
Rainy(1)	1.109	1.092	0.31	3.032	step	25.743	5	0
Snowy(2)	1.467	1.119	0.19	4.335	Block	25.743	5	0
Foggy(3)	4.683	1.025	0	108.095	Model	25.743	5	0
Traffic Facility					Hosmer and Lemeshow tests			
T-Marks/lines	-1.004	0.718	0.162	0.367		Chi-square	df	Sig.
Central IS	1.534	0.709	0.031	4.635		1.817	2	0.403
Constant	-0.44	1.24	0.723	0.644				

Table 7. Coefficients Estimates of Same-direction Crash Model

Variable	Coef.B	Std.Err	Wals.Sig.	Odd Ratio	Tests of Model			
Traffic Facility					Ominbus Tests of Coefficients			
T-Marks/lines	-1.61	0.522	0.002	0.2		Chi-square	df	Sig.
Central IS	-0.278	0.559	0.619	0.757	step	14.878	3	0.002
Side IS	2.019	0.688	0.003	7.533	Block	14.878	3	0.002
Constant	-3.431	0.436	0	0.032	Model	14.878	3	0.002
					Hosmer and Lemeshow tests			
						Chi-square	df	Sig.
						0.328	2	0.849

To begin with the analysis, it is clear that in each crash model in TABLE 2~7 the results of Ominbus Tests of model coefficients remain the same and significant level is beyond 0.05, which indicate each crash model has been set up with at least one variable with favorable statistical significance. At the same time, in the Hosmer and Lemeshowtests, the significance in each test is greater than 0.05, illustrating a prospective goodness of fit. With reference to each crash model table, the odd ratios represent the comparative impact degree between variables in the model. An odd ratio greater than 1 indicates an accelerating effect on accident rate and system instability while it less than 1 indicate that the variable is a protective factor in the operation of system. The value of odd ratio represents the degree of the variable's impact on the accident rate in the system, compared with other factors in the model.

Firstly, with regard to factors of road environment, weather plays a significant role in sideswipe and fixture crash with significance less than 0.05. Especially in fixture crash, odd ratio shows that the foggy weather weighs a lot more than other factors. At the same time, lighting condition markedly impact the sideswipe crash with odd ratio of 1.458.A poor lighting condition such as night without light exerts accelerating effect on accident. Otherwise, the road surface condition doesn't show observably impact on the accident because the odd ratios of each different condition are all approaching 1.

Secondly, according to the models, traffic facility influential factors are all significant in the system. While location is an indicator of where accident happens which is not in the analysis of this part, the traffic signals are one of the most major factors in Head-on crash with the odd ratio of 1.565, indicating that the head-on crash is largely attribute to signals. Furthermore, traffic marks and lines weigh a lot in all such crashes except sideswipe crash. It has great and protective impact degree in rolling crash with odd ratio of 0.353, in rear-end crash with odd ratio of 0.482, in fixture crash with odd ratio of 0.162 and in same-direction scratch with odd ratio of 0.2. In except of its negative influence in sideswipe crash with odd ratio of 1.579, traffic marks and lines are important to serve as a protective factor to keep traffic system stable. Apart from these two factors, central and side isolated strip are also significant factors according to the table. Central isolated strip act as a protective factor in head-on crash and same-direction scratch with odd ratio less than 1 but negative factors in sideswipe, rear-end crash and fixture crash with all odd ratio greater than 1, and most of its influence are larger than other factors. It's a complicated factor and will be deeply analyzed in the next part of this article. At last, the impact of side isolated strip mainly reflect on its influence on same-direction scratch, which weigh much more than comparable factors with the odd ratio of 7.533.

Thirdly, regarding to the vehicle, there are eight types of vehicles as variables in the six models. The vehicle types are significant in the regression. From the result it is obvious that in the comparison of eight types of vehicles, car is most likely to cause a head-on crash. Large truck has greater possibility to cause sideswipe crash. Middle truck is more likely to cause rolling crash. Rear-end crash has greater probability to be caused by large truck. Thus, the type of vehicles running in the traffic system has observable different impact to the system instability.

Influential Factor Analysis in the Traffic System

Base on the logistic regression result of each crash model, influential factors' impact on the traffic system are illustrated by odd ratios. It is necessary to draw a conclusive description of the factors' function in the system in order to get further understanding of system instability.

Considering road environment factors, the most influential factors to the system are weather condition and lighting condition. When it is rainy or foggy, accident rate increases in an extent. At the same time, the drop of light condition also leads to an increase of accident rate. Night with no light is especially easy to result in a disorder of system. The main reason of this attributes to the heavily decline on visibility. While drivers lack a necessary level of vision condition in foggy day or darkness, they are more likely to make mistakes during driving. That's why fixture crash or sideswipe crash is likely to happen in such circumstance. In comparison to weather and lighting, the road surface condition does not weigh much in traffic system. It might be that the poor road surface conditions in Urumqi such as muddy, snow and ice road are not severe, but the commonly slower and more carefully driving behavior in bad road conditions are also the reason why road surface factors are not that significant in statistic. It is an important factor affect the system as well.

With regard to traffic facilities, it is obvious that factors within facilities are influential on the system after scrutinizing the result in TABLE 2~7.In the first place, each factor within facilities has its influence on the system. Traffic marks and lines play an important role to prevent rolling crash, rear-end crash, fixture crash and same-direction scratch. The set of central and side isolated strip mostly help protect the traffic safety. They eliminate the possibility of confliction. But sometimes the set of central isolate leads to accidents. Drivers are more likely to hit the central isolate strip in poor visibility. At the same time, the sideswipe crash is accelerated by central isolated strip. The factors' effect to the system attribute to its placement and function. However, in most situations, accidents are not only attributing to several single factors but also caused by the combining effects of traffic facilities. For example, it is illustrated in the head-on crash model that most head on crash happened at intersections and affected by the disorder of signals, marks and lines. The invalid of traffic control and flawed guidelines cause the misjudgment and unreasonable driving behaviors, which most likely to result in accident.

Vehicles in the traffic system are also influential factors to stability. Different types of vehicles have different characteristics. Ina certain traffic system, if all the cars are replaced to trucks, the stability will change as well. According to TABLE 2~5, the large truck is mostly likely to cause sideswipe crash and rear-end crash. In contrast, there is rarely a light van in such type of accident. The control of large truck in brake, turning direction and evade confliction are much worse than other vehicles. The medium truck is more possible to cause rolling crash, and medium bus is more likely to cause rear-end crash. The result illustrates that compared with other vehicle, the large and medium vehicles are more instable in traffic system.

System Stability Analysis Based on Multinomial Logistic Model

After reaching some results on factors impact on traffic system stability, it is accessible that such influential factors are not separately affecting that system stability but correlate with each other and show a combined effect to the system. In the multinomial crash model, the distribution of accident probability is illustrated under each typical condition of traffic system.

Table8. Prediction of Accident Probability Distribution from Multi-Regression

Value of variable in System 1		Crash probability distribution of System 1					
Environment		**Crash Type**	Frequency			Probability	
Weather	0 (Clear/Cloudy)		Observed	Predicted	Pearson	Observed	Predicted
Lighting	1 (Night with L))	Rolling(3)	1	2.32	-0.871	0.43%	0.99%
Road surface	0 (Dry)	Head-on(1)	182	173.869	1.209	77.45%	73.99%
Traffic Facilities		Sideswipe(2)	29	34.83	-1.07	12.34%	14.82%
Location	2 (Normal Road)	Rear-end(4)	6	8.282	-0.807	2.55%	3.52%
Signal	0 (No)	S-Scratch(6)	0	1.898	-1.383	0.00%	0.81%
Traffic M/L	1 (Yes)	Fixture(5)	1	0.937	0.065	0.43%	0.40%
Central IS	0 (No)	Other(7)	16	12.864	1.345	6.81%	5.48%
Side IS	0 (No)						
Vehicle Type	2 (Medium Bus)						

Value of variable in System 2		Crash probability distribution of System 2					
Environment		**Crash Type**	Frequency			Probability	
Weather	0 (Clear/Cloudy)		Observed	Predicted	Pearson	Observed	Predicted
Lighting	0 (Day)	Rolling(3)	1	3.022	-1.172	0.48%	1.45%
Road surface	0 (Dry)	Head-on(1)	164	156.172	1.255	78.85%	75.08%
Traffic Facilities		Sideswipe(2)	33	29.781	0.637	15.87%	14.32%
Location	2 (Normal Road)	Rear-end(4)	4	7.008	-1.156	1.92%	3.37%
Signal	0 (No)	S-Scratch(6)	0	2.097	-1.455	0.00%	1.01%
Traffic M/L	1 (Yes)	Fixture(5)	0	0.956	-0.98	0.00%	0.46%
Central IS	0 (No)	Other(7)	6	8.964	-1.888	2.88%	4.31%
Side IS	0 (No)						
Vehicle Type	2 (Medium Bus)						

Value of variable in System 3		Crash probability distribution of System 3					
Environment		**Crash Type**	Frequency			Probability	
Weather	0 (Clear/Cloudy)		Observed	Predicted	Pearson	Observed	Predicted
Lighting	0 (Day)	Rolling(3)	4	2.306	1.143	8.33%	4.80%
Road surface	0 (Dry)	Head-on(1)	34	33.573	0.134	70.83%	69.94%
Traffic Facilities		Sideswipe(2)	7	7.035	-0.014	14.58%	14.66%
Location	2 (Normal Road)	Rear-end(4)	2	1.556	0.362	4.17%	3.24%
Signal	0 (No)	S-Scratch(6)	0	0.357	-0.6	0.00%	0.74%
Traffic M/L	1 (Yes)	Fixture(5)	0	0.199	-0.448	0.00%	0.42%

Central IS	0 (No)	Other(7)	1	2.972	-2.197	2.08%	6.19%
Side IS	0 (No)						
Vehicle Type	0 (Motorcycle)						

Value of variable in System 4		**Crash probability distribution of System 4**					
Road Environment		**Crash Type**	Frequency			Probability	
Weather	0 (Clear/Cloudy)		Observed	Predicted	Pearson	Observed	Predicted
Lighting	0 (Day)	Rolling(3)	0	1.016	-1.023	0.00%	2.80%
Road surface	0 (Dry)	Head-on(1)	27	23.276	1.298	75.00%	64.70%
Traffic Facilities		Sideswipe(2)	5	6.173	-0.519	13.90%	17.10%
Location	2 (Normal Road)	Rear-end(4)	2	2.195	-0.136	5.60%	6.10%
Signal	0 (No)	S-Scratch(6)	1	0.885	0.124	2.80%	2.50%
Traffic M/L	0 (No)	Fixture(5)	0	0.801	-0.905	0.00%	2.20%
Central IS	1 (Yes)	Other(7)	1	1.655	-0.976	2.80%	4.60%
Side IS	0 (No)						
Vehicle Type	2 (Medium Bus)						

From all the multinomial logistic regression result, several results with representativeness are selected to analyze and listed in TABLE 8. The data in TABLE8 describes the predicted probabilities distribution of several typical combinations of factors in Urumqi's traffic system. From the data it is illustrated that the absolute value of Parson residual are all beyond 2, most of which are less than 1, indicting a favorable goodness of fit. Obviously, the predicted probability distribution is in accordance with the observed probability distribution. In regard to probability distribution of crash types, head on crash (crash type 1) has highest probability surpasses 60%. The second highest is sideswipe crash (crash type 2) surpassing 10% while the third one (crash type 3, rear-end crash) has got it over 3%. Other group of crash types takes predicted probabilities mostly between 0% and 2%.

Furthermore, with regard to the inside of four systems, such different system instability and accident probability distribution attribute to different combinations of factors in the system. Firstly, in these four systems the environments are almost the same with favorable weather and lighting conditions. The visibility is fine to drivers. There are also no bad road surface conditions in such spots. Thus, the effect of environment is stable and positive to traffic safety. Secondly, however, under the condition that the positions are all in normal road, there is no interference of traffic signal, the factors of traffic mark& lines, isolated strips and vehicles become the major cause of the difference of system instability. As we can see, the central isolated strip is a very important factor in traffic system. The system with traffic marks and lines working together with central isolated strips takes much less possibility to cause accident. In the first two systems there are large numbers of head-on crash, which mostly happened because of no central isolated strips. Without isolated strips, there will be more head-on confliction potential which will easily lead to crash. In System No.3, with the central isolated strip the accident potential of head-on crash drops by about 10%. Even in System No.4, without the traffic marks and lines, the system is more stable than others just with central isolated strips. However, in System No.4, it is obvious that the rear-end crash potential is almost twice higher than other systems due to there are no traffic marks and lines. Accordingly, the key factors to maintain a stable system is largely depend on whether there are favorable settlement of traffic marks & lines and central isolated strips. Otherwise, with regard to vehicles, the medium bus has the most frequency as an accident vehicle. In the traffic system in Urumqi, a medium bus with faster speed than large trucks and buses but worse controllability than cars and motorcycles becomes an instable factor.

In brief, as an example of the four systems in TABLE 8 a poor system with no traffic marks & lines, and no central isolated strip, but with medium buses is the most instable traffic system with high accident potential. In contrast, a stable system which has a high level of traffic safety requires the reasonable set of marks & lines and isolated strips, and vehicles with limited speed and favorable controllability.

CONCLUSION

In most of the three-class cities like Urumqi, the access of traffic system presents disorder and instability under the condition of large number of vehicles and great amount of traffic demand, which also affect traffic safety. By means of analyze the impact degree of each factors in traffic system and investigate the main factors causes the system instability, it is accessible to research an understanding on a solutions to the safety of traffic system accessing. Eventually, the result in this thesis will beneficial to the improvement and development of traffic system construction in third-tier cities in China.

Firstly, six crash types and several factors with observable frequency are selected into investigation. The factors attribute to road environment, traffic facilities and vehicles, which are regard as major influence on urban traffic safety. According to each crash type, a binary logistic model is built to analyze the impact degree of factors in the system and their correlation with each crash. In the investigation it is accessible to take a view that: (I) Traffic signal, traffic marks and guide lines, central isolated strips are the three major influential factors to the head on crash. Especially, the central isolated strips are mostly responsible for the head on crash. (II) The sideswipe crash is largely affected by the central isolated strip. In addition, such environment factors as weather and lighting condition are also related to the sideswipe crash rate. (III) The main reason lead to rolling crash is the defect of traffic planning on marks and guide lines. This type of crash often happens on medium vehicles, because of the specific running performance on medium trucks or buses.

(IV) Referring to the rear-end crash, traffic marks and guide lines, and vehicle types, play an important role on the crash rate. Without a clear guidance of traffic guide lines and marks, it is more likely to happen a rear-end crash when the traffic volume exploding in peak hours. (V) Some crash type rate is determined on road environment, especially fixture crash, influenced by a narrow range of visibility in bad weather condition and mainly happens on a road with central isolated strip. (VI) With reference to same-direction scratch, such traffic facilities as marks, guide lines and side isolated strip are key factors.

Secondly, a multinomial logistic model is built to describe the accident probability distribution in different traffic system, indicating the instability of each traffic system. All the factors and crash type are put into the model. There are four typical traffic system in Urumqi are selected into investigation. According to the comparison between each system, it is approachable to conclude that: (I) Road environment, including weather, lighting and road surface has no significant influence on the urban traffic safety in most of the roadways because road environment are favorable to drivers in most conditions. (II) In major accident spots on normal roadways, the traffic system is flawed with the combination of traffic marks & lines, isolated strips especially and vehicles in the operation. The three factors are the most influential factors leading to a disorder in traffic system. Thus, the key solution to urban traffic safety lies on the settlement of traffic marks, lines and isolated strips in some important spot and restriction on medium buses.

In conclusion, it is applicable to analyze the accident causes with a view of system stability. The factors in the traffic system are not singly taking effect on traffic safety but correlated with each other. Consequently, using the analyzing methodon system instability to measure the degree of safety is far more effective than normal methods. However, there are still some flaws in the process of investigate: (I) Lack depth and fineness of statistical data because of the flawed record in the accident record system in China. (II) The investigation on multinomial logistic regress is not comprehensive enough to describe the whole traffic system situation in Urumqi. (III) Because of the flaw in data about the factors such facilities and vehicles, the thesis failed to reach some comprehensive conclusion on the microscopic causes of each crash and how to improve the situation by changing specific facility. Since the major flaws in this thesis attribute to the statistical data, the train of thought and method are applicable and deserve to be furthermore developed.

ACKNOWLEDGEMENT

The authors acknowledge the support from National Science Foundation of China (51078270, 51238008) and Ministry of Education of China (20120072110019).

REFERENCE

Anastasopoulos, P. C., Mannering, F. L., Shankar, V. N. (2012) “A Study Of Factors

Affecting Highway Accident Rates Using The Random-Parameters Tobit Model." *Accident Analysis & Prevention*, Vol. 45, March, 628-633.

Anastasopoulos, P. C., Mannering, F. L., Tarko, A. P. (2011) "A Note on Vehicle Crash-injury Severity Analysis with the Mixed Logit Model." *Transportation Research Board 90th Annual Meeting,* Washington, D.C.

Aron, M., Seidowsky, R., Nicolas, D. (2013) "Traffic indicators and accidents: the case of a motorway network in the South of France." *Transportation Research Board 92nd Annual Meeting*, Washington, D.C.

Baidu Baike. Introduction of Urumqi. http://baike.baidu.com/view/2154.htm.Undated

Bocarejo, J. P., Diaz, C. (2011) "Characterization of Fatal road traffic accidents using k-means clustering – A case study of Bogotá." *Transportation Research Board 90th Annual Meeting,* Washington, D.C.

Harb, R., Radwan, E., Yan, X. D. (2008) "Freeway Work-zone Crash Analysis and Risk Identification Using Multiple and Conditional Logistic Regression."*Journal of Transportation Engineering-ASCE*.2008.134-5.203-214

Jiang, X., Huang, B., Zaretzki, R. L., Richards, S., Yan, X., Zhang, H. (2013) "Investigating the influence of curbs on single-vehicle crash injury severity utilizing zero-inflated ordered probit models."*Accident analysis and prevention*.2013, 57:55-66.

Kaplan, S., Prato C. G. (2012) "Associating Crash Avoidance Maneuvers with Driver Attributes and Accident Characteristics: A Mixed Logit Model Approach." *Transportation Research Board 91st Annual Meeting,* Washington, D.C.

Lao, Y., Zhang, G., Wang, Y., Milton, J. (2014) *Generalized nonlinear models for rear-end crash risk analysis. Accident; analysis and prevention*.2014, 62:9-16.

Liu, W. (2009) "Fuzzy Diagnosis Method of Dangerous Factors in Road Accident-Prone Section."*Journal of Chongqing Jiaotong University (Natural Science)*, Vol. 28, No. 4, 733-735.

Martensen, H., Dupont, E. (2013) "Comparing single vehicle and multivehicle fatal road crashes: a joint analysis of road conditions, time variables and driver characteristics." *Accident analysis and prevention.* 2013, 60:466-71.

Nowakowska, M. (2010) "*Logistic Models in the Crash Severity Classification on the Basis of Chosen Road Characteristics.*" DVD Compendium. Transportation Research Board 89th Annual Meeting, Washington, D.C.

Pan Z. J., Guo X. C., Sheng Y. G. (2008) "Cause Analysis on the Formation of the Highway Traffic Black-spot with the Gray Association Degree Method." *Journal of Transportation Engineering and Information,* Vol. 6, No. 3, 96-101.

Pei Y. L., Ding J. M. (2005) "Outstanding Factor Method to Black Spot Differentiation." *China Journal of Highway and Transport*, Vol.18, No.3, 99-103.

Qin, X., Wang, K., Cutler, C. E. (2013) "Modeling Large Truck Safety Using Logistic Regression Models." *Transportation Research Board 92nd Annual Meeting*, Washington, D.C.

Romo, A., Hernandez, S., Cheu, R. L. (2013) "Identifying Pre-Crash Factors between Cars and Trucks on Interstate Highways: Mixed Logit Model Approach." *Transportation Research Board 92nd Annual Meeting,* Washington, D.C.

Tay, R., Rifaat, S.M., Chin, H.C. (2008) "A logistic model of the effects of roadway, environmental, vehicle, crash and driver characteristics on hit-and-run crashes."*Accident analysis and prevention.* 2008:40(4):1330-6.

Thompson, J. P., Baldock, M. R., Mathias J. L., Wundersitz L N. (2013) "An examination of the environmental, driver and vehicle factors associated with the serious and fatal crashes of older rural drivers." *Accident analysis and prevention.* 2013, 50:768-75.

Wang, J., Bryden, G., Cheng, W. (2012) "An Examination of the Endogeneity of Speed Limits and Accident Counts in Crash Models."*Transportation Research Board 91st Annual Meeting,* Washington, D.C.

Xiao, S., Guo, X.C., Xu J. D. (2002) "Application Fuzzy Clustering Method in the Highway Traffic Black-Spot Cause of Formation Analysis."*Journal of Transportation Systems Engineering and Information Technology,* Vol. 2, No. 3, 40-43.

Yao, Z. S., Shao, C. F., Long D. L. (2005) "Cause Analysis of Traffic Accidents on Road Block Spots Based on Rough Set Theory."*China Safety Science Journal*, Vol. 15, No. 12, 107-110.

Ye, F., Lord, D. (2011) "Investigation of Effects of Underreporting Crash Data on Three Commonly Used Traffic Crash Severity Models."*Transportation Research Record: Journal of the Transportation Research Board, No.1930, Transportation Research Board of the National Academies,* Washington, D.C., 51-58

Yu, R. J., Abdel-Aty, M., Ahmed, M. (2010) "Crash Type Propensity Analysis with Bayesian Models using Microscopic Traffic and Weather Data." *Transportation Research Board 89th Annual Meeting,* Washington, D.C.

Zhang, Y., Bigham, J., Li, Z. (2013) "Associations between Road Network Structure and Pedestrian-Bicyclist Accidents."*Transportation Research Board 92nd Annual Meeting,* Washington, D.C.

Zhao, J. B., Deng, W., Wang, J. (2011) "Bayesian Network-Based Urban Road Traffic Accidents Analysis."*Journal of Southeast University (Natural Science Edition)*, Vol.41, No.6, 1300-1366.

Analysis of Moped Rider Violation Behavior Characteristics at Mixed Traffic Intersection

Jiayan Zhang[1] and Lanfang Zhang[2]

[1]School of Transportation Engineering, Tongji University, Shanghai 201804, China. E-mail: zhangjiayan0819@163.com
[2]School of Transportation Engineering, Tongji University, Shanghai 201804, China (corresponding author). E-mail: zlf2276@163.com

Abstract
As a kind of environmentally friendly transport tool, mopeds are popular among clerks and students. However, mopeds violation behaviors become a serious threat to the safety of vehicles and pedestrians at mixed traffic intersections. This study investigates the nature of violations and observed behavior and characteristics of moped riders from video data and traffic count data collected. One intersection in Wuxi was selected to obtain the videos of peak hours. Results suggested that the violation rate of moped is 35% exclusive of improper behaviors. Occupying vehicle lane is the most common violation behavior, which has a violation rate of 0.16. According the result and the video, some characteristics of moped riders were summarized, such as group psychology, willing to overtake, and so on. Therefore, enhancement of traffic facilities and traffic management, and traffic safety education are necessary in order to reduce such violation behavior, to improve traffic safety in mixed traffic intersection.

INTRODUCTION

With the constantly growing concerning about the problems of the energy and pollution, such transportation like bicycle and new-energy vehicle has been widely spread. Moped has the advantage of low price, speed faster than bicycle, environment friendly and convenient, which made it a good alternative to vehicle. In shanghai, the number of moped among non-motorized vehicle has reached 40% to 60%. While the improving of the technology of moped and the expanding of the usage, the traffic characteristics and safety problems has attracted large attention at the same time. According to the statistics by police department, in recent years, the number of traffic accident and fatality caused by non-motorized vehicle accounted for half of the number of traffic accident and fatality in Shanghai. And moped riders are mainly to blame for this situation.

Many researchers have devoted themselves to analyze the mixed traffic flow. Assume that there are no isolation facilities between vehicles and non-motorized vehicles on road segments, bicycle will leave their own lanes and ride into vehicle lanes on the condition that vehicle traffic is relatively small and bicycle traffic flow is large (Guan et al 2001). There is a mixed traffic model on city road segment, which is called cellular automata model of friction interference. According to this model, vehicle will decelerate when it was interfered by bicycles. Then, carry out some simulations under different vehicle density and bicycle density. The result is that the traffic capacity of vehicle will be reduced by 20% on average in case of interference and speed will be reduced by 18%. This result is consistent with the actual situation (Wei et al 2010) .

Mixed traffic intersections are also worth studying as accident prone locations. The influence on vehicle will increase with the increase of non-motorized vehicle traffic density, and will tend to be stable when non-motorized vehicle flow reaches a certain level. The interference on vehicle will be different obviously at different intersection (Chen et al 2011). According to the traffic characteristics of vehicle and non-motorized vehicle (such as speed, traffic capacity and delay, etc.), some researchers have established traffic control model of mixed traffic intersection, which meet the benefit of every aspect, such as vehicle, non-motorized vehicle, pedestrians, environment and road (Bai et al 2010).

The driving characteristics of moped and the comprehensive traffic characteristics also received the attention of researchers. The traffic characteristic of moped should be researched from the aspect of vehicle characteristics, rider characteristics, characteristics of traffic flow and so on. It is reported that, moped user are mostly young and middle-aged who have middle income. Moped is suitable for commuter travel and short distance travel (10~20 km). The moped traffic accidents have such characteristics: often happen at rush hour, fatality caused by side impact is increasing rapidly, large amount of moped accident happen in motor-driven driveway. Main cause of such accident and fatality are that moped rider fail to yield vehicle, or some illegal road occupation, reverse driving, driving in violation of traffic signal lamp, speeding and so on (Dong 2008). Some researchers also studied psychology and physiology behavior characteristic of moped rider. Moped rider may drive discretely and most of them fear vehicles. They would like to overtake other bicycles, and also have crowd psychology. So the unsafe behavior of moped can be concluded, such as speeding, illegal turn, reverse driving, run the red light, overtaking and so on (Wang 2007).

METHOD

Choice of location

This study was mainly exploratory with an emphasis on unsafe behavior of moped riders. A selection has been made for a location based upon street view information and on-site location video. Criteria for selection include a signal intersection, mixed traffic, heavy moped traffic, possibility to get video material, etc.

According to the criteria mentioned above, this study chooses a location in Wuxi, an intersection between Qian Rong Road and Qian Hu Road (see Fig.1). One camera was mounted at this intersection to recording the traffic and driving behaviors.

- T-junction between two roads.
- Five lanes for vehicles and one lane for non-motorized vehicle.
- No isolate facilities between vehicles and non-motorized vehicles.

Fig.1Video images location in Wuxi

Fig.1 gives an example of the video images at the location. Because the camera was mounted at a distant place, the process of data extraction has much limitation. For example, the age or gender of moped riders can't be distinguished from the video.

Mopeds

The definition of moped in our country is: "A special bike having electric or electric auxiliary function, which use a battery as a supplementary energy and has two wheels." Table.1 lists some function of the moped meeting the international standard.

Table.1 Characteristic of mopeds

Maximum Speed	≤20km/h
Driving mileage per once charging	≥25km
Power consumption per hundreds of kilometers	≤1.2kw·h
The motor power	≤240w
Braking performance at the speed of 20km/h	≤4m （Dry road condition） ≤15m（Wet road condition）

Moped riders

Middle income earners are willing to choose moped as transportation tool for its low price. Survey data suggested that, the age of moped rider in shanghai is relatively dispersed and mainly ranges from 30 to 49, accounts for 62% of all the users (Dong 2008). However, in other areas, mopeds are also popular among youngers (age from 20 to 29). What's more, the proportion of men and women user are about the same and the occupations are more widely distributed, most are clerks, self-employed person, students and so on.

Analysis method

Selection of analysis periods

Two 2-h periods (7:00-9:00h, 16:30-18:30h) of two days of video recordings have been selected for further analysis, so covering the busiest periods. One day is Friday and another is Saturday.

Traffic counts

To get more insight in the exposure and traffic characteristic, traffic counts have been made for mopeds during two 2-h period. From video, human observers counted the number of mopeds in periods of 11 min, and also counted the number of a given type of violation behaviors.

Violation behaviors of moped riders

Some types of violation behavior of non-motorized vehicle have been listed in other study (Wang 2007). Combined the video with characteristic of moped, this study choose the following violation behaviors for further analysis:

- Running the red light. Moped rider passed the stop line and continues driving behavior when the red signal light is on.
- Reverse driving. Moped didn't ride according to normal direction.
- Illegal waiting. When waiting for red light, moped rider didn't obey the stop line but waiting at the crosswalk line or other place.
- Occupying vehicle lane. On road segment, the track of moped leaned to adjacent lane so as to occupy the vehicle lane, which will have effect on vehicles.

Improper behaviors of moped riders

Improper behaviors have some difference with violation behaviors, for traffic regulations do not specify that such behaviors are illegal, but these actions do have influence on traffic safety and may even cause traffic accident. In this paper, the number of moped rider who conduct improper behaviors did not be counted. Just extract some examples of such behaviors and conduct a preliminary analysis.

RESULT

Traffic counts

The video data we got is recorded every ten minutes. For each ten minutes, the result of the traffic counts for the number of mopeds is given in Fig.2 and Fig.3. The maximum number of mopeds/hour appeared to be 1220 in the morning peak (7:00-8:00). It can be seen in figures that the number of mopeds in morning peak is much higher than it in evening peak. One reason may be that there is a tide phenomenon of this part of area; moped riders went to work from one place to another in the morning. Because the moped riders are mostly clerks, the peak hour occurs at about half past seven. The number of mopeds on Saturday is a little lower than that on Friday. It may because the difference of trip purpose, little people need to ride to work at the morning peak hour on weekends.

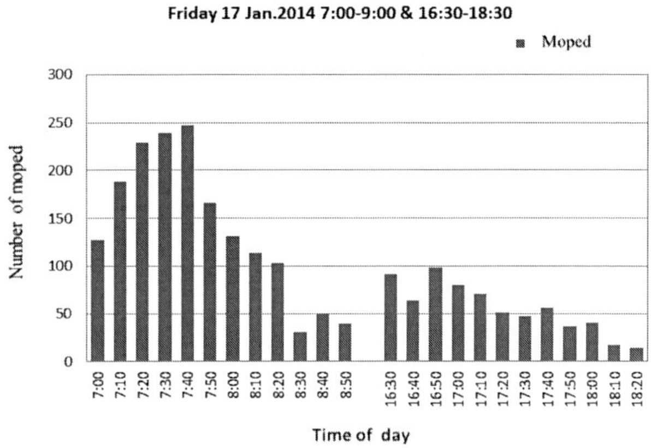

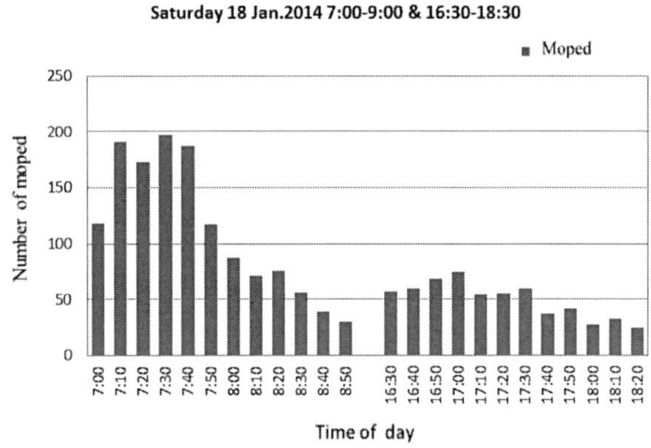

Fig.2 Number of mopes by ten minutes

Violation Behaviors

According to the type of violation behaviors, the numbers of each type of violation behaviors were counted from the video and were shown in Table 2. One thing should be noticed that a moped rider can conduct not only one violation behavior at one time, for example, one may occupied vehicle lane and also ran the red light. Each violation behavior will be counted in this study.

Table.2 Number of different types of violation behaviors

Violation behaviors	Friday	Saturday	TOTAL
Running the red light	23	15	38
Reverse driving	284	241	525
Illegal waiting	98	98	196
Occupying vehicle lane	367	344	711
TOTAL	772	698	1470
Number of mopeds	2609	2171	4255

Fig.3 is a comparison chart of the number of violation behaviors in two days. It suggested that although the number of violation behaviors on Saturday is smaller than that on Friday, the violation rates are a little higher. There may be some effect factors, one is the different traffic control on workdays and weekends, another is the difference of traffic flows and any other reasons.

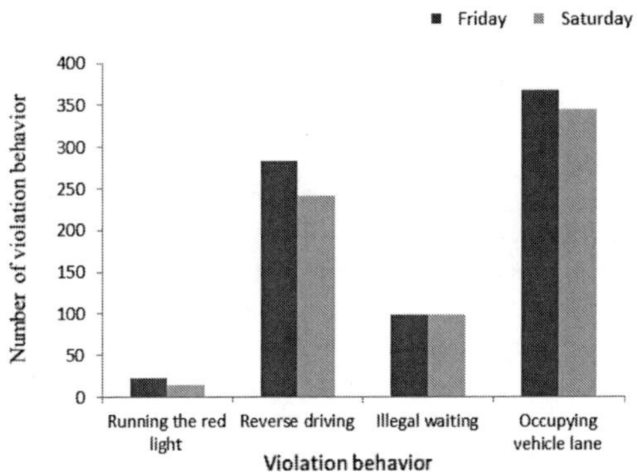

Fig.3 Number of violation behaviors of mopeds

In Fig.3, we can see that occupying vehicle lane is the most common violation behavior, which has violation rate of 0.14and 0.16 for two days respectively. Another common violation behavior is reverse driving, the violation rate for two days are 0.11 and 0.11. Illegal waiting has violation rate of 0.04 and 0.05. The smallest violation rate is running the red light; both are 0.01 for two days. Comparing to other violation behaviors, the number of moped rider who ran the red light is rather small. However, the result it may cause is rather serious, which should arouse the attention of everyone.

Improper behaviors

Expect of such violation behaviors mentioned above, some examples of improper behavior can also be found in video.

The most common improper behavior in video is getting through the intersection casually. There are seven vehicle lanes at this intersection with the width of about twenty six meters which makes it a large intersection. However, problems were caused at the same time. Some moped riders turning left or right along any track they want especially, which will cause many point of conflict. Some moped riders riding on crosswalk line to cross the road but not getting off and push the moped, which may bring great harm to the safety of pedestrians.

Characteristic of moped rider

Some related literatures have studied the influence on the riding behaviors, results suggested that riding behavior is not only affected by moped itself and road environment, but also relate to the experience, knowledge, physiological and psychological function of rider. During the process of watching video, some characteristic of moped rider can be summarized.

Psychological characteristics

(1) Afraid of vehicles

Moped riders fear vehicles, especially large motor vehicles. They are directly affected by the external environment, both in physiological and psychological aspects, and they haven't any security protection. In mixed traffic, motor vehicles and moped get more closer, the psychological pressure of moped rider is larger, so the rider prone to have wrong judgment, which will cause traffic accidents. It can be seen in video that, when a vehicle is approaching or vehicle traffic is large, moped riders are less likely to occupy the vehicle lanes.

(2) Willing to overtake

Regardless of the objective of this travel, time-saving, quick, arrival at the destination safely are the common psychological needs for moped riders. Because of the flexible characteristic of moped, the moped riders like to overtake in order to save time. Sometimes they even dare overtake vehicles, forcing vehicle drivers to slam on the brakes so as to avoid crash. In the video, some examples of such dangerous behavior can be found. What's more, the violation behavior of occupying vehicle lanes is one way to overtake other mopeds.

(3) Group psychology

When counting the number of violation behaviors, it can be found that as long as there is one person conduct such violation behavior, there will be a group of riders follow him, and then more and more. Especially illegal waiting and running the red light, when waiting for the red light, if one rider start early and smoothly crosses through the intersection, the other crowd will follow him to get through the intersection. This phenomenon can not only cause traffic disorder, easy to cause the congestion of vehicles, but also can cause traffic accidents.

Behavior characteristics

(1) Get through gaps

In order to get more convenience, or fear others influence his way, moped riders often ride in the gaps between cars, looking for a shortcut. A lot of examples can be found in video.

(2) Ride at wherever has less traffic

On morning or evening peak hours, there are a large number of non-motorized vehicles on their lane at once. In order not to be affected by others, moped riders often occupy vehicle lanes to keep speed.

(3) Start earlier before the end of red light

Because the flexible characteristic and higher instantaneous startup speed than the vehicle, moped riders often start at the moment that lights just transform from red into green. They may also ride according to the shortest path and moving forward.

DISCUSSION

Nowadays, there is a common consciousness that the moped rider can conduct violation behaviors luckily and escape punishment. Such wrong consciousness was cultivated in the current traffic conditions for a long time. Moped riders should learn the driving methods and rules before they ride on road, constant propaganda and education from media is need to strengthen the cognition of traffic rules. However, lack of public opinion and problems of traffic law enforcement, are the key factor to such serious violation behaviors of moped riders.

Some studies have the conclusion that gender is an effect factor, male riders are more likely to run the red light and have more violation rate. Traffic facilities and environment such as special hard isolation for non-motor vehicle can effectively regulate the moped waiting behavior, and it is also suitable for mixed intersection with large number of non-motor vehicle traffic flow. Law enforcement and management also have effect on the violation rate, study suggested that the violation rate is significantly lower when there are traffic polices (Xu et al 2011).

There are several areas in which related future work would be beneficial. More videos for different locations should be collected, and can compare the violation rate at different place. For instance, the type of intersection, numbers of vehicle lanes, signal phase and the width of lanes. Then we may get more comprehensive result for the characteristics of moped riders and violation behaviors.

CONCLUSIONS

This paper used the method of video to analyze the dangerous condition of an intersection between Qian Hong Road and Qian Hu Road. The number of mopeds and different type of violation behaviors were counted. Then the violation rate was calculated and the characteristic of moped riders was analyzed. In conclusion, there are several findings.

(1) The peak hour of the mopeds is about half past seven in the morning, for the reason of the jobs of moped riders. And there is a significant difference between the number of mopeds in morning peak hour and evening peak hour.

(2) Occupying vehicle lane is the most common violation behavior in this study, with the violation rate of 0.16. The construction of isolation facilities can effectively prevent such behavior.

(3) The propaganda of traffic safety and traffic enforcement should be strengthened to reduce the number of serious violation behaviors such as running the red light. Improper behaviors should also be controlled by traffic police or any other traffic facilities.

REFERENCES

A. Richard A. van der Horsta, Maartje de Goedea, Stefanie de Hair-Buijssen, and Rob Methorstc (2013). "Traffic Conflicts on Bicycle Paths: A Systematic Observation of Behaviour from Video." *Accident Analysis and Prevention.*

Bai, L.T., Bai, H., Bai, Y., and Zhao, L. (2010). "Research on the Method of Traffic Management Control about Motor and Non-motor Vehicles mixed Junction." *Science &Technology Information.*

Cherryya, C., and Cerverob, R. (2007). "Use characteristics and mode choice behavior of electric bike users in China." *Transport Policy*, 14(3): 247—257.

Chen, Z.Q., Sheng, C.Y., Liu, J.F., and He, T. (2011). "Study on the Level and Classification of Mixed Traffics at Urban Road Intersections." *Journal of Transportation Systems Engineering and Information Technology*, 11 (1).

Dong, B.J. (2008) . "The Study of Characteristics of Electric Bicycle." Master Degree Thesis, Tongji University. (in Chinese)

Eleni, I., George, Y., and John.C. Golias (2014). "Detecting powered-two-wheeler incidents from high resolution naturalistic data." *Transportation Research Part F*, 22: 86–95.

Guan, H.Z., Chen, Y.Y., Liu, X.M., and Ren, F.T. (2001). "An Analysis Model for Mixed Traffic Flow on Street." *Journal of Beijing Polytechnic University*, 27(1). (in Chinese)

Guan, H.Z., Wang, M.W. and Chi, H.B. (2005). "Study on Road Traffic Delay Caused by Bicycle for Mixed Traffic Stream." *Journal of Beijing Polytechnic University*, 31(3).

Jia, S., Peng, H., Guo, J., and Chen, H. (2008). "Quantitative Analysis of impact of Bicycles on Vehicles in Urban Mixed Traffic." *Journal of Transportation, Systems Engineering and Information Technology*, 8 (2).

Steg L. Annemarie vail Bmssel (2009) . "Accidents, Aberrant Behaviors, and Speeding of Young Moped Riders." *Transportation Research Part F,* 12(6): 503—511.

Wang, M.L.(2007). "Safety Analysis of Electric Bicycle on Urban Road." Master Degree Thesis, Southwest Jiaotong University. (in Chinese)

Wei, L.Y., and Ying, L.T. (2010). "Frication Interference in Mixed Traffic Based on Cellular Automaton." *Systems Engineering-Theory &Practice*, 30 (10).

Xu. Y.L., Wang, X.S. and Yang, D.Y.(2011). "Analysis of Moped Rider Violation Behavior Characteristics at Urban Signalized Intersections." *China Safety Science Journal*, 21(10). (in Chinese)

Yan, X.P., Ma, M., Huang, H.L., and Wu, C.Z. (2011). "Motor Vehicle–bicycle Crashes in Beijing: Irregular maneuvers, crash patterns, and injury severity." *Accident Analysis and Prevention* , 43: 1751-1758.

Zhang, S.C., He, K.K., and Yang, R.F. (2013). "A Probability Model of Bicycles Crossing Vehicle-bicycle Separation Lines." *Journal of Transportation Systems Engineering and Information Technology*, 13(6).

Evaluation Indexes of Expressway Operating Environment for Safety Design and Operation

Donglan Su[1]; Zhongyin Guo[2]; and Benmin Liu[3]

[1]The Key Laboratory of Road and Traffic Engineering of the Ministry of Education School of Transportation Engineering, Tongji University, No. 4800 Cao'an Road Shanghai 201804, P. R. China. E-mail: 2011sudonglan@tongji.edu.cn
[2]The Key Laboratory of Road and Traffic Engineering of the Ministry of Education, School of Transportation Engineering, Tongji University, No. 4800 Cao'an Road, Shanghai 201804, P. R. China (corresponding author). E-mail: zhongyin@tongji.edu.cn
[3]The Key Laboratory of Road and Traffic Engineering of the Ministry of Education School of Transportation Engineering, Tongji University, No. 4800 Cao'an Road Shanghai 201804, P. R. China. E-mail: liubenming@mail.tongji.edu.cn

Abstract

From the point of expressway operating safety and effectiveness, expressway operating environment is defined here as the system composing road, traffic, climate, and humanistic environment. All traffic activities would be operated inside the environmental space. Therefore, accurate description and reasonable evaluation of operating environment and operation safety seem to be very important. Currently, there is lack of a set of scientific and operable technical indexes to describe and evaluate expressway operating environment for safety design and operation. According to the recent research results, a set of evaluation indexes of expressway operating environment for safety design and operation has been proposed in this paper, where the expressway geometric alignment, road and traffic infrastructures environment, and the traffic flow operation status are considered. This index system could be used as a tool for safety design, safety evaluation, and operation management.

1. INTRODUCTION

The design and operation safety evaluation of expressway operating environment involves road and traffic facilities, traffic flow and other aspects. Many foreign scholars have conducted studies on road alignment safety design. D.Žilionienė analyzed the relationship between the horizontal/ vertical alignment and running safety of low traffic volume roads after re-paved (Žilionienė 2011). Francisco J. presented a design consistency model with V_{85} and R which based on two-lane rural road segments and crash data(Francisco 2012).

However, "Design Specification for Highway Alignment(JTG D20-2006)" has few consideration on traffic safety, and also no the evaluation index. In order to carry out more scientific and effective safety evaluation of alignment, many scholars have conducted relevant researches. Zheng Ke analyzed and evaluated expressway alignment design according to the correlation between the driver's health mental reflection and vehicle speed(Zheng 2003). Wang Xueli established reliability model for expressway alignment design index, analyzed the reliability of some alignment design indexes(Wang 2012)

Guo Zhongyin professor research team has been carried out a series of researches on expressway geometric alignment safety evaluation model and evaluation indexes and proposed the safety evaluation model and index of expressway alignment based on consistency of operating speed and alignment(Guo 2008a). The models has put forward the alignment design evaluation index of tunnel entrance and exit sections based on operating speed (Guo 2012a; Guo 2008b). Expressway operating environment is the foundation of transportation system. Guo Zhongyin professor research team has accumulated some research results through several scientific research projects. A description model has been established to describe and evaluate the characteristics of road division and operating environment attributes (Guo 2011).

Guo Zhongyin group has also carried out researches on traffic flow operation safety evaluation and operation management in event status and proposed collisions avoidance constraints and safety evaluation index during vehicle's lane-changing, established lane volume criterion of lane-changing restriction for deteriorative weather, and proposed rapid assessment of emergency events and traffic organization technology of mountain area highway under severe weather conditions(Guo 2012b).

In summary, the recent domestic and foreign researches on expressway geometric alignment design safety evaluation, operation safety evaluation, and traffic flow status safety evaluation haven't formed widely accepted and operable index system. Through accumulating certain our research outcomes about evaluation of expressway geometric alignment safety, operating environment, operation safety and traffic flow evaluation, , the evaluation indexes of expressway operating environment for safety design and operation in this paper has been developed.

2. Evaluation Index hierarchy of Expressway Operating Environment

2.1 Why Need Establishing Evaluation Indexes of Expressway Operating Environment for Safety Design and Operation

Establishing evaluation indexes of expressway operating environment for safety design and operation are very important work considering the following facts:

(1) Safety evaluation on expressway geometric alignment design is one of effective measures to find out potential safety problems and provide the basis for transformation traffic safety during expressway planning and design.

(2) Evaluation on expressway operating environment is the basis work for the maintenance and sustainable improvement of expressway operating environment.

(3) Evaluation on traffic flow operation status safety is necessary for the decision-making of real time management and control of traffic flow.

2.2 Expressway Safety Evaluation Index Hierarchy

Through reviewing and summarizing Guo team's research results about road traffic safety during the past decade, expressway operating environment for road safety design and operation could be defined as a system composing of road and traffic facilities, climate and humanistic environment and traffic flow which plays a crucial role in the entire people-vehicle-road system. Expressway operating environment system and its sub-components are shown in figure 1.

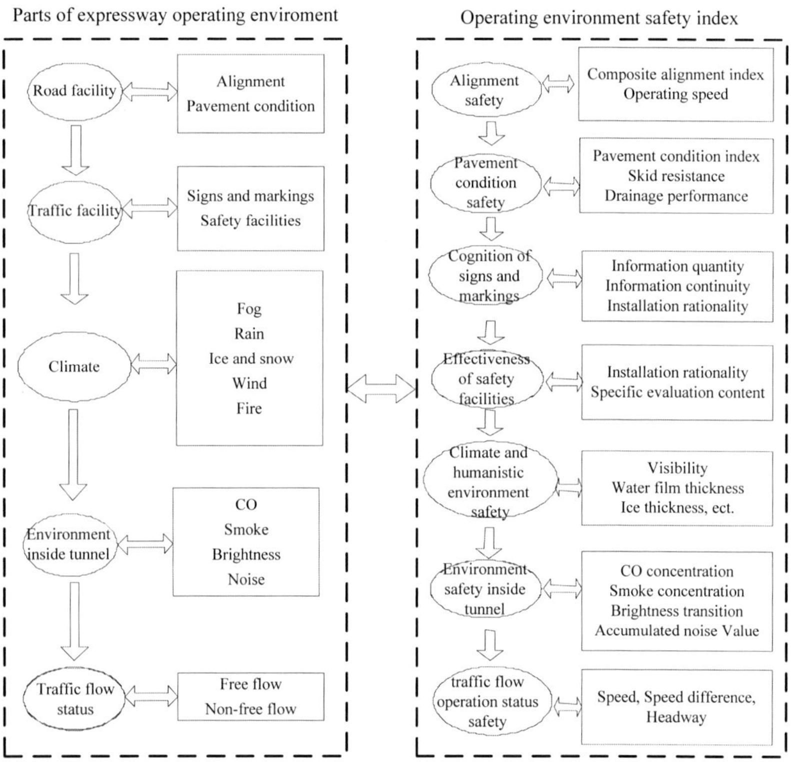

Fig. 1. Analysis Structure of Expressway Operating Environment.

3. SUB-EVALUATION INDEX OF EXPRESSWAY OPERATING ENVIRONMENT

3.1 The Idea of Evaluation Index Hierarchy

The evaluation indexes would be developed based on three aspect of factors related with accident rate, traffic flow and traffic facilities. The evaluation index hierarchy is shown in figure 2.

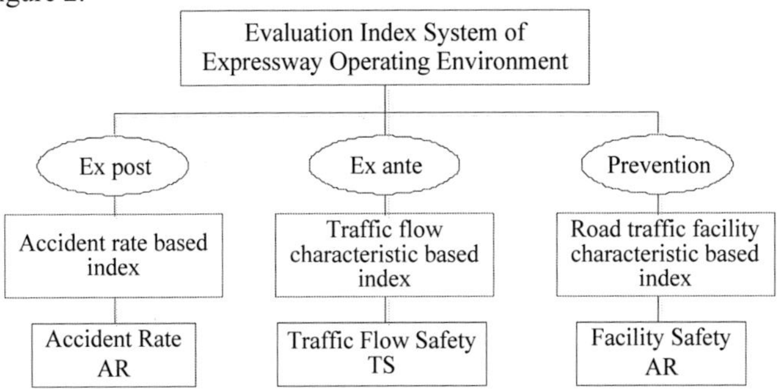

Fig. 2. Evaluation Index Hierarchy of Expressway Operating Environment.

Accident rate is a direct indicator of expressway traffic safety status after certain period time operation, which was belong to ex post index and mostly used for identifying highway accident-prone sections and improving design. Dynamic traffic flow characteristics are parameters reflecting the real-time traffic flow risk that is related with safety status of expressway operating environment. The higher accumulation risk of traffic flow is, the lower expressway operating environment safety performance is. The potential safety hazard of expressway operating environment could be identified before accidents using traffic flow characteristics. So that control and management measures could be improved. However, it's better to carry out reasonable safety design and planning during construction or maintenance phase of expressways. Road and traffic facility characteristic based index is the ex ante index mainly applied for it.

3.2 Accident Rate based Evaluation Index

Accident rate is the most directly index reflecting the expressway environment operating safety. Thus, Evaluation index was established based accident rate and shown in table 1.

Table 1. Accident Rate Level.

Accident Rate Level	Qualitative Description
AR - I	The accident rate is very low and the potential possibility for further reducing accident rate is small.
AR - II	The accident rate is lower than expected level. , Appropriate measures need to be taken to keep the safety level.
AR -III	The accident is higher than expected level and the potential possibility for further reducing accident rate is great and some improvement measures should be implemented
AR -IV	The accident rate is very high and some activities must be implemented for further reducing accident rate.

3.3 Traffic Flow Characteristic based Evaluation Index

Accurate traffic flow operation status safety evaluation can provide the basis for the selection of operation safety management and traffic management measures. The qualitative description of traffic flow safety level was developed based traffic flow risk and shown in table 2.

Table 2. The Qualitative Description of Traffic Flow Safety Level.

Traffic Flow Safety level	Qualitative Description
TS- I	The traffic flow risk is very low and could be acceptable.
TS- II	The traffic flow risk is low and could be acceptable conditionally.
TS-III	The traffic flow risk in is undesirably high.
TS-IV	The traffic flow risk is very high and unacceptable.

Traffic flow safety depends on the following three traffic flow risk, which is freely driving risk, car-following risk, lane-changing risk. These three traffic flow risk are defined and classified as in shown in table 3.

Table 3. Evaluation Criteria of Traffic Flow Safety in Common Road Sections

Safety Grade		Grade I	Grade II	Grade III	Grade IV
Freely driving	Speed(km/h)	80~100	100~110	120~140	>140
	Speed difference $V_j^{max} - V_j^s$(km/h)	—	10~20	0~5	<0
Car-following	Headway $h_{j\_i}^{min} - h_{j\_i}^{s}$ (s)	4~5	3~4	1~2	<1
Lane-changing	Headway of front vehicle $h_{j\_n-1}^{min} - h_{j\_n-1}^{s}$(s)	4~5	3~4	1~2	<1
	Headway of behind vehicle $h_{n+1\_j}^{min} - h_{n+1\_j}^{s}$(s)	4~5	3~4	1~2	<1

*Note: ① V_j^{max} = maximum critical safe speed of the *j-th* vehicle, $V_j^{max} \le \min\{V_{ch}^{yx}, V_{ch}^{qf}, V_{cf}^{yx}, V_{cf}^{qf}\}$; V_j^s = actual speed of the *j-th* vehicle. Safe speed V^s is selected from the smaller one between critical safe speed $V_{critical}$ and the limited speed, there exists $V_{safe} \le \min\{V_{critical}, V_{limited}\}$.

② Car-following risk: Headway= $h_{j\_i}^{min} - h_{j\_i}^{s}$ Δh(s).

$h_{j\_i}^{min}$ s = critical minimum headway between behind vehicle *j-th* and the front vehicle *i-th*. $h_{j\_i}^{s}$ = safety headway between behind vehicle *j-th* and the front vehicle *i-th*.

③ Lane-changing risk: There are two kinds of headway: $h_{n\_n-1}^{min}$ =headway between the *n-th* vehicle and the *(n-1)-th* vehicle, and $h_{n+1\_n}^{min}$ =headway between the *n-th* vehicle and the *(n+1)-th* vehicle.

3.4 Road and Traffic Facility Characteristic based Evaluation Indexes

Road and traffic facility characteristics denote alignment, pavement surface, traffic facilities and related climate conditions. The relative sub evaluation indexes are introduced as follows.

3.4.1 Operating Speed based Alignment Evaluation Index

The expressway geometric alignment characteristics are fundamental factors of expressway and the effects on traffic safety could be indicated using operating speed characteristics. So in order to develop alignment evaluation index, First of all, the operating speed characteristics were set up. Then, the relation model between operating speed characteristics and accident rate was investigated.

The operating speed characteristics and relative models have been developed through the research projects such as Highway Alignment Quality Evaluating Model Based on Design Consistency, Research on Highway Safety Improvement by Application of Operating Speed Theory, etc. The operating speed statistic characteristics such as variation coefficient of speed, speed difference between truck and passenger car, speed dispersion degree, road sectional speed difference and consistent evaluation criteria of expressway alignment have been studied and the definition are shown in table 4.

Table 4. Operating Speed Characteristics.

Operating Speed Characteristics	Calculation Equations
Variation coefficient of speed C_V	$C_V = \frac{\sigma}{\mu}$
	C_V represents dispersion degree of speed, σ -- standard deviation of speed; μ -- the average speed, km/h.
Speed difference between truck and passenger car mb	$mb = \frac{\max\limits_{1\le i\le n}(m_i) - \min\limits_{1\le i\le n}(m_i)}{\overline{m}}$
	mb —range ratio of speed difference between truck and passenger car; m_i —speed difference between truck and passenger car, km/h; $\overline{m}$ —the average value of speed difference between truck and passenger car, km/h.
Speed dispersion degree S	$S = \frac{\sum_{i=1}^{n} \int_{l_i}^{l_{i+1}} \Delta V_i dl}{L}$
	ΔV_i —the difference between 85th percentile speed of i geometric element and the average speed, which means velocity gradient of per unit length(km/h); l_i —starting point stake number of i geometric element(m); l_{i+1} —end point stake number of $i+1$ geometric element(m); L —mileage length(m); n—the total numbers of geometric elements within mileage L.

Speed difference of cross-section ΔV_{85}	In driving into curve direction: $\Delta V_{85} = V_{85(\text{middle curve})} - V_{85(\text{point of tangent to spiral})}$ In driving away from curve direction: $\Delta V_{85} = V_{85(\text{point of tangent to spiral})} - V_{85(\text{middle curve})}$
Speed reduction coefficient SRC	In driving into curve direction: $SRC = \frac{V_{85(\text{middle curve})}}{V_{85(\text{point of tangent to spiral})}}$ In driving away from curve direction: $SRC = \frac{V_{85(\text{point of tangent to spiral})}}{V_{85(\text{middle curve})}}$

The relation between accident rate and operating speed have been investigated through surveying several expressways (Shendan Expressway, Taijiu Expressway, Heda Expressway, etc.). Regression analysis about the relationship between operating speed parameters t and accident rate has been made and five relationship models have been obtained as being shown in table 5.

Table 5. Relationship Between Operating Speed and Accident Rate.

Operating speed characteristics	Relationship Model
Variation coefficient of speed C_V	$I = e^{0.216-0.027/C_V}$
Speed difference between truck and passenger car mb	$I = 2.426mb^2 - 0.151mb + 0.175$
Speed dispersion degree S	$I = e^{(-0.081-0.331/S)}$
Speed difference of cross-section ΔV_{85}	$I = \log\left(0.014\Delta V_{85}{}^2 - 0.016\Delta V_{85} + 2.87\right)$
Speed reduction coefficient SRC	$I = \ln(54.95SRC^2 - 109.56SRC + 56.009)$

*note: I = accident rate (times/mvk)

Combining with expressway safety situation, construction and maintenance cost, evaluation criteria of alignment design consistency based on operating speed has been proposed evaluating alignment consistency as being shown in table 6.

Table 6. Operating Speed based Evaluation Criteria of Alignment Design Consistency.

Safety Level / Speed Index	Good	Adequate	Poor
C_V	$C_V \leq 0.031$	$0.031 < C_V \leq 0.046$	$C_V > 0.046$
mb	$mb \leq 0.46$	$0.46 < mb \leq 0.54$	$mb > 0.54$
S	$S \leq 0.95$	$0.95 < S \leq 2.34$	$S > 2.34$
ΔV_{85}(km/h)	$10 \leq \Delta V_{85} \leq 11$	$-15 \leq \Delta V_{85} < -10$ $11 < \Delta V_{85} \leq 16$	$\Delta V_{85} < -15$ $\Delta V_{85} > 16$
SRC	$0.90 \leq SRC \leq 1.095$	$0.87 \leq SRC < 0.90$ $1.095 < SRC \leq 1.12$	$SRC < 0.87$ $SRC > 1.12$

3.4.2 Other Facilities

(1) Safety index of pavement condition

Pavement conditions can be evaluated with Pavement Condition Index PCI, Sideway Force Coefficient SFC and Skid-resistance transitional index in tunnel sections.

(2) Effective index of traffic safety facilities

(a) Traffic signs and markings

Effectiveness of signs and markings can be mainly assessed with correctness of sign meaning, information quantity, information continuity, setting methods, Sign board size, character size, retroreflective coefficient。

(b) Traffic safety facilities

Traffic safety facilities including anti-collision equipment and separation facilities, sight induced facilities, anti-glare facilities, light facilities and emergency rescue facilities. Due to limited space, the evaluation contents of traffic facilities weren't shown in paper.

(3) Safety index of climate and humanistic environment

Climate factors are mainly described with the parameters: visibility, water film thickness, snow thickness, snow density, ice thickness, temperature, and wind pressure.

(4) Safety index of special environmental in Tunnel

Environmental safety index in tunnel is evaluated by four aspects: CO concentration, smoke concentration, brightness transition safety, and equivalent continuous noise level.

3.4.3 Fuzzy Evaluation Criteria

According to the above analysis on road and traffic facilities, Expressway operating environment safety related with road and traffic facilities is divided into four grades using fuzzy method and the relative evaluation criteria of safety grades are shown in table 7.

Table 7. Fuzzy Evaluation Criteria of Expressway Operating Environment Safety Index.

Evaluation index of expressway operating environment			Expressway operating environment safety grades			
			I	II	III	IV
Road facility	Alignment safety	Alignment transition index	Excellent	Good	Medium	Poor
	Pavement behavior safety	Pavement condition index	100~80	80~60	60~40	40~0
		Pavement skid resistance	≥0.50	0.50~0.40	0.40~0.30	<0.30
		Pavement drainage performance	Excellent	Good	Fair	Poor
	Road capacity	Service Level	Excellent	Good	Fair	Poor

Traffic facility	Cognition of Signs and markings		100~80	80~60	60~40	40~0
	Effectiveness of security facilities		100~80	80~60	60~40	40~0
Climate and humanistic	Climate and humanistic safety	Visibility(m)	>1000	500～1000	100～500	<100
		Rainfall(mm)	<10	10～25	25～50	>50
		Snow thickness(cm)	0～5	5～10	10～15	>15
		Icing thickness(mm)	<2	2～4	4～6	>6
		Wind speed(km/h)	<15	15～20	20～25	>25
Environment inside tunnel	Safety of environment inside tunnel	CO concentration [CO_{Hb}](%)	≤0.7%	0.7%~2.5%	2.5%~6.0 %	>6.0%
		Smoke concentration [smoke]	≤0.0035	0.0035~0.0070	0.0070~0.1200	>0.1200
		Brightness transition safety: entrance	2.40~2.59	2.59~2.94	2.94~3.29	>3.29
		Brightness transition safety: exit	1~1.95	1.95~2.30	2.30~2.65	>2.65
		Accumulated value of continuous noise pollution level [OL_{NP}]	≤48000	48000~54000	54000~60000	>60000

According to evaluation contents of traffic facilities, expressway facility safety leve could be finally obtained (FS) through scoring, the qualitative description as bein shown in table 8.

Table 8. Safety Level of Traffic Facilities.

Safety Level	Qualitative description
FS - I	Road facilities meet the requirements of operating safety and also have forgiven capacity.
FS - II	Road facilities basically meet the requirements of operating safety. However, There is potential need for further improvements.
FS -III	Road facility characteristic don't meet the requirements of operating safety and, should be improved, while it depends.
FS -IV	Road facility characteristic don't meet the requirements of operating safety and must be improved.

4. EXPRESSWAY OPERATING ENVIRONMENT SAFETY INDEX

Expressway operating environment safety is defined as OSC_i, which dependes o accident rate(AR), traffic flow safety(TS) and facility safety(FS) as being shown i equation (1).

$$OSC_i = w_1 f(AR_i) + w_2 f(TS_i) + w_3 f(FS_i) \quad (1$$

In the formula, AR_i, TS_i, FS_i =Actual value of accident rate, traffic safety and facilit safety in *i-th* highway section;

OSCi=Expressway operating environment safety index of certain section i o expressway.

$f(AR_i), f(TS_i), f(FS_i)$ =functions of accident rate, traffic safety and facility safet in *i-th* highway section after dimensionless processing;

w_1, w_2, w_3=constants determined by analytic hierarchy process.

{0.68, 0.20, 0.12} could be experienced value.

The higher comprehensive evaluation index *OSC* of expressway operation environment, the better operation safety. One expressway is composed of many sections with various characteristic and expressway network contains various expressways.

(1)For a single expressway

Expressway operation safety comprehensive value(*OSC*) can be calculated with equation (2)

$$OSC = \frac{\sum_{i=1}^{n} OSC_i \times L_i}{L} \tag{2}$$

Where: *OSC*=Expressway operation safety comprehensive value(*OSC*);

L_i=Length of *i-th* expressway section;

N=Total number of the expressway sections;

L=Total length of the expressway.

(2)For expressway network

Expressway operation safety comprehensive value(*OSC*) of expressway network could be calculated considering the distribution proportion of different expressways with equation (3).

$$OSC = \sum_{j=1}^{m} \frac{\sum_{i=1}^{n} OSC_{ij} \times L_{ij}}{L_j} \times K_j \tag{3}$$

Where: OSC_{ij}=Expressway operation safety comprehensive value(*OSC*) of *i-th* section in *j-th* expressway;

L_{ij}=Length of *i-th* expressway section;

L_j=Total number of *j-th* expressway;

K_j=The proportion of *j-th* expressway length in network.

Based on the value of OSC_i, expressway operating environment safety can be classified into 4 Levels.

Level I: Operating environment satisfies the operational safety requirements and does not need improvement.

Level Ⅱ: Operating environment basically meet the safety requirements, but has the potential need to be improved.

Level III: Operating environment does not meet the safety requirements and should be improved depending on maintenance and investments.

Level IV: Operating environment does not meet the safety requirements and must be improved.

5. REMARKS

The evaluation indexes of expressway operating environment for safety design and operation have been proposed in this paper and also have been applied to evaluate some projects, such as Jing-Zhu-Ao expressway, Zeng-Cong expressway, Yun-San expressway. The application indicated that the evaluation indexes of expressway operating environment for safety design and operation are effective indexes for evaluating expressway operating environment safety.

REFERENCES

Camacho-Torregrosa FJ, Pérez-Zuriaga AM, Campoy-Ungría JM, García-García A (2012). "New Geometric Design Consistency Model Based on Operating Spee Profiles For Road Safety Evaluation". *Accident Analysis and Prevention*. Vol.10 1-10.

Guo, Z.Y. (2008a). "Research on Highway Safety Improvement by Application o Operating speed Theory". *Research Report*. Tongji University. Shanghai, China. (i Chinese)

Guo, Z.Y. (2008b). "Research on Operation Safety of Tunnel Import and Exit". *Researc Report*. Tongji University. Shanghai, China. (in Chinese)

Guo, Z.Y. (2009). "Highway Alignment Quality Evaluating Model Based on Desig Consistency". *Research Report*. Tongji University. Shanghai, China. (in Chinese)

Guo, Z.Y. (2011). "Research on Major Traffic Infrastructure Operation Safety unde Event". *Research Report*. Tongji University. Shanghai, China. (in Chinese)

Guo, Z.Y. (2012a). "Research on Vehicle Operational Security Technology of Highwa tunnel(s)". *Research Report*. Tongji University. Shanghai, China. (in Chinese)

Guo, Z.Y. (2012b). "Rapid Assessment of Emergency Events and Traffic Organizatio Technology of Mountain Area Highway under Severe Weather Conditions". *Researc Report*. Tongji University. Shanghai, China. (in Chinese)

Wang, X. L. (2012). "Safety Analysis of Highway Alignment Based on Reliabilit Theory". *Master Dissertation*. AnHui University of Science and Technology. An Hui China. (in Chinese)

Zheng, K. (2003). "Freeway Alignment Research based on Driver's Psychological an Psychological Reaction". *PhD Dissertation*. Beijing University of Technology Beijing. (in Chinese)

Žilionienė D, Vorobjovas V. (2011). "Correspondence of Horizontal and Vertica Alignment with Safe Driving Conditions on Low-Volume Roads". *Transportatio Research Record: Journal of the Transportation Research Board*. Vol.2203 Transportation Research Board of the National Academies, Washington, D.C. pp.49-56.

Modeling the Safety Benefits of Converting Signalized Y Shaped Junctions to Signalized T Junctions

Weijie Lu[1]; Linjun Lu[2]; Hezheng Bi[3]; and Jian Lu[4]

[1]Transportation Research Center, School of Naval, Architecture, Ocean and Civil Engineering, Shanghai Jiao Tong University, 800 Dongchuan Road, Mulan Building A906, Shanghai 200240, China. E-mail: zjnblwj520520@sina.com
[2]Transportation Research Center, School of Naval, Architecture, Ocean and Civil Engineering, Shanghai Jiao Tong University, 800 Dongchuan Road, Mulan Building A906, Shanghai 200240, China. E-mail: linjun.lu@yahoo.com
[3]Transportation Research Center School of Naval, Architecture, Ocean and Civil Engineering, Shanghai Jiao Tong University, 800 Dongchuan Road, Mulan Building A906, Shanghai 200240, China. E-mail: bihezheng@126.com
[4]Transportation Research Center School of Naval, Architecture, Ocean and Civil Engineering, Shanghai Jiao Tong University, 800 Dongchuan Road, Mulan Building A906, Shanghai 200240, China. E-mail: jianjohnlu@sina.com

Abstract
Junctions are basic joints of road network and their models affect traffic operation and safety greatly. There are quite number of Y shaped junctions at the highway. The existence of such abnormal junctions lowers operation efficiency and triggers traffic accidents. In the rebuilding constructions, T shaped junctions may be new builds but often are converted from Y shaped junctions. For more engineering applications, there is a need to estimate the safety effects of conversions. Several studies have estimated reductions in crashes and severity; however, these results pertain mainly to conversions from Y junctions without signal control. Results for conversions from signalized Y junctions have been less conclusive. Our study aimed to fill this void by estimating the safety performance of converting signalized Y shaped junctions to signalized T shaped junctions. Several cities in Jiangsu province along the national highway G205 helped to identify in the recent past. The empirical Bayes method was used to estimate the safety effects of conversations. The results indicated a safety benefit for converting signalized Y shaped junctions to T shaped junctions. There were reductions in both total and injury crashes, with a larger benefit for non-vehicle injury crashes. Further analysis indicated that the safety benefit of T shaped junctions for total crashes decreased as traffic volumes increase, a result that suggests the need

for the development of a crash modification function (CMF), a task for which more data would be required.

INTRODUCTION

Massachusetts Department of Transportation (DOT) redesigned the intersection of Old County Road and State Road in West Tisbury recently. The new design was a more T-shaped junction rather than the Y shape of the present junction. Until 1984, the chapters on at-grade intersections in the AASHO Policy on Geometric Design of Rural Highways of 1954 and 1965 included Y intersections along with T intersections in discussions of three-legged intersections. It was noted that simple unchannelized three-legged intersections are appropriate for junctions of minor or local roads if the skew is not too great. In rural areas this type usually is used in conjunction with two-lane roads carrying light traffic. In suburban or urban areas it may be satisfactory for higher volumes and multilane roads, AASHTO. (2010).

Traffic accidents statistical of G205jiangsu section, China ,shows that between years 2007–2011 at the national highway, traffic accidents taken place at "three legged junctions" account for 27%, followed by "median openings" (25.2%), "crossroads" (26.7%), "interchange"（9%）and "roundabout" (6.1%) , and other type6%. In terms of mileages travelled and injury severity, three-legged shaped junctions pose the greatest risks to non-vehicles such as electrical bicycles. So analyzing the conflicts of three-legged junctions and further studying transformation design of Y shaped junctions is of great significance for correctly and reasonable junction designs.

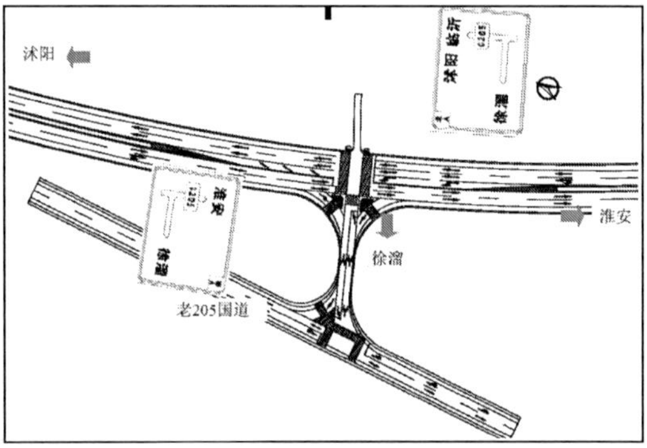

Fig. 1. Typical T Junction Conversation at the National Highway of G205.

The common problems Y shaped junctions exiting in the highway of china is as follows: 1) the area of Y shaped junctions is mainly too small and do not have enough turning vehicle guide lines that may cause the vehicles heavily mixing; 2) the turning radius of Y shaped junctions is too small, causing some of the longer vehicles difficult to turn, some occupy the non-motorized road, increase the collision risk of non-vehicles and pedestrians.

A study of three-legged intersection by MDOT district, United States contained before-after comparisons of uncontrolled Y junction conversions that occurred between 1994 and 1999. The study found that: for the two-lane T shaped junctions locations, there was a significant reduction in the total crash rate (1.37 crashes per million entering vehicles (MEV) before to 0.35 crashes per MEV after) and the injury crash rate (0.34 crashes per MEV to 0.12 crashes per MEV). A comparison of predicted crash rates for conventional junctions with T junctions at the highway in Jiangsu province revealed safety benefits for T junctions at lower traffic volumes (less than 15,000 entering vehicles per day). For main lanes entering volumes of 13,000 vehicles per day, the crash rate was 23 percent lower for T junctions than for Y uncontrolled junctions in suburban areas and 41 percent lower in rural areas. For main lanes entering volumes of 18,000 vehicles per day, the crash rate was 36 percent lower for T junctions than for uncontrolled Y junctions. The safety performance of T shaped junctions and Y uncontrolled junctions was relatively comparable at higher volume. For the two-lane one direction T junction conversions, there was a general reduction in both total and injury crashes, although one location had a 25 percent increase in total crash frequency.

A before-after study of T shaped junctions conversions in Jiangsu province employed the empirical Bayes (EB) methodology to control for regression to the mean and other trends in crash occurrence. The analyses used data from four cities where a total of 26 Y junctions were converted to T junctions between 2005 and 2012. Of the 26 junctions, 10 were previously stop-controlled, and 16 were signal-controlled. For the signalized junctions, the EB procedure estimated highly significant reductions of 26 percent for all crashes and 46 percent for injury crashes. A later study, applying the same methodology and using these same four sites with an additional five converted junctions, found highly significant reductions of 38 percent for all crashes and 57 percent for injury crashes. When considering area type, four suburban sites had a 52 percent reduction in all crashes, but no results could be obtained for injury crashes due to a small sample size. The five urban sites had a statistically insignificant one percent reduction for all crashes and a 43 percent reduction in injury crashes.

To sum up on this literature review, there is not enough consistent knowledge on the safety effects of conversion of signalized Y junctions to signalized T junctions and little or no knowledge on the circumstances under which such conversion will be more or less safety effective. This void provided the motivation for our study, for which the primary objective was to use a substantially larger database than earlier studies to support estimates of the safety performance and to better identify circumstances (e.g. traffic conditions, land use) under which conversion of signals to T junctions may be more safety effective. Target crash types included: (1) total crashes (all crash types and severities), (2) property damage only crashes (any crash resulting in property damage only), and (3) fatal/injury crashes (any crash resulting in a fatality or injury). It should be noted that study sites were limited to new installations and relatively recent signal to T junctions conversions (i.e., since 2003) and data were provided by state and local agencies on program research.

EMPIRICAL BAYES METHOD

Before–after analysis

The Empirical Bayes (EB) method for before-and-after studies goes further by introducing an estimate for the mean crash frequency of similar sites that is used to adjust the crash record of the site for regression to the mean. The mean crash frequency of similar sites is usually estimated from a Safety Performance Function (SPF) calibrated from untreated "reference" site based on the AADT, and sometimes on other characteristics of the site. This SPF also accounts for traffic volume changes and those from other factors unrelated to the treatment. The result is a true estimate of crashes expected without the treatment, and ultimately, a true safety effect of the treatment, Hauer, 1997.

Using the SPFs and the annual SPF multipliers to account for changes in AADT and time trends, (expected crashes in the after period if the treatment had not been implemented) was estimated as the product of the EB expected number of crashes in the before period and the sum of the annual SPF predictions for the after period divided by the sum of these predictions for the before period (for each treatment site).

The empirical Bayes method was developed to account for the regression to the mean (RTM) effect. The EB method makes joint use of two clues to account for the RTM effect, i.e., the observed accident record and the predicted accident frequency at similar entities, shown in equation 1. The equation form is followed the "Highway Safety Manual" (AASHTO, 2010).

$$N_{expected,B} = \omega_{i,B} N_{predicted,B} + (1 - \omega_{i,B}) N_{observed,B} \quad (1)$$

Where

$N_{expected,B}$ = Expected average crash frequency at site i for the before period
$N_{observed,B}$ = Observed crash frequency at site i for the before period
$\omega_{i,B}$ = Weighting factor

The SPF is an equation giving an estimate of average accident per year on a site, as a function of some explanatory values (e.g., daily traffic, site area, etc.) and SPF is developed from the crash data of the reference group. It should be noted that the concept of 'predicted' differs from that of 'expected'. The number of predicted crashes is estimated by the safety performance function (SPF) while the number of expected crashes is estimated by the EB procedure. A weighting factor is obtained by equation 2, where, k is an over-dispersion parameter from a negative binomial regression model with the use of a maximum likelihood procedure described by Washington et al. 2003. The weight $\omega_{i,B}$ for each site i, is determined as:

$$\omega_{i,B} = \frac{1}{1 + k \sum_{\substack{Before \\ years}} N_{predicted}} \quad (2)$$

The next step, we estimate the expected number of crashes in the after-period using the conversions and the $N_{predicted,B}$ adjustment factor r_i. $N_{predicted,A}$ is the output value of SPF using the average annual daily traffic (AADT) of the after period. The adjustment factor r_i , equation 3 reflects the changes in crash frequency. The variance of can be estimated approximately from equation 5. The CMF can be estimated by equation 6 ,AASHTO, 2010.

$$r_i = \frac{\sum_{\substack{\text{After} \\ \text{years}}} N_{\text{predicted,A}}}{\sum_{\substack{\text{Before} \\ \text{years}}} N_{\text{predicted,B}}} \quad (3)$$

$$N_{\text{expected,A}} = N_{\text{expected,B}} \times r_i \quad (4)$$

$$OR_i = \frac{N_{\text{observed,A}}}{N_{\text{expected,A}}} \quad (5)$$

$$OR_i{}' = \frac{\sum_{\text{All sites}} N_{\text{observed,A}}}{\sum_{\text{All sites}} N_{\text{expected,A}}} \quad (6)$$

Equation 7 shows a more precise estimate of the CMF and equation 8 shows the variance. The standard error can be obtained by taking the square root of the variance. By applying equation 1 through 8, the CMF by the EB method can be obtained, which accounts for RTM effect of the treatment site and changes in the traffic volume.

$$CMF(OR) = \frac{OR'}{1 + \frac{Var\left(\sum_{\text{All sites}} N_{\text{expected,A}}\right)}{\left(\sum_{\text{All sites}} N_{\text{expected,A}}\right)^2}} \quad (7)$$

$$Var\left(\sum_{\text{All sites}} N_{\text{expected,A}}\right) = \sum_{\text{All sites}} \left[(r_i)^2 \times N_{\text{expected,B}} \times (1 - \omega_{i,B})\right] \quad (8)$$

An observational before/after evaluation can be conducted for a single project at a specific site to determine its effectiveness in reducing crash frequency or severity. However, results from the evaluation of a single site will not be very accurate and, with only one site available, the precision and statistical significance of the evaluation results cannot be assessed.

Cross-Sectional Safety Evaluation Method

A cross-sectional analysis was employed to supplement the results of the EB analysis. This was necessary because the general sample size that the conversations from Y to T was relatively limited, The national highway of Jiangsu section has many group segments of rural two-lane Y junctions with skew angle (in degrees) in a certain range, such as 0-30, 30-60, 60-90degrees .These three groups of rural two-lane road segments could be used in a cross-sectional study. Data are collected for a specific time period for both groups. The crash estimation based on the accident frequencies for one group is compared with the crash estimation of the other group. It is, however, very difficult to adjust for differences in the various relevant conditions between the two groups.

Because heterogeneity effects are assumed to be constant for given cross-sectional units or for different cross-sectional units during one time period, they are absorbed by the intercept term as a means to account for individual or time heterogeneity, Hsiao, 1986. More formally, a cross-sectional regression is written as:

$$\gamma_{it} = \alpha + X'_{it}\beta + u_{it}, i=1, \cdots n; , t=1, \cdots, T$$

(9)

where i refers to the cross-sectional units (individuals, counties, states, etc.), t refers to the time periods, α is a scalar, β is a vector, andX'_{it} is the,i th observation on the Pth explanatory variable.

A negative binomial regression model was applied in this evaluation to estimate the safety effects of T junctions compared to Y signalized junctions. Spss software was used to estimate the models, using a clustered robust standard error to adjust for intra-group correlation when multiple years of data were included for the same site. Once the models were estimated, the coefficient for the T junction indicator was used to estimate a CMF. The following additional variables were considered in the model development.

- Number of Y junctions approaches (3-legged/4-legged indicator).
- Number of approach or T junctions lanes (single lane/multilane).
- Y shaped junctions skew angle (in degrees).
- Signalized junctions phasing (permissive/protected/protected permissive indicator).
- Junctions class defines (Private, locals, collectors and minor arterial / private, locals, collectors and minor arterial).

Development of CMF

Data for signalized Y junctions, similar to those converted to signalized T junctions, were sought for use in developing the SPFs. But such data were difficult to obtain for all states in which treatment sites were identified. For all other locations, the SPFs previously used in NCHRP Project were applied. These SPFs were recalibrated for use in the specific jurisdictions using data for the sample of T intersection conversions for the period immediately before conversion. Only the data in the 1-year prior to T intersection construction were used for this purpose. This guarded against the possibility that a randomly high crash count in earlier years may have prompted the decision to install the roundabout, which would have resulted in SPFs that would overestimate safety improvements after conversion. Examination of annual crash trends in the before periods indicated that this decision was justified. The functional form of the SPF is given by Eq. (10). The coefficients of a, b and c and the over-dispersion parameter (k) are parameter estimates from negative binomial regression model. AADT = annual average daily traffic (vehicles per day) for junctions i.

$$\text{Crash/year} = e^{a}(MajAADT)^{b}(MinAADT)^{c} \qquad (10)$$

For a three-leg junctions, road AADT are determined as follows: $MajAADT$ =the major road AADT;$MinAADT$ =the minor if the AADT.

Table1.Intersection Safety Performance Function for National Highway.

Crash Type	a (s.e.)	b (s.e.)	c (s.e.)	Dispersion parameter
Total	-10.7132 (1.7507)	0.4945 (0.1744)	0.8458 (0.1268)	0.0494
Injury	-6.8523 (1.7905)	0.6028 (0.1706)	0.4091 (0.1018)	0.3373
Left Turn	-12.3749 (2.3291)	0.8226 (0.2199)	0.5448 (0.1349)	0.5717
LTOPP	-11.2308 (2.5444)	0.9466 (0.2803)	0.3423 (0.1676)	0.7710
Rear-End	-11.2095 (2.2141)	0.8905 (0.1890)	0.6178 (0.1115)	0.4583

DATE COLLECTION

Geometric, traffic and crash data were obtained for two-lane one direction highway segments in National highway G205 Jiangsu section from 2007 to2011. The data were obtained in two parts: (1) a crash inventory database extracted from the crash reporting system and (2) a roadway inventory file. Crash data were available for each year of the study period.

The crash data were merged with the geometric data using the three unique identifying features (i.e., county, route number and segment number) and separated by year. The crash inventory data include all reportable crashes for mid-block locations (i.e., non-junctions crashes). Reportable crashes are defined as those in which at least one vehicle is towed from the scene. This dataset does not contain crashes occurring at or near junctions and any data from "phantom" or "hit-and-run" crashes are excluded. The dataset includes state roads only. Again, this study investigates specific crash types that tend to be influenced by lane and shoulder width. The "related" crash types include head-on, run-off-road, opposite direction sideswipe, and same direction sideswipe crashes. In this dataset, the target crashes rep-resent approximately 67 percent of total crashes. The roadway inventory data include information regarding the location (i.e., county, route, and segment number) as well as the geometric and traffic characteristics of each roadway segment. The segmentation is based on the roadway characteristics and each segment is homogeneous with respect to the roadway characteristics.

Table2.Summary Statistics of Signal Y to Signal T Junction Conversions

Variable	Minimum	Maximum	Average
Years before	1.00	7.00	4.03
Years after	1.00	7.00	3.67
Major road AADT before	9,489	21,990	15,253

Major road AADT after	10,504	30,697	21,094
Minor road AADT before	5,466	7,323	6,209
Minor road AADT after	6,487	8,190	7,253
Total crashes/year before	51.2	69.0	26.09
Total crashes/year after	47.0	64.2	25.7
Injury crashes/year before	32.1	43.4	39.0
Injury crashes/year after	24.1	28.3	26.5

Crash data were collected from archived data sources maintained by the police department. One issue with the crash data collected from police records is that not all accidents are reported and not all reported accidents are correctly recorded. Many crashes that do not involve injury are unreported. This leads to overrepresentation of fatal and injury accidents in the crash database. It was seldom easy to trace the exact location of individual crashes from the database. This can affect the accuracy of crash counts of road sections, especially in the case of short sections. The driver-vehicle unit level variables include driver age, gender, vehicle type and severity for each crash. The crash database also lacked details of the victims and only the details of the accused were available for analysis. Reliable estimates of safety can be produced only with the help of comprehensive and accurate crash, geometric, traffic and operational data. In India there is an immediate need for reasonably reliable data to be easily and readily available for analysis and future research in road safety.

EVALUATION RESULTS AND DISCUSSION

Table 3 presents the results for total and injury crashes, including the sum of the expected crashes that would have occurred in the after period without the treatment ($\mathbf{N_{expected,B}}$), sum of the variance (**Var**), sum of crashes during the after period ($\mathbf{N_{observed,A}}$), CMF, and standard deviation of the CMF. Over all 26 conversions it is seen from the first two rows of results that the safety benefit for total crashes (CMF = 0.872) and injury crashes (CMF = 0.841) is statistically significant at the three percent significance level and the reduction in injury crashes is much larger than the effect for all crashes combined.

This general indication confirms the results of a recent study that was based on only nine conversions (four suburban and five urban conversions). The safety benefit for suburban area conversions is larger than for urban conversions. The results also indicate that the CMFs for intersections with three approaches are larger than for intersections with four approaches. While this trend holds for both total and injury crashes, the confidence intervals overlap when comparing the CMFs for injury crashes (i.e., the difference is not statistically significant). The CMFs for 2 lanes versus multi-lane T junctions are also not significantly different.

Table3. Intersection Level Evaluation Results for Conversions.

Crash Type	Grouping	Sites	$N_{expected,B}$	Var	$N_{observed,A}$	CMF	Standard error
	All sites	26	511	338.56	435	0.872	0.033
ALL	2 treated	18	404	30.27	359	0.889	0.022

	approach						
	>2 treated approach	8	107	378.3	76	0.891	0.042
Injury and Fatal	All sites	17	253	71.34	219	0.841	0.039
	2 treated approach	11	136	212.28	109	0.801	0.035
	>2treated approach	6	117	159.1	110	0.940	0.053
Suburban	Total	15	123	85.13	137	0.655	0.041
Suburban	Injury	15	18	34.44	12	0.667	0.036
Urban	Total	13	278	176.61	145	0.521	0.081
Urban	Injury	13	34	28.23	21	0.617	0.098

CONCLUSIONS

Our research investigated the safety effects of converting signalized Y junctions to signal T junctions. The EB methodology was employed in an observational before-after study to estimate CMFs for total and injury crashes. Three potential confounding factors were included in the analysis, including traffic volume, area type, and number of approach lanes.

The analysis suggested that the safety benefit is larger for suburban than for urban conversions, about 5-10persent higher. It is possible that changes in land development increased traffic, altered traffic patterns, and increased accidents at a location that previously and not exhibit safety problems. In any event, an urban unsignalized Y junction that exceeds any of the thresholds given above should be given priority in safety improvement programs.

Traffic volume is a strong predictor of crash frequency and may also influence the effect of T junctions. The results from the before-after analysis indicated a change in effect for different volumes. Specifically, the safety benefit of T junctions appeared to decrease as traffic volumes increase, at least with respect to total crashes. A similar analysis of injury crashes did not show a trend (i.e., the safety benefit was relatively consistent across the range in AADT).

The examination of the standard deviation of the mean value of the CMFs and of the distribution of CMFs showed that the standard deviation of the distribution was larger than the standard deviation of the mean. This indicates that there are differences among the CMFs from different sites. This difference was substantial with respect to the CMFs for total crashes, but less so for the injury-related CMFs. These substantial differences for the CMFs for total crashes emphasizes the need for further research into the development of crash modification functions instead of crash modification factors, particularly for total crashes.

REFERENCES

AASHTO. (2010). Highway Safety Manual. Retrieved 2010, from *http://www.highwaysafetymanual.org*

AASHTO,A Policy on Geometric Design of Highways and Streets, American

Association of State Highway and Transportation Officials, Washington, DC, 1990.

Bahar, G. (2009). Methodology for the Development and Inclusion of Accident Modification Factors (AMFs) in the *First Edition of the Highway Safety Manual (Draft Version).*

Belluz, L., & Forbes, G. (2003). Synthesis of Safety For Traffic Operations. Annual Conference of the Transportation Association of Canada. St. John's Newfoundland and Labrador: *Transportation Association of Canada.*

Brehmer, C.R., K.C. Kacir, D.A. Noyce, and M.P. Manser.(2003). Evaluation of Traffic Signal 17 Displays for Protected/Permissive Left turn Control, *NCHRP Report 493.*

Brase, C. H., and C. P. Brase.(1995). Understandable Statistics: Concepts and Methods. D.C. *Heath and Company, Lexington, MA.*

David, N.A. and Norman, J.R.(1989). Motor Vehicle Accidents in Relation to Geometric and Traffic Features of Highway Intersections, Stanford Research Institute, prepared for FHWA and NHTSA, July.Press, New York.

Space Properties of the Road Alignment and Impact on the Traffic Accident

Xiaohuan Zhou[1]; Zhongyin Guo[1]; Benmin Liu[1]; and Zixuan Song[2]

[1]Key Laboratory of Road and Traffic Engineering of the Ministry of Education, Tongji University, Shanghai, ST 201804, China. E-mail: zzxxhhzxh@gmail.com
[2]Shaanxi Academy of Urban and Rural Planning and Design, Xianan, ST 710075, China. E-mail: zixuan926@hotmail.com

Abstract
Based on numerous research works on road traffic safety, the space curvature index of expressway alignments was developed considering road alignment space geometric properties and its impact on traffic accident. This index was modified to reflect the influence of the curvature on highway operating speed. The road safety performance and operating speed of several expressways were investigated and the regression relationship between single-vehicle accident rate and space curvature index of highway alignments was studied. So that the space curvature index of highway alignments could be used as a prediction factor of traffic accident in highway alignments during highway alignment safety design and road safety analysis.

Keywords: Highway safety; Road alignment; Single-vehicle accident analysis; Space curvature

INTRODUCTION

The issue of road traffic safety has been getting more and more consideration. Out of these road safety research topics, the effects of road alignment on safety are being studied at home and abroad. A number of various models and quantitative assessment methods of the road alignment (Fambro et al. 2000) have been developed. Alignment consistency in terms of vehicle speed change was used by Leisch (1977) as a representative feature of highway alignment continuity. According to the "Highway Safety Design and Operations Guide" (Ministry of Communications of P. R. China 2004) highway design consistency is defined as "the avoidance of abrupt changes in geometric features for contiguous highway elements and the use of design elements in combinations that meet driver expectations". In developing them to the road safety audit guide, researchers studied the features of highway alignments and operating speed related highway traffic safety.

Quite a few research works have been conducted on predicting operating speeds. Many studies took horizontal curves and vertical sections as factors (Bella 2005; Abdul-Mawjoud and Sofia 2008), and these kinds of models established the linear consistency evaluation based on the difference of operating speed. (IHSDM 2010)

In this paper, we describe research relating with highway alignments and safety performance using an index of space geometric properties. Highway alignments can play the function guiding driver's operating and restricting traffic movement. The highway alignment (also the vehicle movement trajectory) is assumed to be a three dimensional curve. The space curvature of the three dimensional curve was adopted as the main alignment index combining the restrictions of highway alignments. Then the relationship between the index and single-vehicle crash rate was developed by investigating the road safety performance and operating speed.

METHOD

Curvature is an expression index of the curve degree of curve line generally, expressed as K. Usually the curvature of a curve can be calculated with the general curve equations as shown by equation (1):

$$\kappa(t) = \frac{|r'(t) \times r''(t)|}{|r'(t)|^3} \tag{1}$$

Throughout the highway alignment design, the horizontal alignment features are the priority ones to be determined. Any complex route is composed of straight segment, circular curve segment and the transition curve segment which are combined with various forms. In measuring a highway alignment in a plane coordinate system, plane coordinates (x,y) can be expressed in a parametric equation (2):

$$r(x,y) = (F_x(l), F_y(l)) \tag{2}$$

Where, l is the station (mileage), m. This parameter equation is a piecewise function. $F_x(l)$ and $F_y(l)$ have different expressions respectively for straight segment, circular curve segments and transition curve segments.

Any curve element can by expressed by the curvature radius at the start point of and the end point of the curve, R_o, R_e and the length of the curve, s. The coordinate (x,y) any point of a horizontal curve can be determined with the station of the horizontal start point. Let the distance of the calculation point p is $l=L_p-L_o$, curvature is κ, declination angle is β, the distance of the curve segments $S=l_e-l_o$.

A clothoid is applied in the design of an expressway's horizontal curve, which connecting a circular curve and a straight line, whose alignment is spiral. The radius of the convolute curve changes from radius R_o to R_e. When designing the convolute curve, the general formula for calculating coordinates taking the coordinates of any point as the origin coordinate, the calculation may be greatly simplified.[Tang and Wu 2006] The curvature of the convolute curve changes with distance linearly. The curvature at any point of the convolute curve $k_i=1/R_i=k_o+l(k_e-k_o)/S$. After the temporary coordinate system *X'O'Y'* is established as shown in Figure 1(a), the convolute curve can computed.

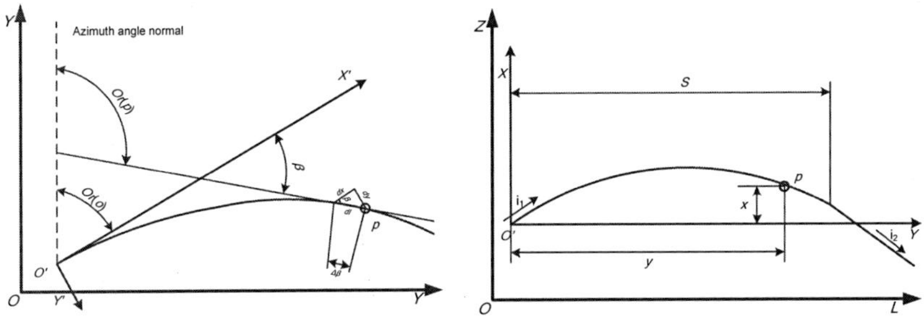

(a)Horizontal Curve **(b)Vertical Curve**

Fig.1 Coordinate Calculation of Road Alignment

Taking the differential distance $dl=R_i d\beta \Rightarrow d\beta=k_i dl$, equation (3) can be deduced.

$$\beta=\int_0^l k_i dl = k_o l + \frac{k_e - k_o}{2S} l^2 \tag{3}$$

According to the computing formula of the temporary coordinate system, let the curve length at the calculated point *p*, equal to *l*, the point coordinates in the temporary coordinate system are:

$$\begin{cases} x=\int_0^l \cos\beta dl \\ y=\int_0^l \sin\beta dl \end{cases}, \text{which, } \beta=\int_0^l k_i dl = k_o l + \frac{k_e - k_o}{2S} l^2 \tag{4}$$

Taking straight lines and circular curves be special forms of the convolute curve, the formulas also can be used to compute straight line segments and circular curve segments.

The vertical alignment plays a longitudinal transition, ensuring a smooth ride and adequate sight distance. The parabola or circular curve with large radius is used generally as a vertical curve. The vertical curve length at the calculated point p is $l=l_i-l_o$, the altitude at starting point is z_o, slope difference $\omega=i_1-i_2$ and the curve segment length $S=l_e-l_o$. Where the quadratic parabola is used for vertical curve, straight line segments connect to the parabola at each ends and the entire slope is constituted. After the temporary coordinate system *XO'Y* is established as shown in Figure 1(b), the quadratic parabola as a vertical curve can be computed.

The function for quadratic parabola curve within the temporary coordinate system can be expressed by equation(5).

$$x = ay^2 + by, i = \frac{dx}{dy} = 2ay + b \tag{5}$$

The quadratic parabolic formula of vertical curve can obtained putting the coordinates value at the end of the vertical curve, as shown in equation(6):

$$z = z_o + \frac{\omega}{2S} l^2 + i_1 l \tag{6}$$

As talking about horizontal curve above, the calculation formula of vertical curve segment can be applied to computing straight line segments.

The projected curve length t (mileage, l) at *XOY* plane is as the only parameter while develop the three-dimensional mathematical model of expressway alignment in this paper. Combining the results previous derived, the space curve of expressway alignment can be expressed as:

$$r(t) = \left(\int_0^t \cos(k_o t + \frac{k_e - k_o}{2S} t^2)dt, \int_0^t \sin(k_o t + \frac{k_e - k_o}{2S} t^2)dt, z_0 + \frac{i_e - i_o}{2S} t^2 + i_o t\right) \quad (7)$$

The start and end point of the alignment segment are defined in accordance with horizontal and vertical alignment respectively. By putting expressway alignment equation into formula (1) and making simplification, the space curvature of expressway alignment can be computed using the following expression (8):

$$\kappa(t) = \frac{\sqrt{(\frac{i_e - i_o}{S})^2 + (\frac{i_e - i_o}{S} t + i_o)^2 (k_o + \frac{k_e - k_o}{S} t)^2 + (k_o + \frac{k_e - k_o}{S} t)^2}}{(1 + (\frac{i_e - i_o}{S} t + i_o)^2)^{3/2}} \quad (8)$$

Where, $(i_e\text{-}i_o)/S{*}t{+}i_o$ is the linear interpolation of the vertical curve slope inside the computed alignment segment. Because the quadratic parabola is used for vertical curves, the values were equal to the slope, i_t; $k_o+(k_e\text{-}k_o)/S$, is the linear interpolation of horizontal curve curvature k of the computed alignment segment. The values is equal to the curvature k_t of the point; $(i_e\text{-}i_o)/S$ is the linear change rate of vertical curve slope i of the computed alignment segment, so let it be di_t. It has to be recognized that even though i in the z direction has great influence on operating speed, its magnitude is small compared with the influence of horizontal curve curvature. Vertical curve curvature and horizontal curve curvature have different influences with different way. The influence of curvature is produced through the action of alignment on vehicle operation. The impact of vertical slope is generated by the gravity in Z direction.

Therefore, i_t may be modified a correction factor B where computing three-dimensional curvature. The parameters B used to reflect that the action in the z direction is different from that in other two directions. So the original formula is modified to be equation (9):

$$\kappa(B) = \frac{\sqrt{di_t^2 - Bi_t |i_t| k_t^2 + k_t^2}}{(1 - Bi_t |i_t|)^{3/2}} \quad (9)$$

DATA COLLECTION

The researchers collected crash data and measure operating speed about 200 sections of three expressways in the north east parts of China. The investigated data of expressways are shown in Table 1. The traffic-related crash data were collected from department of the police office, which includes the year of vehicle data collecting and two years before. The crash data was treated according to the types of crashes. The three-year accident data are shown in Table 2.

Table 1 Investigated Expressways

Expressways	Lane (Two-way)	Total Long (km)	Design Speed (km/h)
I	8	348	120
II	4	222	100

III	6	361	120

Table 2 Investigated Expressways and Accident Data

Expressways	Accident data					
	Accident type	Single-vehicle accident	Rear-end collision	Collision	Other	Total
Expressway I	Accident No	669	182	10	34	895
	Percent (%)	74.7	20.3	1.1	3.9	100
Expressway II	Accident No	1810	612	61	120	2603
	Percent (%)	69.5	23.5	2.3	4.7	100
Expressway III	Accident No	1142	45	1095	244	2526
	Percent (%)	45.2	1.8	43.3	9.7	100

It can be observed from the statistics of these 3 expressways that single-vehicle crash takes a large proportion in China, which is 45-75%. Vehicle information was collected by setting up the roadside detection units in the daytime during 9:00 and 17:00 under clear weather condition. The data of vehicle speeds and axles on the sections of more than 200 straight, transition and horizontal lines were collected. All the vehicles under 6s gap to the front vehicle were removed, to ensure free flow conditions. There are at least 400 vehicle records measured in each section. These records were sorted and the 85th speed was selected as the representative speed.

Correction parameters were obtained by analyzing expressway operating speed. The operating speeds were derived from sample exceeding 400 vehicles per section for a total of about 200 sections along several expressways.

RESULT

Through analyzing the effects of highway alignment features on road safety and operating speed, the space curvature index was modified regarding the influence of the curvature on highway operating speed. The regression relationship between single-vehicle accident rate and the index of highway alignments was developed. Through non-linear regression analysis and considering the expressway alignment features, the calculated value of B for A-Type vehicle is 134.041 with R^2=0.730 and 250.772 for D-Type vehicle with R^2 = 0.665. The space curvature index of expressways is computed using the following equations:

$$\kappa_a = \kappa(134) \quad ; \quad \kappa_d = \kappa(250) \tag{10}$$

The section of 10 meters expressway alignment can be taken as a calculation step while the space curvature index is computed for an expressway segment for the passage car and heavy truck separately. According to research results [Miaou and Lum 1993] and preliminary analysis of some expressway sections, the following typical feature parameters are selected to be studied: the standard deviation of the space curvature of the computed segment, σ_κ; the mean value of space curvature of the computed segment, $avg\kappa$; the coefficient of variation space curvature of the computed segment, C_κ; the difference between the space curvature for passenger car and that for heavy truck, Δm_κ and the relative difference ratio, M_κ. The detail definitions of these parameters are described below:

$$\sigma_\kappa = \sqrt{\frac{OP \times \sum_{1 \le i \le n} (\kappa_{ai} - avg\kappa_a)^2 + (1-OP) \times \sum_{1 \le i \le n} (\kappa_{di} - avg\kappa_d)^2}{n-1}} \tag{11}$$

$$C_\kappa = \frac{\sigma_\kappa}{avg\kappa} \tag{12}$$

$$avg\kappa = (OP \times \sum_{1 \le i \le n} \kappa_{ai} + (1-OP) \times \sum_{1 \le i \le n} \kappa_{di}) / n \tag{13}$$

Taking into account that the difference between the space curvature for passenger car and that for heavy truck is not well indicated the discrete of the difference Δm within the computed segment, the relative difference ratio, M_κ is introduced as a parameter to reduce the impact of limited sample data:

$$M_\kappa = (\max \Delta m - \min \Delta m) / avg\Delta m \tag{14}$$

Where in Formula 11-14: κ_a is the space curvature for passenger car, m^{-1}; κ_d is the space curvature for passenger car, m^{-1}; OP (occupancy of passenger cars); $avg\Delta m$ is the average value of difference between the space curvature for passenger car and that for heavy truck Δm, m^{-1}.

The single-vehicle accident rate (accidents per million vehicle kilometers) of 20 expressway segments (more than more than 3,000 cross sections, OP is about 0.691) and the relative alignment features indicated by σ_κ, avgκ, C_κ and M_κ were investigated. The collected and analyzed results are shown in Table 3 and Table 4.

DISCUSSION

The correlation analysis results in Table 3 indicate that there is a good correlation between accident rate and typical parameters of space curvature of expressway alignments. The regression correlations between accident rate and σ_κ and M_κ are very good with level of 0.05.

Regression analysis between single-vehicle accident rate and σ_κ, avgκ, C_κ and M_κ were carried out with quadratic polynomial regression models. The regression results are shown in Figure 2. In these figures, the abscissa is the space curvature index and the vertical axis is the single-vehicle accident rate (accidents per million vehicle kilometers):

Table.3 Single-vehicle Accident Ratio and Indexes of Space Curvature in Different Road Section

Section Number	Accident Rate	Standard Deviation	Mean Value	Coefficient of Variation	Range	Range Ratio
1	0.091	0.000965	0.000722	0.749	0.000422	2.19
2	0.181	0.001135	0.000493	0.435	0.000422	3.13
3	0.181	0.001537	0.001095	0.712	0.000596	2.68
...	...	...	...	...	...	...
20	0.263	0.000947	0.000379	0.400	0.000085	2.10
21	0.214	0.001357	0.000360	0.265	0.000079	2.05

Table.4 Correlation Test of Single-vehicle Accident Ratio and Indexes of Space Curvature

		Accident Rate	Standard Deviation	Mean value	Coefficient of Variation	Range	Range Ratio
Accident	Pearson Correlation	1	.773**	.445*	-.424	.837**	.831**

Rate	Significant (both sides)	.000	.043	.056	.000	.000

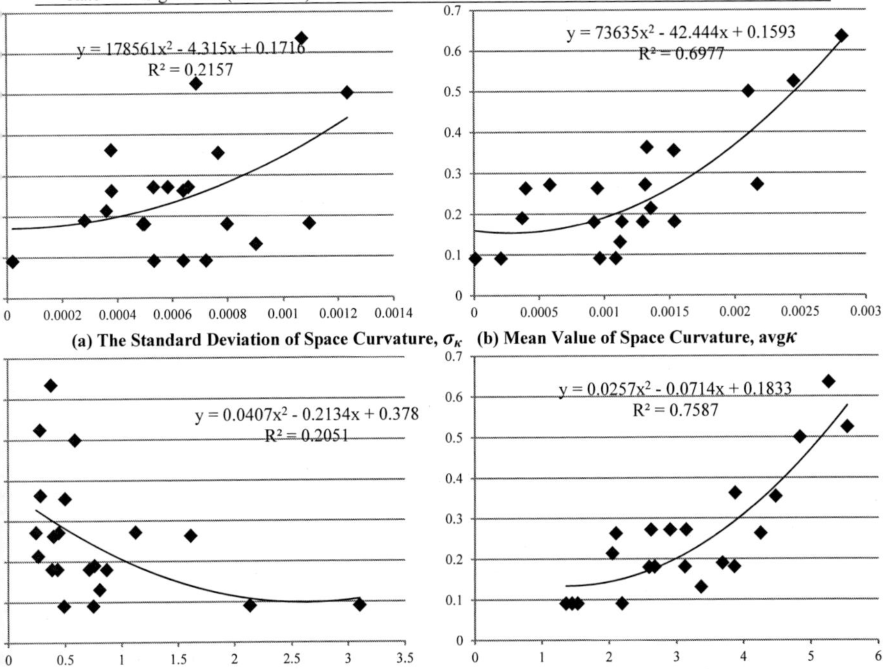

(a) The Standard Deviation of Space Curvature, σ_κ (b) Mean Value of Space Curvature, avgκ

(c) Coefficient of Variation Space Curvature, C_κ (d) Relative Difference Ratio, M_κ

Fig.2 Regression Result of Single-vehicle Accident Ratio and the Alignment Indexes

The quadratic polynomial regression analysis results in Figures 4 show that: (1) there is a low correlation coefficient between single-vehicle accident rate and the standard deviation of the space curvature, and the correlation coefficient between the accident rate and the mean of the space curvature correlation is high up to 0.7. (2) the coefficient of relationship between the accident rate and the coefficient of variation space curvature C_κ is very low, only about 0.2. So the relationship between the accident rate and the coefficient of variation space curvature C_κ is not significant.

Compared with the traditional analysis method of single index and independently multiple indexes, this paper was focused on the single-vehicle accident, which is based on the highway geometric characteristics. The space curvature of the geometric design was established as the base of the safety evaluation index, and this series of evaluation indexes have good regression result (0.75).

CONCLUSION

The road safety issue is now getting more and more attention in China. The research work about the impact of highway alignment on road safety has been carried out for many years. In this paper, space features of highway alignment for the evaluation of road safety were developed and focused on the single-vehicle accident.

Unlike relatively independent alignment feature parameters which were used to indicate the road safety performance of highway alignment, the relative integrated alignment feature parameters based on the space curvature of highway alignment were studied. This could be used to evaluate the road safety performance of expressway and optimize the linear design of expressway.

ACKNOWLEDGEMENT

This work was supported by Liaoning Communication Department of the project No.062089 (Study on Application of Operating Speed Theory for Safety Improvement of Expressway) and by Ministry of Communications of the P. R. China of Western Transportation Construction project No.2009318791034 (Risk Assessment and Tactic of the Operation in Yunnan Highway Based on The Subjective – Objective Safety).

REFERENCE

Abdul Mawjoud, A. A., Sofia, G. G. (2008). "Development of models for predicting speed on horizontal curves for two-lane rural highways." *Arabian J. Sci. Eng.*, 33(2B), 365-377.

Bella, F. (2005). "The evaluation of design consistency: Predicting models of operating speed on three-dimensional alignment from tests on driving simulator." *3rd Int. Symp. on Highway Geometric Design*, Transportation Research Board, Washington, DC.

Fambro, D.B., Collings, J.C., Vrdova, R.L., Leisch, J.P., Mason, J. J. M. (2000). "Geometric Design: Past, Present, and Future." *Transportation in the New Millennium,* 2000: A2A02: 1-7.

Hassan,Y., Sayed T., Tabernero V. (2001). "Establishing Practical Approach for Design consistency Evaluation." *Journal of Transportation Engineering*, 127(4): 295-302.

Leisch, J. E., Leisch, J. P. (1997). "New Concepts in Design Speed Application. " *Transportation Research Record* 631, 1977: 4-14.

Miaou,S.P., Lum, H. "Modeling Vehicle Accidents and Highway Geometric Design Relationships." *Accident Analysis & Prevention*, 1993, 25(6): 689-709.

Ministry of Communications of P. R. China (2004). Guide Lines for Safety Audit of Highway JTG/T B05-2004.

Tang, Z.Q., Wu, Z.M. (2006). "Research on United Calculation Methods for 3D Coordinate of Road Curve." *Bulletin Of Surveying And Mapping*, 2006, 12: 20-23.

The U.S. Federal Highway Administration(2010). *Interactive Highway Safety Design*

A Car Following Method to Relief Road Access Intervals

Jinyu Sun[1], Junhua Wang*[2]

[1]School of Transportation Engineering, Tongji University, No. 4800 Cao'an Highway, Shanghai 201804, China. E-mail: littlex2yu@gmail.com
[2]Key Laboratory of Road and Traffic Engineering of the Ministry of Education, Tongji University, No. 4800 Cao'an Highway, Shanghai 201804, China. E-mail: benwjh@163.com
* Corresponding author

ABSTRACT: Access of relief road links traffic with facilities on the roadside that can be terminal of drivers. Nevertheless, the intervals between accesses are often not taken into consideration, or only in very rough ways. This paper aims at demonstrating a modeling approach to determine appropriate access intervals, which is based on car following theory. Vehicles always change lane from the relief road into the main stream at a quite lower speed, as a conflict to the main stream, which would lead to a temporary decline to traffic capacity. Utilizing the model of car following and the design capacity of freeway correspondingly, we find the expected maximum length it will affect. The proposed length is assumed not to reach the next access in order to avoid negative interaction, and therefore determine it as the minimum interval between accesses. We expect that this study can make available appropriate relief road access intervals.

INTRODUCTION

Traffic has developed at an immensely high speed recent years, especially in metropolises and mega cities of China. Today, one of the most negative results of developing modern transportation systems is congestion. So, a reasonable design for smooth operation is the most critical issue in agencies' transportation strategy. Congestion may appear when some vehicles are driven slowly. As is practiced, a relief road has been proved to be a very important and comprehensive management tool for the problem (Dong, 2012). Relief roads can separate rapid traffic from non-motorized traffic and links traffic with facilities on the roadside that can be terminal of drivers.

Vehicles change lanes between relief road and main road through accesses when needed. In this way, access intervals can be another issue to be considered. Long access interval can be a problem for divers to get to their terminal, while short one makes it easier. On the other hand, short access intervals mean more frequent lane changing in a certain road section and the influence of lane changing brings to the main road may be overlapped and amplified. Thus a proper access interval should be determined to ensure the relief road will work with efficiency.

BACKGROUND

Relief road is generally defined as a byroad that lies on one or both sides of main road, for vehicles which cannot get into the main road and going to reach the roadside (Chen and Jiang 2012). Urban expressway is usually constructed with its relief road. The main road of the urban expressway has a high operation velocity and volume level while the relief road has a low speed and is aimed at collecting and distributing. As for the distinct position of relief road, its various functions are listed below.

- Vehicles change lane from to the main road when needed;
- Buses and taxis may park temporarily on;
- Non-motorized traffic operate on;
- Undertake mixed traffic flow before entering intersection.

There are three types of connection between relief road and main road: paralleling type, separation type and large green belt type. Of the three types, paralleling type is widely considered in China for less occupation of land, more simple construction and more convenience (Shao-hong, 2004),and this is the type the research focuses on. Relief roads lie on the both side of main road, and railing or simple blocks are arranged to distinguish them.

In this way, the main road is parallel with relief road, which is the reason why it is called paralleling type, and it has similar characters such as alignment and curve. Access between and relief road and main road is just an opening of the railing while interchange ramp is a length of road. The traffic flow around the access is also quite different from in the intersection. Vehicles have the same orientation before get into the main land and there will be only one or two instead of several vehicles enter the main road at a single access once a time, which means the traffic flow is discrete rather than consecutive. Therefore, the method for intersection intervals is not suitable for this research.

A relief road access minimum interval is the minimum length between two adjacent relief road accesses when traffic situation and service level are guaranteed. There is a great gap of speed (about 20 km/h-50 km/h), so improper access will obviously disturb the operation of main road.

Car following models describe the processes that drivers follow each other in the traffic stream in a mathematical approach. The car following models focuses on the driving behavior in a constraint circumstance (Zou and Yang 2001). There have been plenty of scientific researches since 1950s. The models can be classified into four categories: Gazis–Herman–Rothery (GHR) model, safety distance or collision avoidance models, Psychophysical or action point models and Fuzzy logic-based models (Wang, et al. 2006)

The psychophysical or action point model is short for AP model. The first discussion of the underlying factors that would eventually lead to the construction of AP the models was given by Michaels (1963), who raised the concept that drivers are initially able to tell they were approaching the front vehicle due to changes in the apparent size of the vehicle, and by perceiving relative velocity through changes on the visual angle. The formula of the model is:

$$d\theta/dt = -w\Delta v/R^2 \tag{1}$$

Wiedemann (1974) put up the theory that the car following status can be divided by perception threshold and set up a perception threshold model based on this theory. Psychophysical or action point models nowadays are all based on this model with different modification. Perception threshold model takes driver's psychological and physiological characters and the various driving behavior that will lead into consideration, which make the model more reasonable, but too many parameters is a problem and they are also difficult to be calibrated and validated (Owsley, C. and McGwin Jr, G. 1999).

The traffic flow model in VISSIM is a discrete, stochastic, time step based, and microscopic model with driver-vehicle-units as single entities (PTV, A. 2009). The model uses the psycho-physical driver behavior model for longitudinal vehicle movements and a rule-based algorithm for lateral movements. The model is based on a series of work by Wiedemann (1991).

The fundamental concept of this model is that when the driver of a faster moving vehicle reaches his individual perception threshold to a slower one, he is beginning to decelerate. Then, his speed will fall below that lower vehicle's speed because he cannot exactly determine the speed of that vehicle, until he starts to slightly accelerate again after the distance reaching his another perception threshold (see Fig. 1). Overall, in the model, the car following behavior is an iterative process of acceleration and deceleration, and in the iterative process, driving behavior can be classified four modes as free driving, approaching, following and braking.

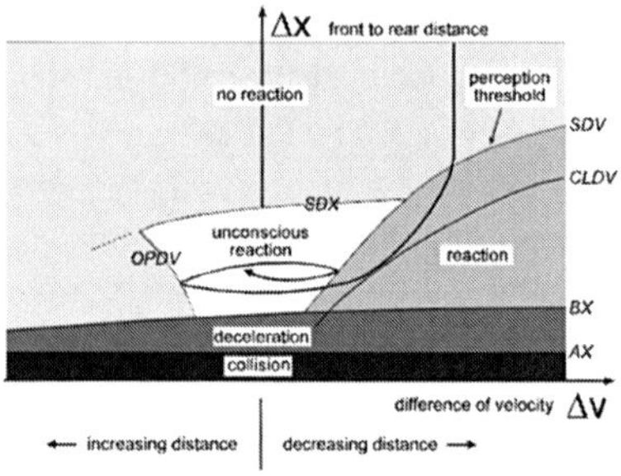

Fig.1. Car Following Logic

METHODOLOGY

In order to simplify the complicated traffic component and status, the research starts with two hypotheses:

(a). main road and relief road are both single lane road;

(b). vehicle stopped when preparing to enter the main road. After getting into the main road, vehicle accelerates to the average speed of main road.

When main road has a high volume, vehicles on the right lane may not be able to change lane to the rapid lane, and the hypotheses can be used to simulate the status. Under this circumstance, drivers has to stop the vehicle before enter the main road to observe, and then, when the driver begin to change lane, vehicles in the main road have to decelerate even stop to wait. With these two hypotheses, the research simulate in VISSIM, to determine the relief road access minimum intervals.

Modeling

As driving environment and behavior vary in different situation, parameter calibration is needed before traffic simulation. Characters of facilities such as the size of vehicle and road width doesn't differ much thus does not need much justification. Adjustment is made when parameter refers to driving behavior. Researches have shown that adjustment of some parameters lead to a 35% - 45% distinction in volume per second and density.

The model used in VISSIM is an improved version of Wiedemann's 1974 car following model. The following parameters are available in this model:

Average standstill distance (*ax*): It defines the average desired distance between stopped cars with a fixed variation of ±1 m;

Additive part of desired safety distance (*bx_add*) and multiple part of desired safety distance (*bx_mult*) :they together affect the computation of the safety distance.

The distance *d* between two vehicles is computed using this formula:

$$d = ax + bx \tag{2}$$

where *ax* is the standstill distance.

$$bx = (bx\_add + bx\_mult * z) * v \tag{3}$$

where *v* is the vehicle speed [m/s]

z is a value of range [0,1] which is normal distributed around 0.5 with a standard deviation of 0.15.

When taken into practice, the two parameters matter much to the result. Parameter *bx_add* and *bx_mult* is focused on for they partly determine the parameter *bx*, Research (YANG, H., HAN, S., and CHEN, X. 2006) has pointed out the advised value of *bx_mult* is 3.5-3.75, and *bx_add* 2.5-2.75. If the two parameter value is lower than the advised, the capacity of the road is quite sensible to operation velocity while lack of sensitivity otherwise. In this simulation, *bx_mult* is 3.75, and *bx_add* is 2.75 (see Fig. 2).

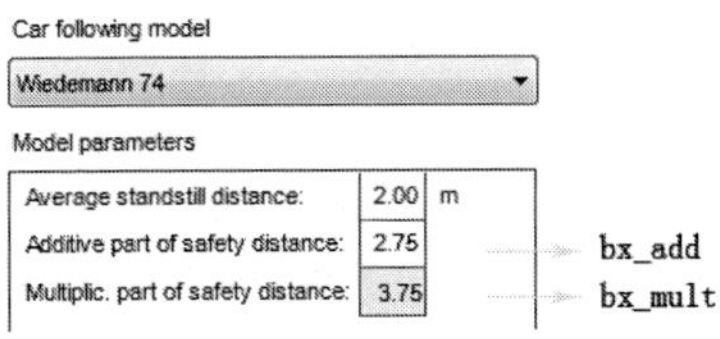

Fig.2. Value of *bx_mult* and *bx_add*

In order to get the relatively conservative minimum value, the ratio of small vehicle in the link is set to be 100%, as for the longer acceleration procedure and greater volume of truck comparing with small vehicle. The average speed of vehicles to be 80 km/h and to be linear distribution from 75 km/h to 85 km/h and the volume is at an average level to be 1000 pcu. When the traffic is heavy in the main road, vehicles in the relief road have to decelerate to a quite low level or even stop waiting to enter the main road. To simplify the process of calculation, we set the speed of the vehicle preparing to enter the main road from the relief road to be 0.

Time interval is long enough between the adjacent vehicles enter the main road to match the actual operation status of the relief road, and thus guarantees the impact made by one vehicle has dissipated when the next set off. Vehicles accelerate to the average speed of main road after it enters. Because the influence length is the only factor taken into consideration, the conflict rule is set to be relief road first to ensure enough vehicles may get into the main road, with little negative impact on the accuracy. In VISSIM, the main road and the relief road are both 3.5 m wide and arranged parallel. The main road is long enough to ensure the affected area can be covered. Time, number, velocity and relative coordinate on the main road of all the vehicles will be recorded in the simulation, and time interval of data collection is set to be 0.2 s.

Data

Situation is different as traffic flows, thus there is no distinct feature for the vehicle to change lane. Firstly select the number of vehicle that is affected manually. If there is only one vehicle that decelerate for the interrupt, the data. Circumstance that two or more vehicles are affected will be used for further analysis while ones that only one vehicle decelerating for the interrupt are filtrated. Vehicle line that is affected is written as *i*, while there are *n* vehicle lines in total in the simulation, and *m* vehicles are in the line.

The distance D_{ij} is computed using this formula:

$$D_{ij} = l_{i1} - l_{ij} \tag{4}$$

where l_{i1} is the coordinate that the first vehicle of line *i* located on the link when begin to decelerate;

l_{ij} is the coordinate that the *j*th vehicle of line *i* located on the link when begin to decelerate;

D_{ij} means the distance between *j*th vehicle and location at which first vehicle begin to decelerate when the *j*th vehicle begin to decelerate.

define

$$D_{imax}=max_{1\leq j\leq m}(D_{ij}) \tag{5}$$

and D_{imax} (see Fig. 3 and Table. 1) is the maximum length that affected in line *i*.

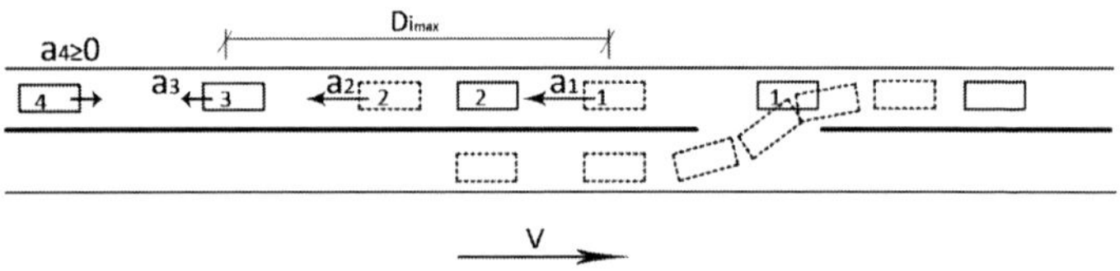

Fig.3. Affected Maximum Length

Table 1. D_{imax} in Different Vehicle Line

i	1	2	3	4	5	6	7	8
Distance (m)	110.9	32.9	31	72.5	2.6	29.3	4.9	28
i	9	10	11	12	13	14	15	16

Distance (m)	32.1	29.5	43.1	103.3	192	36.9	183.4	161.5
i	17	18	19	20	21	22	23	24
Distance (m)	143.8	50.6	187.1	36.4	60.8	171.3	85.8	183.8
i	25	26	27	28	29	30	31	
Distance (m)	203.6	233.3	75.1	116.9	173	138.9	81.4	

The scatter diagram of D_{imax} is shown below (see Fig. 4).

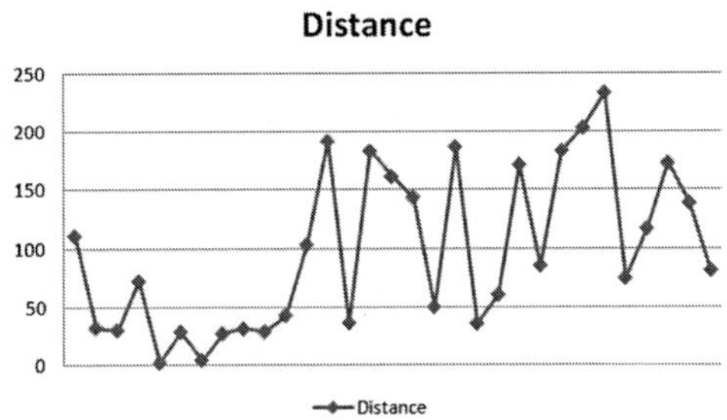

Fig.4. Scatter Diagram of D_{imax}

Considering the great fluctuation of the distance in the scatter diagram, we choose the 85th percentile as the relief road access minimum interval D_{min} instead of the average value from a statistical perspective (Mosteller, F., and Tukey, J. W. 1977), and D_{min}= 183.6 m. In other words, the distance between two relief road access in urban roads should not be less than 183.6 m to assure that the influence area will not reach the adjacent access.

CONCLUSION

Relief road access minimum intervals in urban roads is determined by the influence length on main road when the vehicles change lane from relief road to main road. In the calculation of relief road access minimum intervals, we should take account of various factors, such as the performance and size of vehicles. Meanwhile, the timing of the changing lane also plays a vital role in the determination of the minimum interval. The influence length will be relatively long when the vehicles turn into a dense traffic and will be short otherwise. Based on certain assumptions, this research obtained a series of data in different conditions with the help of traffic simulation software and chose the 85th percentile of these data 183.6 m as the relief road access minimum intervals.

There are several problems unsolved in the study. Although we can use single lane with relief roads to simulate two lanes with relief roads when the vehicles can't change lane into the main road due to the heavy traffic, it is not a common case after all. Moreover, the value we chose for D_{ij} tends to be smaller than the exact one. Because vehicles are discrete entities, we consider the influence is discrete, namely the influence will stop spreading after it reach a certain vehicle. This spreading process is similar to the influence that shock wave model makes in the micro-level; the terminate position of the influence might locates between two adjacent vehicles in real circumstances. The calculated influence length is therefore smaller than the actual one, and the relief road access minimum intervals D_{min} is conservative.

ACKNOWLEDGMENTS

The authors appreciate the support of the International Engineering Foundation.

REFERENCES

Brackstone, M., and McDonald, M. (1999). "Car-following: a historical review." *Transportation Research Part F: Traffic Psychology and Behaviour*, *2*(4), 181-196.

Chandler, R. E., Herman, R., and Montroll, E. W. (1958). "Traffic dynamics: studies in car following." *Operations research*, *6*(2), 165-184.

Chen, D. and Jiang, J. (2012). " *Optimization design method for auxiliary road of urban arteries*." *KEJI ZHIFU XIANGDAO*, (9).

Dong, A. (2012). "Design and application of urban auxiliary road." Shanxi Architecture, *38*(11), 153-154.

Golob, T. F., Recker, W. W., and Leonard, J. D. (1987). "An analysis of the severity and incident duration of truck-involved freeway accidents." *Accident Analysis and Prevention*, *19*(5), 375-395.

Mathew, T. V., and Radhakrishnan, P. (2010). "Calibration of microsimulation models for nonlane-based heterogeneous traffic at signalized intersections." *Journal of Urban Planning and Development*, *136*(1), 59-66.

Michaels, R. M. (1963, June). "Perceptual factors in car following." In *Proceedings of the 2nd International Symposium on the Theory of Road Traffic Flow (London, England), OECD*.

Mosteller, F., and Tukey, J. W. (1977). "Data analysis and regression: a second course in statistics." *Addison-Wesley Series in Behavioral Science: Quantitative Methods*.

Owsley, C., and McGwin Jr, G. (1999). "Vision impairment and driving." *Survey of ophthalmology*, *43*(6), 535-550.

PTV, A. (2009). "VISSIM 5.20 user manual." Karlsruhe, Germany.

Wang, X. Y., Yang, X. Y., Shan, G., and Wang, F. Q. (2006). "Review of the simulation model of driving behavior." *In Machine Learning and Cybernetics, 2006 International Conference on* (pp. 911-918). IEEE.

Wiedemann, R. (1974). "Simulation des Straßenverkehrsflusses. Schriftenreihe Heft 8." *Institute for Transportation Science, University of Karlsruhe, Germany.*

Wiedemann, R. (1991). "Modelling of RTI-Elements on multi-lane roads." *In DRIVE CONFERENCE (1991: BRUSSELS, BELGIUM)* (Vol. 2).

Yang, H., Han, S., and Chen, X. (2006). "Parameter calibration and application for the Vissim simulation model." *Urban Transport of China*, *6*, 004.

Zou,Z., and Yang,D. (2001). "A Comprehensive Review of Road Traffic Simulation Research." *Journal of Traffic and Transportation Engineering*, *1*(2), 88-91.

A Study of the Opening Size of Auxiliary Lanes on the Driving Behavior–Based Analysis

Yangyu Liu[1] and Lanfang Zhang[2]
[1]Department of Transportation Engineering, Tongji University, Cao'an Highway 4800, Shanghai 201804, China.
[2]Department of Transportation Engineering, Tongji University, Cao'an Highway 4800, Shanghai 201804, China (corresponding author).

Abstract

In China, with the auxiliary lane construction acting as an important means of access control management, more and more auxiliary lanes are constructed on the high-grade roads, but the opening size of those auxiliary lanes lacks definite rules and theoretical basis. In this paper, the characteristics of driving behaviors of entering and exiting the auxiliary lanes are considered, and a relative kinematics model is established on the basis of lane changing theory. The Minim Opening Safe Spacing (MOSS) is presented and acts as the safety index of the opening at certain speed. A final driving simulation test proves the good applicability of the model.

INTRODUCTION

In China, more and more auxiliary lanes are constructed on the cross section of urban arterial roads and high-grade highways to ease the impact of road access on the main traffic flow. The auxiliary lane construction has been acting as an important means of access control management.

The auxiliary lane refers to the road constructed for the vehicles to enter and exit the high-grade roads, and for the vehicles not permitted into the high-grade roads. The vehicles to enter and exit the high-grade road shall cross the auxiliary lane as a transition, and then enters or exits at the designated opening.

Generally, the auxiliary lane is constructed on both sides of the main lain, and its designing speed is lower than that of the main lane. The common auxiliary lanes mainly exist in two types: integral type and segregated type as shown in the figure below. The integral type has advantages over the segregated type like less land occupation, more convenient access to the main lane and so on. Therefore, the former is more widely applied in actual practice.

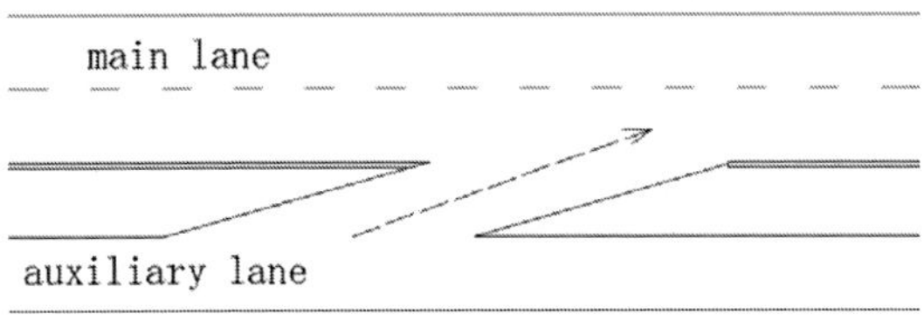

Fig.1.Integral auxiliary lane

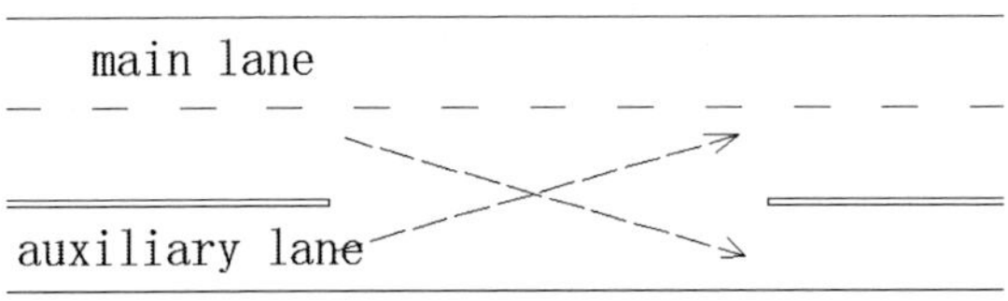

Fig.2. Segregated auxiliary lane.

The main function of the auxiliary lane is to transversely isolate the fast traffic flow from the slow traffic flow in space. Before reaching the opening of the road, the vehicle to exit the main lane shall slow down to make sure that it can turn safely and stability, and then exit. While the vehicle to enter the main lane shall slow down to give way or stop to wait for the traffic gaps at the open of the road, and then speed up to enter the main lane. Therefore, taking the auxiliary lane as the speed transition can improve the traffic speed of the main lane and the traffic safety of the whole road environment.

The domestic researches in the auxiliary lane present in the several following directions: (1) the Model of the Length of the Urban Freeway Ground Weaving Area (Li 2012), on the basis of experimental data from VISSIM simulation; (2) Traffic Thermodynamics Entropy Model (Wang et al. 2010),through the contrast between the two concepts of traffic flow in the auxiliary lane and the fluid in fluid mechanics, which provided a basis for the simulation system to describe traffic flow; (3) the impact on the Traffic Flow of Urban Expressway Exit from the auxiliary lane (Wang et al. 2008) ,analyzed through simulated test. In general, the researches relevant to auxiliary lanes mostly focus on the theoretical models of the traffic flow in the auxiliary lane, whereas still lack guidance on the practical detailed design of the auxiliary lane.

However, in regard of the auxiliary lane's design, there still remain many detailed problems to be settled, especially the determination of the opening size of the auxiliary lane. *Specification for Design of Urban Expressway CJJ129-2009,* as the design basis, merely regulates the requirement on the transverse isolation between the main lane and the auxiliary lane, i.e. the isolation shall be 0.5m distance from the lateral side strip of the main lane and 0.25 of the auxiliary lane. *Design Specification for Highway Route* however suggests that the auxiliary lane can also be constructed into the integral type by connecting the lane separator openings, in the circumstance of land tension.

Many research have been focused on lane changing of driving behavior. The relationship between driver awareness information and vehicles on different speeds ,different lane width and different driving direction(VanWinsum 1999), was researched through driving simulating; The region difference was found between different countries on lane changing behavior (Ioannis 2001) ; behavior characteristics during the process of lane changing was researched with a fixed driving simulator (Hiroshi 2001); vans and passenger cars drivers' different viewpoint features was described (Louis 2005).

If the auxiliary lane opening is unreasonably constructed, the moving status of vehicles entering or exiting the auxiliary lane and vehicles on the main lane may be affected, with the most direct reaction that whether the vehicles can enter or exit at the expected speed and not affect the traffic flow in the adjacent lane.

The text, by virtue of analysis on the driving behaviors of entering and exiting the auxiliary lanes, combining with lane changing theory, establishes a kinematics model and determines the opening size of auxiliary lanes.

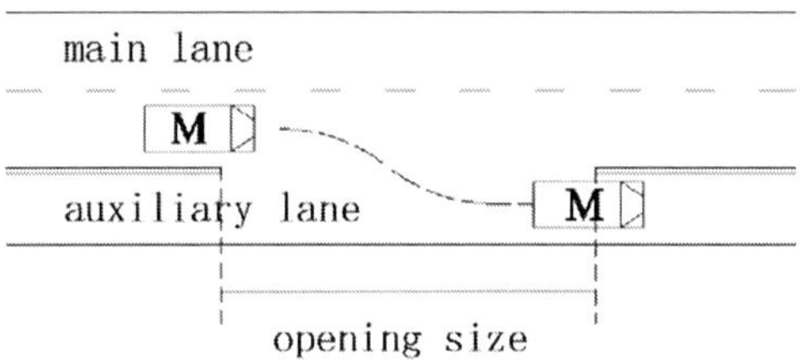

Fig.3.The diagrammatic sketch of the auxiliary lane opening.

Analysis on the driving behaviors

The driving behaviors exist in three phases as shown in the figure below: (1) information perception phase, in which the driver starts to percept the input external environment information; (2) decision making phase, in which the driver makes decisions according to his own intentions and interests; (3) vehicle handling phase, in which the driver handles the vehicle and achieves the driving expectations. The upstream module is the environment information, while the downstream module is the vehicle moving status. The environment information works as a basis for decision making. The vehicle moving status is a reaction to the vehicle handling and affects the environment information in next phase, which is then delivered to the driver. The overall process moves in circles.

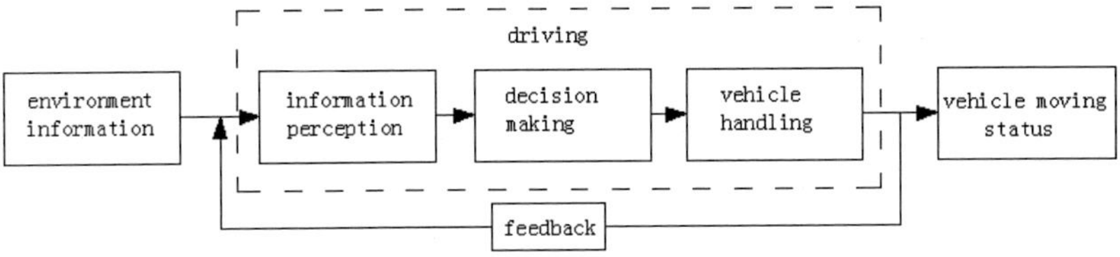

Fig.4.Thediagram of driving behaviors.

In terms of the driving behaviors of entering and exiting the auxiliary lanes, the driver firstly acquires and percepts the information like the location and the size of auxiliary lane opening as the decision basis. And then the driver makes the decision to enter or exit the auxiliary lane or not, according to the requirements of driving routes or self-willingness. The handling behaviors, including adjusting the speed by stomping on the brake pedal or accelerator pedal, turning the steering wheel, passing the opening area of the auxiliary lane and driving into the target lane, may change the moving status of the vehicle at the next moment.

The lane changing behavior describes the characteristic of the driver changing lanes. It is an integrated process combined by perception of environment information including the surrounding traffic environment i.e. the vehicles' speed and gaps and road environment, making the driving judgment and decisions, and adjustment to achieve the self-driving targets. According to the difference in the motivations of interests pursuit, the lane changing behaviors can be divided into Mandatory Lane Changing and Selective Lane Changing. Mandatory Lane Changing refers to lane changing behavior which must be implemented within certain areas with the definite target lanes, such as the diverging behavior the merging behavior of vehicles at the ramps or weaving areas. Selective Lane Changing refers to the lane changing behavior in pursuit of faster speed and more free driving space, when encountering slower vehicles ahead. Therefore, the driving behaviors of entering and exiting the auxiliary lanes belong to Mandatory Lane Changing. The driver shall abide by the traffic rules and travel plans and have to change lanes to achieve the normal driving targets, otherwise he may suffer a great time loss because of his ignorance.

According to the driving behaviors of entering and exiting the auxiliary lanes and the vehicles' moving status, the whole process of entering and exiting the auxiliary lanes can be divided into two phases: decelerating phase and turning phase. Decelerating phase refers to the process from stomping on the brake pedal when noticing the auxiliary lane opening ahead to loosening the brake pedal. Turning phase refers to the process from turning the steering wheel towards the target lane to returning the steering wheel and adjusting the vehicle to the target lane.

In the research, the major indexes of environment information are the vehicle speed and spatial distance, and most sensitive variables affecting the driving decision making are the speed of vehicles changing lanes and the limited adjustable spatial distance i.e. the opening size range of auxiliary lanes. The indexes indicating the vehicles' moving status are the vehicle speed and the vehicles' vertical and horizontal acceleration. Therefore, the above variables are the significant factors affecting the safety and efficiency of entering and exiting the auxiliary lanes.

MODEL SPECIFICATION

Scenario Analysis

Wu Xiaorui and Yang Hongxu proposed the Lane Changing Model to consider the car-following behavior(Wu ,Yang 2013). They added more space constraints to make the model more suitable for the actual road scene with a complex traffic environment. The figure below shows the model's structure. Vehicle M is the vehicle to change lane, and vehicle A and vehicle B are the following vehicle and the leading vehicle in the same lane with vehicle M, and vehicle C and vehicle D are the following vehicle and the leading vehicle in the target lane. Vehicle M shall change from the lane space between vehicle A and vehicle B to the lane space between vehicle C and vehicle D. Hajjaji pointed out that vehicle A made little impact on the lane changing process of vehicle M (Hajjaji, Ouladsine 2012). However, the distance between and speed difference vehicle B, vehicle C, vehicle D and vehicle M and speed variation remain as the major kinematics factors.

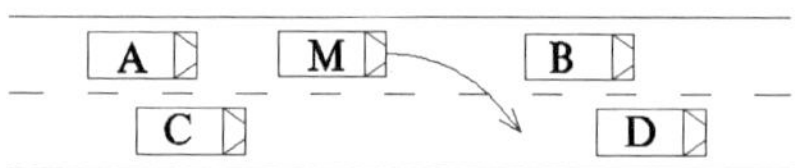

Fig.5. Lane Changing Model to consider the car-following behavior.

We may as well regard the upstream and downstream auxiliary lane fences as two still vehicles S1 and S2, so we can get a similar lane changing model.

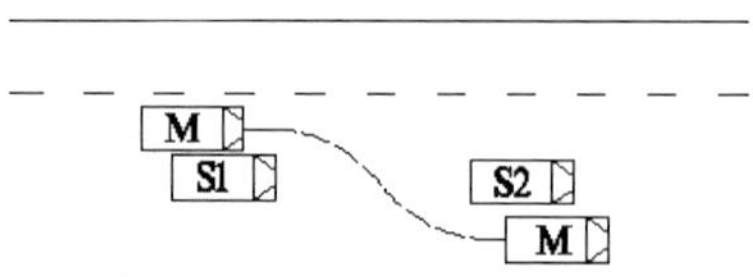

Fig.6. a similar lane changing model.

Vehicle M accelerates to gain the speed advantage and overtakes Vehicle S1, and it turns the steering wheel towards the target lane after gaining some certain distance advantage. Before entering the target lane and adjusting its direction, vehicle M shall control the speed difference and distance between itself and vehicle S2 to make sure that it does not crash into vehicle S2. Actually, the speed advantage of vehicle Mover vehicle S1 i.e. the speed difference is just the speed of vehicle M. Likewise, the speed difference between vehicle M and vehicle S2 is just the speed of vehicle M. Therefore, vehicle M has an obvious speed advantage. Vehicle M usually does not need to accelerate, but it needs to slow down ahead of time to control the speed difference between itself and vehicle S2. In the whole process, the effect of vehicle S1 on vehicle M is equivalent to the effect of vehicle C on vehicle M in the origin model, and vehicle S2 acts the role of vehicle B. therefore, the actual meaning of the distance between vehicle S1 and vehicle S2 is providing the opening size of auxiliary lane.

Model specification

Referring to the lane changing theory, the driving process of entering and exiting the auxiliary lane can be divided into two parts. The vehicle's speed is adjusted in part one, and the vehicle moves into the target lane in part two. If the vehicle enters the opening area at a too fast speed and the opening area is too narrow, it may be very urgent for the driver to turn the steering wheel. The hasty turning of the steering wheel may lead to the instability of the vehicle's moving, which breeds some security risk. On the contrary, the too low speed may leads to rear-end collision between the subsequent vehicles on the origin lane. Therefore,, the focus must be on the guidance and control of the vehicle's speed.

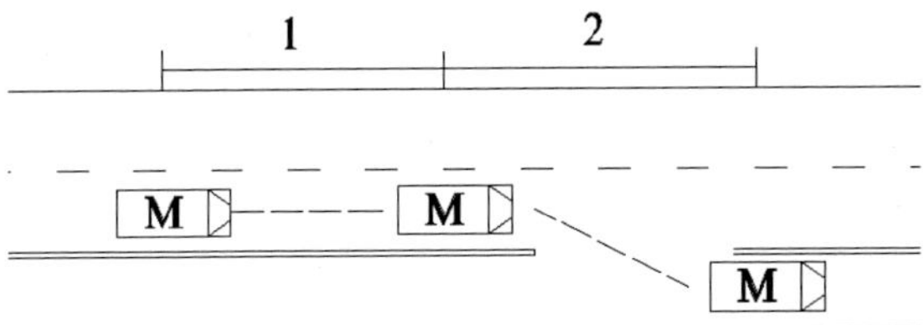

Fig.7.The diagrammatic sketch of the process

In the actual vehicle moving process, the driver may adjust the speed i.e. slow down before entering the opening areas of the auxiliary lane, and then turns the steering wheel and drives into the target lane at a constant speed, and finally drives normally on the target lane. In part one, the driver only needs to percept the location of the auxiliary lane opening and to make the decision of slowing down, let alone the size of the auxiliary lane opening. In part two, the vehicle needs to use the longitudinal space of the auxiliary lane opening to accomplish changing lanes. Therefore, we need only to consider the second part and establish the model relevant to the opening size.

The model of entering and exiting the auxiliary lanes is proposed on the kinematics analysis, and abides by the following assumptions:

In the process of lane changing, the transverse motion does not affect the longitudinal motion, the speeds of the two-direction- motion bear no connection.

The moving speed of vehicles on the main lane is higher than that of vehicles on the auxiliary lane.

According to the geometrical characteristic of the vehicle trajectory, the process of entering and exiting can be divided into two parts as shown in the figure below, with the location at the extension line of the auxiliary lane fences as the center. Part one consists of the Longitudinal Displacement S_{x1} and the Lateral Displacement S_{y1}, and part two consists of the Longitudinal Displacement S_{x2} and the Lateral Displacement S_{y2}.

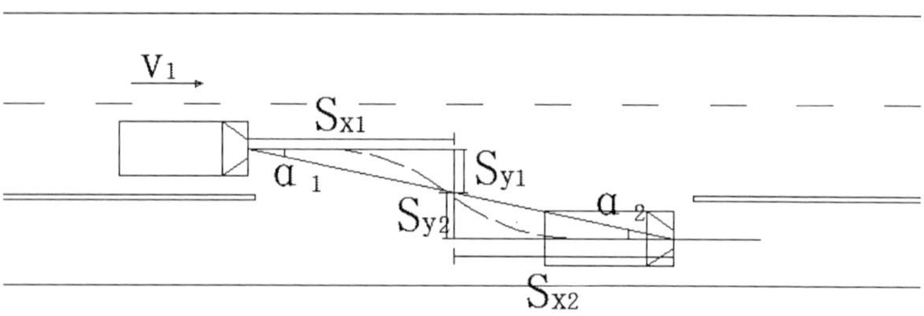

Fig.8. process of entering and exiting

Jose L Bascunana held that the Vertical Deviation Anglesα_1,α_2 were quiet small when the vehicle was changing the lanes (Bascunana 1995). Therefore it can be held that the Longitudinal Velocity v_x of moving in the opening area equals to the Velocity v_1 of entering the opening area of the auxiliary lane, and v_x remains the same. So the Longitudinal Displacements can be described as follows:

$$S_{x1} = v_x t_1 \quad (1)$$

$$S_{x2} = v_x t_2 \quad (2)$$

Note: t_1, $t_2$are the respective time of part one and part two.

The vehicle's maximum velocity of the lateral-direction motion is lower than that of the longitudinal-direction motion, and within less time, so the moving process can be deemed as the uniformly accelerated motion. In order to describe the vehicle's lateral dynamic characteristic, acceleration is selected to calculate lateral displacement.

So the Lateral Displacement can be expressed as following:

$$S_{y1} = \frac{1}{2} a_1 t_1^2 \quad (3)$$

$$S_{y2} = \frac{1}{2} a_2 t_2^2 \quad (4)$$

Note : $a_1$is the Lateral Acceleration of part one, and $a_2$is that of part two.

According to the constraints of lateral movement during the lane changing, the vehicle moves at least a distance of a lane's breadth. So the Expression presents:

$$S_{y1} + S_{y2} \geq W_l \quad (5)$$

Note: W_l is a lane's breadth.

According to symmetry, the following relation can be concluded:

$$a_1 = a_2 = a_y$$

$$t_1 = t_2$$

Therefore, the compilation of (3), (4) and (5) leads to:

$$S_{y1} + S_{y2} = a_y t_1^2 \geq W_l \quad (6)$$

$$t_1 \geq \sqrt{W_l / a_y} \quad (7)$$

In order to make sure that the vehicle's longitudinal motion in the opening area does not lead to the crash into the auxiliary lane fences, the following constraints shall be satisfied:

$$OS \geq S_{x1} + S_{x2} \quad (8)$$

Note : OS is opening size.

Substituting(1), (2) and (7) into (8) leads to:

$$OS \geq 2v_1 \sqrt{W_l / a_y} \quad (9)$$

It can be seemed from the Expression (9) that the opening size of auxiliary lanes is only relevant to three variables: the Velocity $v_1$of entering the opening area of the auxiliary lane, the crossing lane's breadth W_l and the Lateral Acceleration a_y.

Minim Opening Safe Spacing (MOSS)

In the lane changing theory, Safe Spacing works as the technical standard for safety. Safe Space refers to that the vehicle may crash into the leading vehicles in the same lane and the target lane in the process of changing lanes in urban roads, within the microsystem consisting of those vehicles. In order to avoid the occurrence of this situation, the driver has to consider the safe spacing to keep before changing the lanes. Correspondingly, Minim Opening Safe Spacing (MOSS) is used to describe the safety level of the auxiliary lane opening. MOSS refers to the minimum size to be constructed so that vehicles can pass the auxiliary lane opening safely and steadily at a certain speed. Otherwise, traffic accidents like crash into the fences or lateral instability may take place.

According to the definition, Minim Opening Safe Spacing (MOSS)can be expressed as following:

$$MOSS = 2v_1\sqrt{W_l / a_y} \quad (10)$$

The value of the Velocity v_l of entering the opening area of the auxiliary lane shall be related to the designing speed of this road. *Specification for Design of Urban Expressway* regulates that the designing speed of the auxiliary lane shall be 30 to 40 km/h. According to the theory that the operating speed shall be keep in line with the designing speed, the value of v_l and $|\Delta v|$i.e. the difference of the designing speeds in the auxiliary shall satisfy the in equation $|\Delta v| \leq 20km/h$.

Referring to *Code for design of urban road engineering* and *Design Specification for Highway Alignment*, the value of W_l, i.e. the crossing lane's breadth shall be the lane breadth of the high-grade road, namely $W_l = 3.75$m.

The value of a_y i.e. the Lateral Acceleration shall take a full consideration of the stability of the vehicle and comfortableness of the driver and passengers, so that a_y=0.1g=0.98m/s^2.

The design value of MOSS can refer to the table below.

Table.1.The value table of Minim Opening Safe Spacing (MOSS).

Velocity km/h	**20**	**30**	**40**	**50**
calculated value m	21.7	32.60	43.47	54.34
recommended value m	25	35	45	55

Simulation Test

The simulation test operates as follow: Virtools Simulator works as the experimental apparatus; the 1km long straight road presents the experimental scene; the auxiliary lane opening is constructed around 700m; the main lane presents two-way 4-lane, and the auxiliary lane is on the outside of the main lane; the designing speed of the auxiliary lane is 30km/h, and the designing speed of the main lane is 60km/h.

50 people participated in the test, with 30 of more than 10 years' driving experience. Every participating person drove twice, once to accomplish lane changing from the main lane to the auxiliary lane and once from the auxiliary lane to the main lane around

700m.

The test records the average lateral acceleration of entering and exiting the opening area. By virtue of regression analysis of the average lateral acceleration data, the data present significant normality. ($\mu = 0.9936, \sigma = 0.4197$)

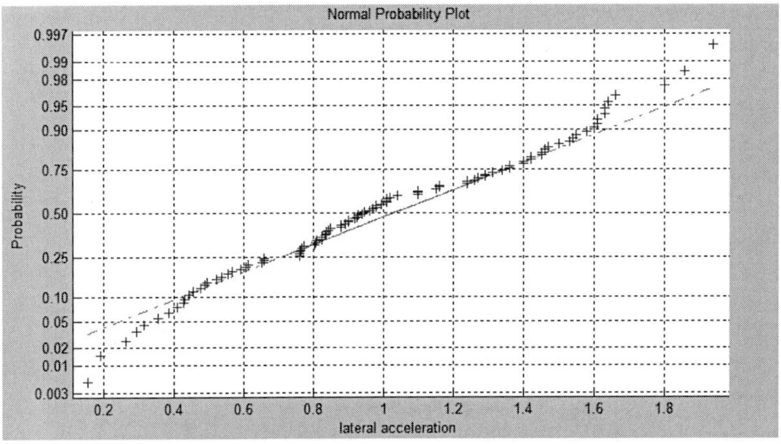

Fig.9.Normal regression of lateral acceleration.

The normality of the test data demonstrate that the model conforms to the driver's driving habits, so the data have certain credibility.

The lower part of lateral acceleration i.e. the left side of 0.7 may be the result of the lower driving speed by some unskilled drivers. While the higher part i.e. the right side of 1.5 may be caused by the hasty handling behaviors of the driver, when noticing the auxiliary lane opening at a rather high speed after running in the straight lane for a long while.

CONCLUSION

In the text, the characteristics of driving behaviors of entering and exiting the auxiliary lanes are considered, and a relative kinematics model is established on the basis of lane changing theory and the analysis on the motion process of entering and exiting the auxiliary lanes. The minim opening safe spacing (MOSS) is presented and acts as the safety index of the opening at certain speed. A final driving simulation test finally proves the good applicability of the model.

However what needs to be further studied is that whether the distinction between oversize vehicles and small vehicles i.e. the vehicles' size affects the design of the auxiliary lane opening, as well as whether the vehicle suffers the variation of moving status under different lighting conditions when running in the auxiliary lane area.

Therefore, the model will be relevantly modified on basis of above further studies.

REFERENCES

Hajjaji, A.E., Ouladsine, M. (2012).“Modeling Human Vehicle Driving by fuzzy logic for standardized ISO double lane change maneuver.”IEEE international workshop Robot and Human interactive Communication.

Code for design of urban road engineering(CJJ37–2012), Beijing, China Architecture and Building Press, 2012.

Design Specification for Highway Alignment (JTGD20—2006), Beijing, China Communications Press, 2012.

Hioshi, O. (2001).“Analysis and Modeling of Human Driving Behaviors Using Adaptive Cruise Control” *Applied Soft Computing*, 37(1),237-243.

Ioannis, G., Mstthew, G.K.(2002).“An International Comparative Study of Self-report Drive Behavior” Transportation Research Part F,32(4),243-246.

Bascunana,J.L. (1995).“Analysis of Lane Change Crash Avoidance” *SAEPaper 951895*. Future Transportation Technology Conference and Exposition, Costa Mesa, CA: Society of Automotive Engineers.

Louis, T. (2005). “Van and Passenger Car Driver Eye Glance Behavior During the Lane Change Decision Phase.” TRB2005 Annual CD-ROM.

Li, Z.B.(2005). “Study on Determination the Length of the Urban Freeway Ground Weaving Area and Improvement Measures”. Chang’an University, 2012.

*Specification for Design of Urban Expressway (CJJ129-2009).*Beijing, China Architecture and Building Press, 2009.

Peng, J.S., Fu, R. et al. “Research of Driver’s Lane Change Decision-making Mechanism” *Journal of Wuhan University of technology* 33(12), 46-50.

Kiefer, R.J. and Hankey, J.M. (2008).“Lane change behavior with a side blind zone alert system”*Accident Analysis and Prevention* 2008 (40), 683-690.

Wang, C.L., Qiang, T.G., Zhu, L. (2010).“Research on Traffic Thermodynamics Entropy Model of Urban Expressway Supporting Road*”, Forest Engineering*, 26(5),63-65.

Wang, Z.B., Gu, N., Li, J.S. “Impact on the Traffic Flow of Urban Expressway Exit from the Auxiliary Lane”*Journal of Transportation Engineering and Information*, 1(6),31-37

VanWinsum,W., de Waard, D., Brookhuis, K.A.(1999). “Lane Change Manoeuvres and Safety Margins.” Transportation Research Part F.,(2),139-149

Wu, X., Yang, H. (2013).“A Lane Change Model with the Consideration of Car Following Behavior.” *Procedia-Social and Behavioral Sciences* 96, 2354-2361

Freeway Access Management Practices in California

Zhongren Wang[1]

[1]California Department of Transportation, 2389 Gateway Oaks Dr., Suite 200, MS-91, Sacramento, CA 95833.

Abstract

Freeway is the type of facility that enjoys the highest level of access control. It serves the main purpose of mobility instead of accessibility. However, due to the ever-increasing travel demand and challenges to building new freeways, the existing freeway facilities are retrofitted with a full range of system management strategies, such as entrance ramp metering and managed lanes, and the consideration of non-motorized travel demand. The implementation of these strategies presented a unique set of access management issues, such as new geometric design standards, and project development policies and procedures. A general introduction of the freeway access management practices in California is presented in this paper, together with the key traffic engineering and design considerations related to ramp metering, managed lanes, and consideration of non-motorized travel demand.

THE GENERAL ACCESS MANAGEMENT PRACTICE

In California, the Streets and Highways Code [California$_a$, 2014] and California Environmental Quality Act (CEQA) [California$_b$, 1970] formed the basis for the access management business. According to the Streets and Highways Code, the Department of Transportation (Caltrans) is the sole owner and operator of the state highway system, including freeways. It is incumbent upon Caltrans to carry out the access management function of the entire state highway system.

In California, the state highway system has a total centerline length of 24, 371 km. The system accounts for less than 10% of the total roadway mileage in California, but serves 15 billion miles of travel per day [Caltrans$_a$, 2013]. In the conventional sense, freeway facility is typically a divided multilane facility with fully controlled accesses. Freeway interchanges are spaced apart no less than 1.6 kilometers.

There is no single dedicated program that administers highway access management in California. Instead, multiple Caltrans Divisions are involved. For facility design, the Division of Design maintains the Caltrans Highway Design Manual (HDM) [Caltrans$_b$, 1999], where all facility design standards such as intersection spacing and freeway design reside. The Access Management (AM) principle of consolidating accesses along major roadways is well incorporated into the HDM and practiced throughout California. In addition, the Division maintains the Project Development

Procedure Manual (PDPM) [Caltrans$_c$, 2014], which prescribes the detailed project development and review processes so that design oversight may be caught.

For facility planning, the Division of Transportation Planning develops a long-term, statewide vision for the transportation system. It ensures the integration of efficient land use and transportation system to provide the international, interstate, and interregional mobility and accessibility needs of people, goods, and services. The Division of Transportation Planning leads the Local Development-Intergovernmental Review (LD-IGR) process, which belongs to the mandated environmental review process under the California Environmental Quality Act (CEQA). LD-IGR is the primary vehicle that Caltrans employs to reflect its stewardship of the state highway system. Through this process, transportation impacts resulting from land use development are supposed to be either eliminated or reduced to a level of insignificance.

A guide for the preparation of Traffic Impact Studies (TIS) was adopted in 2002 [Caltrans$_d$, 2002]. The TIS Guide outlined the necessity and detailed contents of a TIS report. It provided guidance on the selection of analysis scope, methodologies, and mitigation strategies. It also provided a fair share calculation methodology which is widely referenced [NCHRP 2012]. On the basis of fair-share, more than 30 million dollar worth of mitigation projects have been implemented on the state highway system since the year 2000 [Wang, 2007].

Following the TIS process is the encroachment permit process. Any project proposed outside of Caltrans must obtain a permit from Caltrans as long as the project will encroach upon the right-of-way of a state highway. The encroachment permit process is administered by the Division of Traffic Operations following the Caltrans Encroachment Permit Manual [Caltrans$_e$, 2013]. The permittee must abide by the time window, material used, and traffic control requirements. Together with the CEQA and TIS process, all project development activities that affecting the state highway system are reviewed, documented, and managed; and the impacts are also properly mitigated.

In addition to the reviews for the project development activities, truck access that influences the day-to-day operations of the state highway system is also reviewed. Oversize or overweight trucks must obtain special permits, called transportation permits to access the state highway system. According to the California Vehicle Code [DMV 2014], Caltrans has the discretionary authority to issue special permits for the movement of vehicles/loads exceeding statutory limitations on the size, weight, and loading of vehicles. The transportation permit process is administered by the Division of Traffic Operations [Caltrans$_f$, 2011].

THE MOBILITY PYRAMID

Due to the ever-increasing travel demand and challenges of building new freeways, the existing freeway facilities are retrofitted with a full range of traffic management strategies, such as entrance ramp metering and managed lanes, and the consideration of non-motorized travel demand. The purpose of these strategies is to exploit to the maximum extent the productivity of the existing facilities, before investing in any expansion. This concept is best illustrated in the mobility pyramid shown in Figure 1.

The mobility pyramid encompasses almost all functions that the California Department of Transportation (Caltrans) performs. These functions include system monitoring and evaluation, maintenance and preservation, smart land use and demand management, Intelligent Transportation System (ITS) implementations, operational improvements and system completion and expansion [Caltrans$_g$, 2013]. These functions also include highway crash prevention and safety. Examining the contents of the mobility pyramid, one finds that the pyramid uses system monitoring and preservation as its base; ITS implementation as its center; and system expansion as its top. Such a layout reflects the ideology of Caltrans mobility improvement. System preservation and safety is of fundamental importance. Before any significant investment on system completion and expansion, effort must be made to unleash the potential productivity of the existing system by implementing ITS technologies, such as demand management, traffic control, traveler information, and incident management.

Figure 1. The California mobility pyramid.

Such an ideology leads to aggressive planning, design, and implementation of transportation management strategies, such as ramp metering and managed lanes. Other strategies such as shoulder running, variable speed advisories, coordination of ramp meters and street signals are also being planned or experimented. The implementation of these technologies transformed the freeway system and the way it is operated, and thus created a unique set of access management issues. These issues may include new geometric design and system operation standards, and new project development policies and procedures. These may also involve the re-balancing of mobility and accessibility of a freeway facility. The key traffic engineering and design considerations related to Managed Lane, ramp metering, and consideration of non-motorized travel demand are presented in the following Sections.

MANAGED LANE

Managed Lane is a lane-level traffic management strategy by setting vehicle eligibility, or limiting the access points of a designated lane, typically the left-most lane. Vehicle eligibility can be set based on vehicle type or occupancy (the minimum number of people carried). In particular, certain vehicles are exempted from the occupancy requirement to access a managed lane facility. For example, an electric car with a solo driver is allowed to use the facility with a minimum occupancy requirement of two or more. In California, the managed lane facility includes HOV Lanes, Express Lanes and Park and Ride facilities.

More than 40% of the total managed lane miles in the United States exist in California. The managed lane is typically constructed by widening the existing facility to the median, and not by taking an existing lane. Some managed lane facilities may have dedicated on- and off-ramps. But more often than not, the managed lane, due to its retrofitting nature, is accessed by using the existing on- and off-ramps on the far right. The managed lane traffic has to weave through all the existing general purpose lanes to enter or exit. A freeway, retrofitted with a Managed Lane grants operates as a combo system, and its operations include the operations of not only the conventional general purpose lanes, the managed lane, but also their interactions. The access point design, access spacing, and sight distance are just a few topics that have access management connotations.

Mainline access design

Mainline access refers to how vehicles move in and out of the managed lane along the mainline. Two types of access designs are typically used in California, that is the continuous and the limited access design. As shown in Figure 2(L), the continuous design allows vehicles to move in and out of the managed lane continuously along the lane. There is no physical separation between the managed lane and its neighboring general purpose lanes. This type of access is more frequently used in Northern California. On the other hand, as shown in Figure 2(R), the limited access design allows vehicles to access the managed lane only at designated locations, or the ingress/egress locations. This type of access design is more frequently used in Southern California. Studies indicated that the both types of access designs have similar safety performance [Jang et al. 2009].

Figure 2. A HOV lane contiguous with mixed-flow lanes.

Managed lanes typically do not have dedicated ingress/egress ramps (on the left hand side), and they are accessed by weaving through the neighboring general purpose lanes as shown in Figure 3. The primary design standard is the 240 m (800 ft.) per lane change, which is applicable for both types of access designs. Apparently, the 240 m (800 ft.) per lane change standard is a key consideration in selecting the location of the ingress/egress, especially in retrofitting design. For the limited access design shown in Figure 3, the access location may not be appropriate if the on- and off-ramp spacing is less than 1200 m (4000 ft.).

There are various types of designs for the ingress/egress openings. When ingress and egress are collocated as shown in Figure 3, the primary standard is to keep the opening at least 600 m (2000 ft.) long. A weaving lane may be provided to ease the weaving operations at the ingress/egress location if the weaving demand warrants.

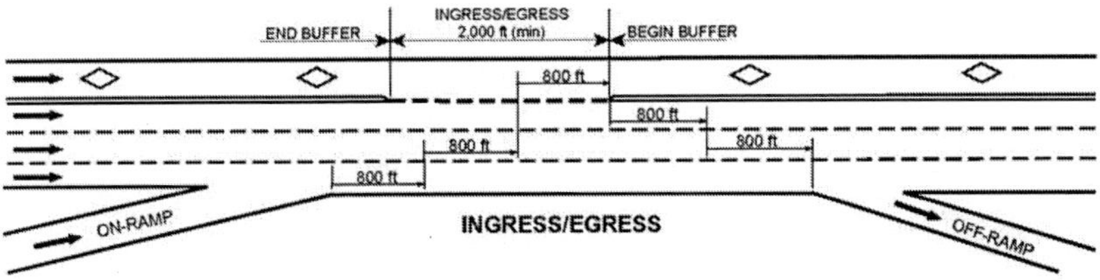

(1 ft. = 0.3048 m)

Figure 3. Mainline limited access design [Caltrans$_h$, 2003, Caltrans$_i$, 2011].

The ingress and egress may also be separated to mitigate the ingress/egress weaving operation problems. A speed change lane may be provided to smooth the ingress/egress operations. The speed change lane design for ingress is schematically shown in Figure 4. The speed change lane is composed of approaching taper, bay taper, and an acceleration distance (the general purpose lane traffic should travel at lower speed). The egress with a speed change lane is simply the mirror image, but the acceleration distance becomes the deceleration distance.

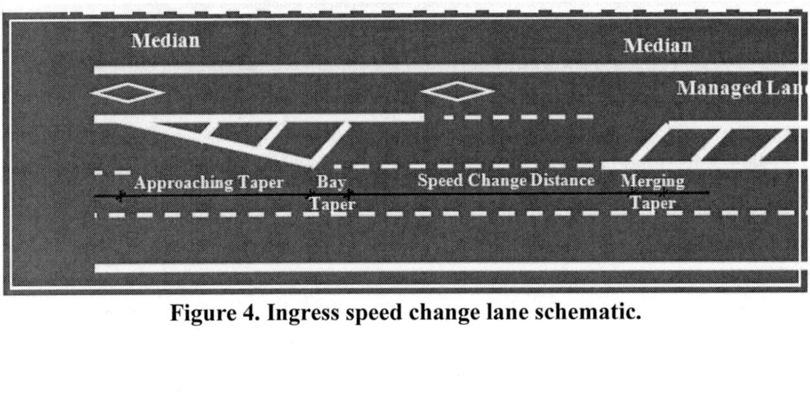

Figure 4. Ingress speed change lane schematic.

Dedicated ingress/egress ramp may be provided for the managed lane facility. As the managed lane facility is typically built next to the median, the dedicated ramps are typically elevated, and thus costly. What is shown in Figure 5 is a dedicated transit ingress/egress serving a transit center, and park and ride lots. Following the arrows shown in Figure 5, transit vehicles can access the manage lane facility through the overcrossing structure, without weaving through the general purpose traffic. Such design may increase transit ridership, and improve transit performance.

Mainline access spacing

For limited ingress/egress design, the minimum separation between mainline access openings is not specified. It is leaving to engineering judgment to determine the most appropriate access location. Many factors play into the selection of an appropriate access spacing. Facility geometry, traffic characteristics and operating conditions for the managed lane, the neighboring general purpose lanes, and the on- and off-ramp traffic are just a few of these factors. Certain commuting demand pattern may also limit the number and location of ingress/egress for a certain section or the freeway. For example, an egress will not be placed at a location with potential local street queue overspills.

Figure 5. Transit dedicated access ramps.

Sight distance

When retrofitting a managed lane facility in the median of an existing freeway, the available right-of-way may not only limit the lane and shoulder width, but also create sight distance issues. As the retrofitted managed lane facility is typically a single-lane facility, obstructions alongside both the left- and right-hand-sides may cause sightline blockage. For example, when the alignment curves to the right as shown in Figure 2L, the queued traffic in the general purpose lane causes the sightline blockage. On the other hand, when the alignment curves to the left, the tall median concrete barrier may also become apparent sight blockage as shown in Figure 2R.

For the continuous access design, the queued traffic in the adjacent general purpose lane will govern the design, as long as the left shoulder width is more than 0.92 m. It is therefore equally necessary to check for sight distance impairment as a result of queued traffic in the adjacent general purpose lanes when the alignment curves to the right [Wang, 2013].

RAMP METERING

A ramp meter is a traffic signal that controls the rate of entry of vehicles from a ramp onto a limited access facility; the signal allows one or two vehicles to enter on each green or green flash [TRB 2010]. Ramp meters cause on-ramp vehicles to stop before accelerating to merge with mainline traffic. Ramp meters changed the way a freeway is conventionally accessed when no such stop is necessary. What is more, freeway-bound traffic is favored in the ramp terminal intersection design. Free-right bypass is typically provided for expedient freeway access. However, with the introduction of ramp meters, together with the consideration of the safety of non-motorized traffic, freeway-bound traffic no longer enjoys such favors. The freeway-bound traffic may not only experience queue overspill issues at on-ramps; but also possible acceleration issues due to the regulatory stop at the ramp meter. To accommodate this new operating mechanism, the on-ramp, the acceleration length, the ramp terminal intersection, and possibly the associated upstream feeding facilities all have to be designed differently.

Storage length

At a signalized intersection, storage length is an essential design element at every approach as part of the intersection influence area. The storage length is estimated based on twice the average number of arrivals for the specific phase in consideration as recommended by the Caltrans Highway Design Manual 405.2 [Caltrans_b, 2013]. Adequate storage length is necessary to temporarily 'store' traffic queue during a red phase. Inadequate storage length may lead to queue overspill.

A ramp meter is nothing different from a regular street signal in terms of the need of storage length. The queue caused by a ramp meter should best be kept within the ramp, so that the operations of upstream feeding facilities will not be affected. It is unfortunate that most California freeways were constructed before the advent of ramp metering. No storage length was thus built in to the on-ramps. Once a limit line is drawn as shown in Figure 6, inadequate storage length and/or acceleration length becomes unavoidable with a given on-ramp design. That is why ramp meter installation is typically associated with on-ramp improvement. In California, any time when a ramp meter is installed, a High Occupancy Vehicle (HOV) preferential lane shall be installed as shown in Figure 6 [Caltrans_j, 2011].

Various methods exist to estimate the storage length [Wang 2013]. In a current effort to update the Caltrans Ramp Metering Design Manual (RMDM) [Caltrans_k, 2000], storage length is recommended to be designed using seven percent of design hour volume. This method agrees with practices at the States of Wisconsin, Minnesota. The practice also agrees with that in Australia. But it might overestimate storage length when demand is high, say at a metered connector, as freeway-to-freeway connectors may be metered in California. An academic research project is underway to further develop the storage length design method. The detailed arrival patterns will be taken into consideration in such a methodology.

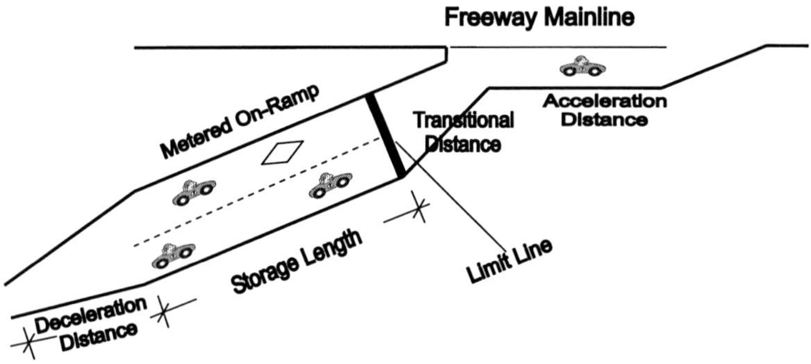

Figure 6. Schematic drawing of a metered on-ramp.

Acceleration distance and auxiliary lanes

At a metered on-ramp, vehicles start from zero speed to accelerate to merging speed. The speed-change process requires adequate accelerating distance to materialize. When an existing on-ramp is not designed as a metered one, improvement is necessary to provide the adequate acceleration distance. As it can be seen in Table 1, auxiliary lane installation has been a standard practice at on-ramps with special geometric or traffic situations, especially at metered on-ramps. A new design standard is being proposed to install auxiliary lane at all on-ramps with a minimum length of 100 m. Auxiliary lane is believed to improve traffic safety.

Table 1. Auxiliary Lane Standards for Metered On-Ramps in California.

Facility		Minimum Length, m	Remarks	Source
Metered On-Ramps	1- Lane	150	Truck demand is 5% or greater on ascending on-ramps to freeways with sustained upgrades exceeding 3%.	HDM 504.3(2)(a)
	2-3 Lanes	300	Truck demand is 5% or greater on ascending on-ramps to freeways with sustained upgrades exceeding 3%.	HDM 504.3(2)(b)
			Ramp volume exceeds 1,500vph	
	All	100		Proposed New Criteria

Sight distance

Ramp meters will create queues along metered on-ramps. The sight distance to the back of queue is a concern. Adequate sight distance must be provided, especially when the back of queue is hidden from the view of the approaching motorists. Together with the queue storage length, and additional acceleration distance needed, such a requirement will further increase the length of a metered on-ramp. When it is not possible to contain the queue within a metered on-ramp, it is time to change the design of upstream feeding facilities. For example, some local agencies (owner and operator of the upstream feeding facilities) agree to use their facilities as queue storage for ramp meters during peak hours. This is sometimes agreed upon in exchange of somewhat smaller interchange footprint.

NON-MOTORIZED TRAFFIC AND AT-GRADE INTERSECTION DESIGN

Elimination of the free-right movement

The 2013 California Household Travel Survey [Caltrans_l, 2013] found that trips made on foot, bike, and transit totaled nearly 23% of all daily trips statewide. Driving solo trips have decreased by 11%, while walking rates doubled from 8.8% to 16.6% for all trips. Non-motorized travel has become an important mode choice for Californians.

In 2009, Caltrans settled a $1.1 Billion worth of lawsuits to improve the accessible conditions of 4000 km of state-controlled sidewalks, cross walks, ramps and 300 park and ride facilities throughout California. Each year in the next 30 years, Caltrans shall continue to improve the accessible conditions of these public right-of-way to meet the requirements of the Americans with Disabilities Act (ADA). It becomes imperative for Caltrans to consider non-motorized traffic demand and build complete streets in its system improvement effort.

In the 2013 revision of the HDM, the free-right type of design for both on- and off-ramps was eliminated. Instead, all turning movements are controlled as shown in Figure 7. Such design is believed to be safer for non-motorized travel needs because all movements are controlled. In addition, with a ramp meter operating downstream, it does not make much sense for the freeway-bound traffic to bypass the signal at the ramp terminal intersection.

At such signalized intersections, dedicated bicycle detection is required per Caltrans policy [Caltrans_m, 2009]. In California, a special type loop detector is used. This type of loop detector is more sensitive than those used for vehicle detection. The minimum pedestrian timing requirement must also be followed. A walking speed of 1.07 meter per second is typically used.

In California, bike lanes are classified into 3 categories [Caltrans_b, 1999]. Category I bike lane has dedicated right-of-way, separating from other traffic. Category II bike lane shares the road with motorized traffic, but with marked lane designation. Category III bike lane has no specific bike lane designation, and shares the right-of-way with motorized traffic. Bicycle traffic is also allowed to ride on freeway shoulders in certain routes. The total mileage reaches 500 km.

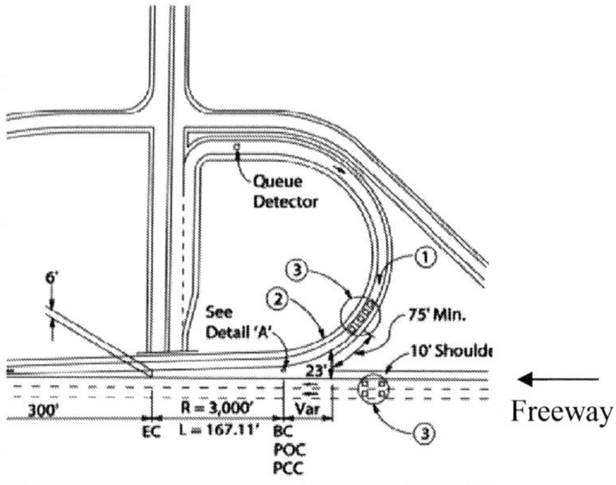

Figure 7. Ramp terminal intersection design considering bicycle and pedestrian traffic [Caltrans$_b$, 1999]**.**

HOV Preferential Lane Access Design

With the introduction of the HOV preferential lanes at metered on-ramps, the non-motorized traffic has to cross one-more lane at the ramp terminal intersection. In order to shorten the crossing distance, the HOV preferential lane access typically employs a bulb-out design as shown in Figure 8. The HOV access flares up downstream of the cross walk. Such an access design also avoids trapping the general purpose traffic to the HOV preferential lane.

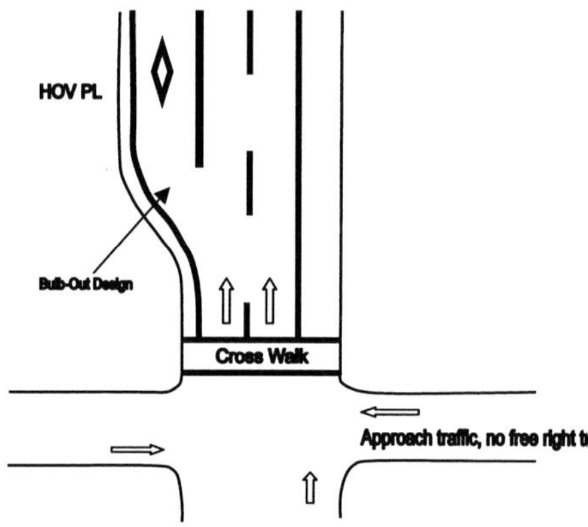

Figure 8. HOV preferential lane access design.

Roundabout

Caltrans started to install modern roundabout in the State Highway System in the early 2000s. Up to now, there are 20 existing roundabouts, with another 60 planned or programmed [Caltrans$_n$, 2012]. Due to its safety and mobility benefits, roundabouts start to be taken as feasible alternative intersection control strategies in California. The year 2013 saw the institution of a Traffic Operations Policy Directive (TOPD), which established an integrated, systematic, and performance-based approach to engineering and investment decision affecting state highway intersections and interchanges [Caltrans$_o$, 2013]. The key process change is that intersection access control must consider all possible strategies, including signal, stop, and yield control (roundabout).

With such a policy, roundabout will be evaluated as an option for any intersection. Whichever option is the best suited for the prevailing and/or expected traffic demands and operating conditions at particular locations will be selected. For example, the multi-legged ramp terminal intersection shown on the left of Figure 9 was to be improved to provide acceptable operation performance for both the intersection and the freeway. This roundabout alternative was determined by Caltrans and the relevant local agency to be the only viable improvement. A partial dual-lane roundabout will a right-turn bypass was finally constructed as shown on the right-hand-side of the same Figure.

Figure 9. Roundabout at a freeway ramp terminal intersection.

A word of caution is that roundabout installation at ramp terminal intersections must coordinate with ramp metering. If a ramp meter is to be installed downstream of a roundabout, it is better to check whether adequate queue storage space has been provided at the on-ramp. The queue caused by the ramp meter may overspill and interfere with the continuous operations of the roundabout. If this may happen, it then makes better sense to place the ramp meter upstream, instead of downstream, of the roundabout.

SUMMARY

In California, the system management goal is to fully exploit the potential of the existing facilities, before building any new. This goal makes system management strategies such as managed lane and ramp metering indispensable. On the other hand, the installation of managed lane and ramp meters onto the existing freeway system fundamentally changes the way how the system is accessed. With managed lane, the freeway system becomes a combo system. With ramp meters, the on-ramps operate like signal-controlled urban streets. With the accommodation of the non-motorized traffic demand, the ramp terminal intersection design introduces new elements. The general design practice for such a freeway system is presented in this paper, together with the policies and procedures.

On a nationwide basis, the studies of system management strategies emphasize more on their congestion-relief benefits, and not on the access management context. In fact, ramp meter may help ease freeway weaving operations, and may have implications on ramp spacing. Ramp meter also increases the influence area of an on-ramp which may interfere with the operations of the ramp terminal intersection. On the other hand, managed lane without dedicated access ramps may require longer ramp spacing. The friction between the managed lane and its neighboring general purpose lane(s) drives the search for better ingress/egress designs. Investigations down this line may well become a new chapter in the Access Management Manual [TRB_b 2013].

ACKNOWLEDGEMENTS

The contents of this paper reflect the views of the author who is solely responsible for the facts and accuracy of the data presented herein. The contents do not necessarily reflect the official views or policies of the State of California. This paper does not constitute a standard, specification, or regulation.

REFERENCES

$California_a$ (2014). *Streets and Highways Code,* Section 90. State of California Sacramento, California.

$California_b$ (1970). *California Environmental Quality Act.* State of California, Sacramento, California.

$Caltrans_a$ (2013). The Vehicle Miles of Travel (VMT) report. California Department of Transportation, Sacramento. Available online: http://traffic-counts.dot.ca.gov/.

$Caltrans_b$ (1999). *Highway design manual, 5th Edition.* California Department of Transportation, Sacramento.

$Caltrans_c$ (2014). *Project development procedures manual.* California Department of Transportation, Sacramento.

$Caltrans_d$ (2002). *Guide for the Preparation of Traffic Impact Studies.* California Department of Transportation, Sacramento.

NCHRP 2012. *State of the practice in highway access management.* National Highway Cooperative Research Program Synthesis Report 404, TRB, Washington, D.C..

Wang, Z. (2007). *"Local development-intergovernmental review: challenges and opportunities."* TRB 86th Annual Meeting Compendium of Papers CD-ROM, 2007.

$Caltrans_e$ (2013). *Encroachment permits manual.* California Department of Transportation, Sacramento. Available online: http:// www.dot.ca.gov/hq/traffops/permits/manual.htm.

DMV (2014). *California vehicle code,* Division 15. California Department of Motor Vehicles (DMV). Sacramento, California.

$Caltrans_f$ (2011). *Transportation permits manual.* California Department of Transportation, Sacramento, California.

$Caltrans_g$ (2013). *Transportation management system master plan.* California Department of Transportation, Sacramento.

Jang, K., Ragland, D., and Chan, C-Y (2009). *A comparative safety study of limited vs. continuous access high occupancy vehicle (HOV) facilities.* Final Report, UCB-ITS-PRR-2009-22, March 2009. Available online: http://www.dot.ca.gov/research/researchreports/reports/2009/final_report_task_0968.pdf.

$Caltrans_h$ (2003). *HOV Guidelines.* California Department of Transportation, Sacramento. Available online: http://onramp.dot.ca.gov/hq/traffops/quicklinks.html.

$Caltrans_i$ (2011). Traffic Operations Policy Directive (TOPD) 11-02: *Updated managed lane design.* California Department of Transportation, Sacramento. Available online at: http://www.dot.ca.gov/hq/traffops/liaisons/ice.html.

Wang, Z. (2013). "The sight distance issues with retrofitted single-lane HOV facilities." *International J. of Transportation Science and Technology,* Vol. 2 (2), 2013, 149 – 157.

TRB_a (2010). *Highway Capacity Manual 2010.* Transportation Research Board. National Research Council, Washington, D.C..

$Caltrans_j$ (2011). Deputy Directive (DD) 35-R1 "Ramp Metering". California Department of Transportation, Sacramento.

Wang, Z. (2013). "Queue storage design for metered on-ramps." *International J. of Transportation Science and Technology,* 2(1), 47-64.

$Caltrans_k$ (2000). *Ramp meter design manual.* Traffic Operations Program, California Department of Transportation, Sacramento.

$Caltrans_l$ (2013). *2013 California Household Travel Survey.* Final Report, 2013. Available online at: http://dot.ca.gov/hq/tsip/otfa/tab/documents /chts_finalreport/Final Report.pdf.

$Caltrans_m$ (2009). Traffic Operations Policy Directive 09-06: *Provide bicycle and motorcycle detection on all new and modified approaches to traffic-actuated signals in the state of California.* California Department of Transportation, Sacramento. Available online at: http://www.dot.ca.gov/hq/traffops/signtech/signdel/policy/09-06.pdf.

$Caltrans_n$ (2012). *California State Highway System Roundabout Inventory.* California Department of Transportation, Sacramento. Available online at http://www.dot.ca.gov /hq/tpp/offices/oasp/rtb_inventory.html.

$Caltrans_o$ (2013). Traffic Operations Policy Directive 13-02: *Intersection Control Evaluation (ICE).* California Department of Transportation, Sacramento. Available online at: http://www.dot.ca.gov/hq/traffops/liaisons/ice.html.

TRB_b (2003). Access Management Manual. Transportation Research Board. National Research Council, Washington, D.C..

Model of Deceleration Lane Length Calculation Based on Quadratic Deceleration

Benmin Liu[1] and Lingzong Kong[1]

[1]Key Laboratory of Road and Traffic Engineering of the Ministry of Education, Tongji University, Shanghai 201804, China. E-mail: 278887478@qq.com

Abstract

Through observation and data analysis of the eight-lane highway deceleration lane existing in China, it finds that large vehicles have lower speed better driving regularity than small ones, and that the deceleration lane can meet the demands of large vehicles. In this study, it focuses on the small vehicles and combines running speed, average speed, and designing speed to study, in which it considers the impact of large vehicles. Considering the way of deceleration lane calculation based on quadratic deceleration, it divides the deceleration lane into three sections which are triangular transition section, the first deceleration section and the second deceleration section, according to the deceleration process and characteristics in each one, and study the related parameters. At last, it proposes the deceleration lane calculation model and the corresponding recommended value.

Keywords: Highway; Quadratic deceleration theory; Deceleration lane length

INTRODUCTION

The highway deceleration lane is the additional lane that is used to make sure of security for shunt vehicle deceleration space. It is composed of the triangle transition section and the deceleration part. The United States, Germany, Canada, Japan and other countries dictate the length of varying velocity lane. Through analysis of the reference value in these specifications, it finds that on one hand, some reference values of the China's "Design Specification for Highway Route" (JTG D20-2006) are smaller than those of the United States, Germany, and Canada, but are larger than those of Japan; On the other hand, the problem that how to determine the length of the deceleration lane corresponding to different velocities of ramps has not been further formulated. Some designers ignored the assumed different velocities of ramps, and employed the values that are equal to or larger than the standard values. Obviously, the designing method is not reasonable and needs further discussion. Through observation and data analysis of the eight-lane highway deceleration lane existing in China, authors propose a calculation method and the recommended length of the deceleration lane based on the theory of quadratic deceleration theory.

FORMS OF DECELERATION LANE

Vehicles distribution state analysis

Through investigating on the spot and statistical analysis, it finds that the traffic volume of the eight-lane highway interchanging area line fell far behind of the corresponding service level. The outermost lane of the main line has light traffic, so the shunt vehicles all can nearly complete the outflow behavior. The speed of every lane presents a multistep change from inner to outer. The vehicles in the inside lane have the highest speed, and the vehicles in the outmost lane have the lowest speed. The vehicle type on the different lanes of eight-lane highway is more obvious than that of four-lane highway. Cars run on the inside of main lane, heavy haul wagons run on the outermost and both run on the middle lane.

Determination of the forms of deceleration lane

The two forms of highway deceleration lane are the direct type and parallel type. The characteristics are as follows:

(1)Parallel-type deceleration lane

The parallel-type deceleration lane is the one that is separated from the main line with a certain gradient rate reaching a lane width which is parallel to the main lane until a deceleration lane outflow nose form (Fig. 1). It can provide striking export region for drivers, but the track of the shunt vehicle is S-shaped which does not match with the deceleration lane. So the driving comfort is poorer, and it is prone to rear-end collision accident.

(2) Direct-type deceleration lane

Direct-type deceleration lane is the one that smoothly connects the main line and ramp in a certain gradient rate (Fig. 2). This type satisfies the drivers' directly driving track. Shunt vehicles can directly drive into the gradual deceleration lane along the triangle transition section. Therefore, it is advantageous for vehicles' quickly and smoothly driving out and can reduce the accidents of the shunt vehicles in the mainline causing by the abrupt deceleration.

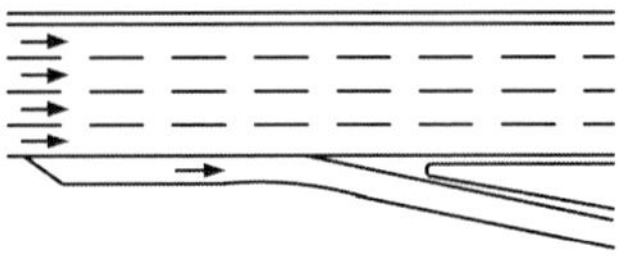

Fig. 1 Parallel-type deceleration lane

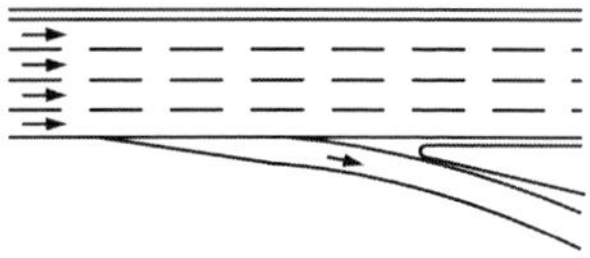

Fig. 2 Direct-type deceleration lane

Through the comprehensive analysis of the eight-lane traffic running state and the form characteristics of deceleration lane, the direct-type deceleration lane is more compatible with the moving track of eight-lane highway vehicles (Yuan et al. 2009). Therefore, this article mainly focused on the design model of the direct-type deceleration lane and the length calculation.

MODEL ESTABLISHMENT

Model description

By actually measuring the running speed of different kinds of vehicles on the four sections of the eight-lane direct-type deceleration lane more than one, the paper got the statistical graph of running vehicles' speed, as is shown in Fig.3.(Data is collected in normal traffic and weather at the daytime).

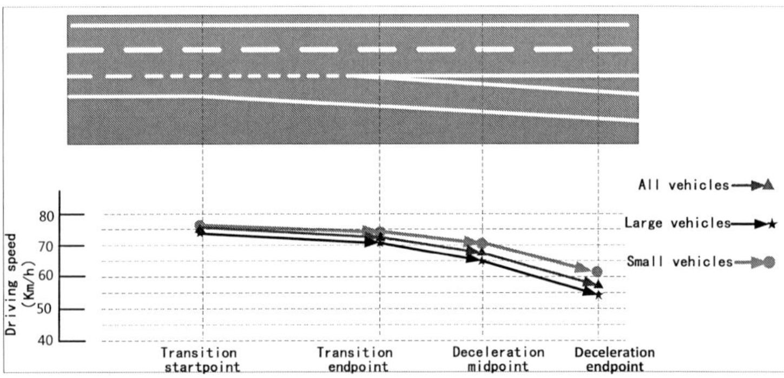

Fig. 3 Running speed of direct-type deceleration lane

The Fig. 3 shows that all the vehicle models in the transition section have nearly the same speed. In the triangle transition section, vehicles are nearly kept at an even speed and the deceleration speed can be seen as zero; from the end of the triangle transition section to the middle of the deceleration lane, the speed is relatively obvious. This is because the outing speed limit signs and the lines of the ramp section are in the driver's visual field and it forces the drivers speed down consciously; from the middle part of the lane to the outflow nose, limited by the linear of the outing ramp, the speed drops greatly for safety.

According to the deceleration process and running characteristics of vehicles, the determination of lane length is divided into three sections: the triangle transition section, the first deceleration section and the second deceleration section (Pan and Kong 2011). In reference to the calculation model of the deceleration lane AASHTO and an eight-lane highway vehicle distribution state, the paper assumed the model of the eight-lane highway deceleration lane as follows: vehicles maintain speed in the triangle transition section, and the speed is related to the designing speed of the mainline; In the first deceleration section, vehicles speed down by the engine and the speed descend a little; In the second deceleration section, vehicles speed down by the brake until they reach the outflow nose and the speed drops a lot.

Model establishment

According to the assumption of the eight-lane highway deceleration lane model, it got an illustration of the lane length (Fig. 4). The length of deceleration lane L is the combination of the three sections. That is,

$$L = L_0 + L_1 + L_2 \tag{1}$$

In which L_0(m) is the triangle transition section length, L_1(m) is the first deceleration length and L_2(m) is the second deceleration length.

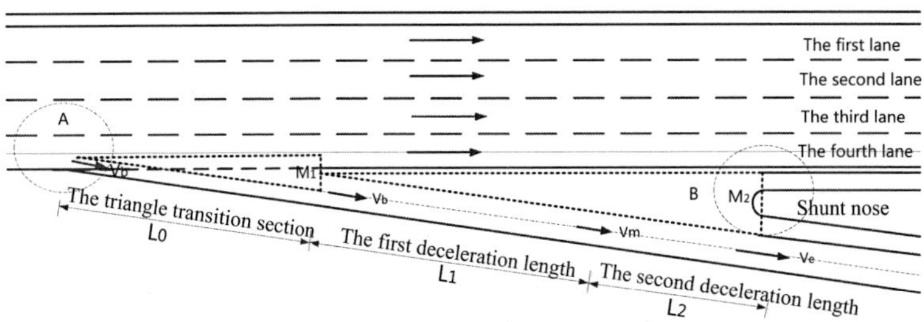

Fig. 4 Schematic deceleration of the lane length calculation

(1)The triangle transition section length

The triangle transition section is the one that the vehicles complete horizontal movement and drive into the deceleration lane with the initial outflow speed. The section's calculation formula is determined according to the geometry relationship:

$$L_0 = \frac{M_1 \cdot (L_1 + L_2)}{M_2} \tag{2}$$

In which M_1(m) is the width of horizontal movement in the triangle transition section, and M_2(m) is the width of horizontal movement in the one-lane deceleration lane section.

(2)The first deceleration length

The first deceleration length is the one that the vehicles pass through the triangle transition section and drives into the section that the vehicle speeds down by engine. The vehicle makes uniformly retarded motion. The initial speed V_b is the outflow speed, and the deceleration a_1 is engine deceleration. According to kinematics, it got the formula for the section:

$$L_1 = \frac{V_b \cdot t_1}{3.6} - \frac{a_1 \cdot t_1^2}{2} \tag{3}$$

In which t_1(s) is the time for first deceleration section and it is usually 3s; V_b(km/h) is the initial speed at outflow point and the values can be found in Table 3.1; a_1 (m/s^2) is the engine deceleration and the values can be found in Table 4.

(3)The second deceleration length

The second deceleration length is from the full width of the ramp sections to end of the deceleration lane that the vehicles use the brake to deceleration. The vehicle makes uniformly retarded motion. The initial velocity V_m is the final velocity of the first deceleration and the deceleration a_2 is brake deceleration. According to kinematics, it got the formula for the section:

$$\begin{cases} V_m = V_b - 3.6 \cdot a_1 \cdot t_1 & \text{if } V_m > V_e \\ V_m = V_b & \text{if } V_m \le V_e \\ L_2 = \dfrac{V_m^2 - V_e^2}{25.92 \cdot a_2} & \end{cases} \tag{4}$$

In which V_m (km/h) is final velocity of the first deceleration; V_e (km/h) is end speed of outflow nose and the values can be found in Table 3.3; a_2 (m/s^2) is brake deceleration and the values can be found in Table 4.

MODEL PARAMETER AND RECOMMENDED VALUE

Parameter selection

(1) The initial velocity of outflow point

Through measuring many eight-lane highways' initial velocity of outflow point, it finds that the shunt vehicles usually run with the speed less than that in the mainline. In the paper, it takes the velocity of the outflow point as the initial velocity, which facilitates the vehicles to outflow fast. What's more, accidents caused by the deceleration on the mainline which is due to the shortness of deceleration lane are less.

The measured outflow velocity of eight-lane highways is close to that of Japan, so the paper uses the value of Japan. The designed velocity of mainline and the corresponding initial velocity of outflow point are in Table 1(Japan highway design essentials 1991).

Table 1 Outflow point initial velocity

Mainline design velocity (km/h)	120	100	80	60	50	40
Outflow point initial velocity (km/h)	90	80	70	60	50	40

(2) The velocity of outflow nose end

China's "specification'' clearly regulates the minimum radius value of ramp at different designed ramp velocity and the minimum radius value of horizontal curve ramp on the outflow nose at different designed velocities of mainline. AASHTO determines the velocity of the outflow nose as shown in Table 2(AASHTO 1988). According to the "specification" in ramp designation in China and the recommended value of AASHTO, the paper uses the value as shown in Table 3.

Table 2 Outflow nose speed of American

Ramp design speed (km/h)	80	70	60	50	40	30	20
End speed of Split nose (km/h)	70	63	51	42	35	28	20

Table 3 End speed of outflow nose

Ramp design speed (km/h)	80	70	60	50	40	35	30
End speed of outflow nose (km/h)	70	63	60	50	40	30	30

(3) Retarded velocity on deceleration lane

The first vehicle deceleration length is from triangle transition section to Full-width section of deceleration lane. Vehicles use the engine to decelerate at this stage. The second deceleration followed the first deceleration and use Automotive Brake to decelerate again. Since the vehicle just runs away from the mainline and the driver needs an adaption time, the designation of retarded velocity should be based on comfort. AASHTO regulates the retarded velocity of engine and brake in terms of "driving comfort" discussed in the two reduction theory. The engine retarded velocity is 1.0~1.5m/s^2 and the brake retarded velocity is 1.5~2.0m/s^2 according to "Driver manual" in China. In this paper, it lists the first and second retarded velocity in Table 4.

Table 4 Retarded velocity

Initial velocity of branch point (km/h)	90	80	70
Engine deceleration (m/s^2)	1	0.9	0.8
Brake deceleration (m/s^2)	2	1.8	1.6

(4)Efflux angle and the length of triangle transition section

At the exit of highway intersection, the direct-type deceleration lane length refers to the longitudinal length from the designed initial point of the ramp to the maximum width. The designed initial point of the ramp should be located on the outermost lane of the mainline. Since the width of deceleration ramp lane is 3.5m and the width of mainline is 3.75m, the connection line of the initial point locates in the point where the distance to the outermost line is 1.7m, as shown in Fig. 5.

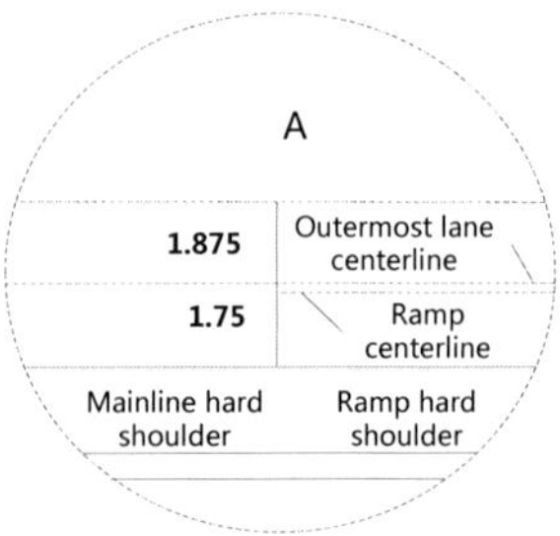

Fig. 5 Detailing of Ramp starting position

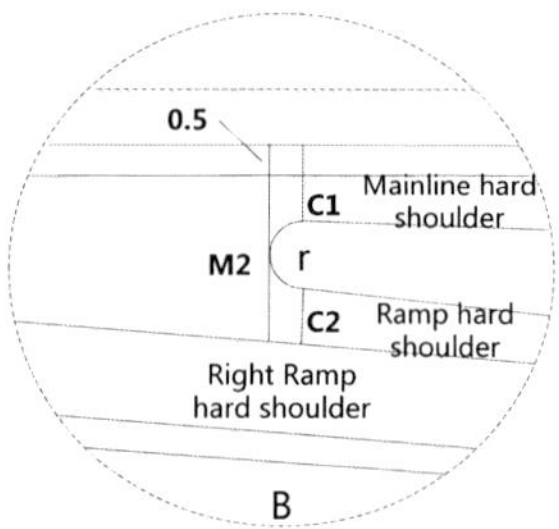

Fig. 6 Detailing of Ramp offset

The "Specification" clearly regulates the nose radius, the width of main line hard shoulder and the left hard shoulder of ramp (Fig. 6), so the lateral offset value and the width of triangle transition section can be calculated, as shown in Table 5.

Table 5 The width of lateral offset and triangle transition section

mainline design speed (km/h)	120	100	80
the nose radius r (m)	0.6	0.6	0.6
the width of main line hard shoulder C1 (m)	3.5	3.0	3.0
the width of left hard shoulder of ramp C2 (m)	0.6	0.8	0.8
the lateral offset value M2 (m)	3.5-0.5+1.2+1.6=5.8	5.5	5.5
the width of triangle transition section M1 (m)	3.5+0.5=4.0	4.0	4.0

The direct-type deceleration lane deviates from the mainline at a fixed outflow angle, and finally reaches the end. In the process, the width of the deceleration lane outflow transverse value is shown in Table 5. The deceleration lane deviating ratio (outflow angle) can be deduced by the length of the two deceleration lane which can be used to determine the triangle transition section length (Sun and Zhu 2003).

Recommended length value

As is shown before, the two deceleration lane length can be calculated respectively by Formula 3 and Formula 4. The paper put the parameter values in Table 1, Table 2 and Table 4 into the formula and can get different recommended deceleration lane length values under different designed speed of mainline and ramp, as is seen in Table 6.

Table 6 Deceleration length recommended values(m)

Mainline design speed /(km/h)	**Initial velocity of outflow /(km/h)**	**Ramp design speed /(km/h)**						
		80	70	60	50	40	35	30
		End speed of nose outflow /(km/h)						
		70	63	60	50	40	35	30
120	90	100	115	120	145	160	170	175
100	80	65	85	90	115	135	140	150
80	70	-	55	60	85	110	120	125

The outflow angle of direct-type deceleration lane can be calculated in Table 7 referring to Table 5 and Table 6.

Table 7 The outflow angle of direct-type deceleration lane

Main-line design	**Initial velocity of outflow / (km/h)**	**Ramp design speed / (km/h)**					
		80	70	60	50	40	30

speed /(km/h)			End speed of nose outflow / （km/h)					
			70	63	60	50	40	30
	90	gradual change ratio	1/17.25	1/19.83	1/20.69	1/25.00	1/27.59	1/30.17
	80	gradual change ratio	1/11.82	1/15.45	1/16.36	1/20.90	1/24.55	1/27.27
	70	gradual change ratio	-	1/10.00	1/10.90	1/15.45	1/20.00	1/22.73

The outflow angle should be less than 1/15~1/20 in Japan's specification (Zhang and Cheng 2012). In China, specifies the outflow angle should be equal to or less than 1/15. In this paper, the maximum outflow angle is 1/15 which is used to rectify the deceleration lane length and get the recommended value, as is shown in Table 8.

Table 8 The recommended length of deceleration lane

Mainline design speed /(km/h)	**Initial velocity of outflow / （km/h)**		**Ramp design speed / （km/h)**					
			80	70	60	50	40	30
			End speed of nose outflow / （km/h)					
			70	63	60	50	40	30
120	90	deceleration length	100	115	120	145	160	175
		gradual change ratio	1/17.5	1/20.0	1/21.0	1/25.0	1/27.5	1/30.0
		transition length	70	80	84	100	110	120
100	80	deceleration length	82.5	85	90	115	135	150
		gradual change ratio	1/15	1/15.5	1/16.5	1/21.0	1/24.5	1/27.5
		transition length	60	62	66	84	98	110
80	70	deceleration length	-	82.5	85	85	110	125
		gradual change ratio	-	1/15	1/15.5	1/15.5	1/20	1/22.8
		transition length	-	60	62	62	80.00	90

CONCLUSIONS

Through the analysis and study of the eight-lane highway deceleration lane, it draws the following main content:

(1)By analyzing the changes of the speed and trajectory of a vehicle traveling at the exit, in order to ensure safety and reduce the difficulty of the operation for drivers, it determines that the eight-lane highway deceleration lane should adopt direct-type deceleration lane.

(2)The paper gets the conclusion that direct-type gradual deceleration lane is made up of triangle transition section, the first deceleration and second deceleration. what is more, it gives the calculation method of each part and the recommended length of deceleration lane which make up the lack of China's “Specification” in respect of deceleration lane length, and it has some reference value in deceleration lane of highway design and planning.

REFERENCES

AASHTO(1988) *Highways and urban roads geometric design.* Xi'an: Northwestern Poly-technical University Press.

AASHTO(2004). *A Policy on Geometric Design of Highways and Streets.* Washington, D. C: AASHTO.

Pan, B. H., Kong, L. C. (2011). “Research on Calculation Model of Minimum Net Distance between Interchanges on Eight-lane Expressway”. *Applied Mechanics and Materials Vols.* 97-98, 121-129.

Shan xi Tourism Press(1991) “Japan highway design essentials.” *Project Management Division of the Ministry of Communications dubbed assemble*, Xi'an: Shan xi Tourism Press, 1991.

Sun, J. S., Zhu, X. B. (2003). *Design Data Sets of Road.* Beijing: China Communications Press， 2003.

Yuan, Z. D., Xue, L., Wang, W. L. (2009). “Length Design of Speed Change Lane at Interchanges of Expressway”. *Journal of Chongqing Jiao tong University: Natural Science Edition*, 2009, 28(4): 689-693.

Zhang, H. J., Cheng, W. (2012). “Length Design of Acceleration and Deceleration Lane at Expressway Interchange”. *Highway Engineering.* 1002-4786, 12-0130-03.

Research on Driving Safety of Urban Interchange Ramp under Crosswind

X. Wen[1]; F. Chen[2]; L.J. Yu[3]; and X.D. Pan[4]

[1]Key Laboratory of Road and Traffic Engineering of the Ministry of Education, Tongji University, 4800 Cao'an Road, Shanghai 201804, China.
E-mail: 1105896004@qq.com
[2]Key Laboratory of Road and Traffic Engineering of the Ministry of Education, Tongji University, 4800 Cao'an Road, Shanghai 201804, China (corresponding author). E-mail: chenfeng.csu@gmail.com
[3]Shanghai Municipal Engineering Design Institute, 901 Zhongshan Road, Shanghai 200092, China. E-mail: 546036697@qq.com
[4]Key Laboratory of Road and Traffic Engineering of the Ministry of Education, Tongji University, 4800 Cao'an Road, Shanghai 201804, China.
E-mail: panxd@163.com

Abstract
During the last decade, urban interchange ramps were widely constructed in China, which also associated with a rapid increasing number of traffic accidents. Among various causal factors to traffic accidents, the environment factor, crosswind, has been proved to be correlated with occurrence of accidents happened on urban interchange ramp. In order to improve the traffic safety for urban interchange ramp, the impacts of crosswind on traffic safety were investigated in this paper. Using hot-wire anemometer, it is demonstrated that the wind speed at traffic accident hot spots is much bigger than that of the nearby weather station at the same time. Then several computational fluid dynamics (CFD) simulation models were developed to illustrate the impacts of crosswind on traffic safety. The results indicate that driving around the windward side on a ramp is the most dangerous situation compared to the other scenarios and the aerodynamic force of automobile can change significantly during overtaking process.

INTRODUCTION

With the rapid development of national economy and the growing traffic demand in China, fundamental traffic facilities has been built in a large scale, such as grade separation interchange. Urban interchange ramps serve as important links of the whole road network, so the traffic safety for ramps has direct and significant impact on the road capacity and the service level of expressway and highway. Through the investigation cooperating with Shanghai Traffic Police Department, it was found that several similar traffic accidents have occurred in the same place unexpectedly. In addition, these accidents usually happened during strong wind period. Therefore, the impacts of crosswind on traffic safety of urban interchange ramp should be paid more attention and further study for its effect on traffic safety is necessary. This paper focus on studying the impacts of crosswind on ramp traffic safety and the most dangerous situation during driving on the ramp is found.

LITERATURE REVIEW

There are some researches about the traffic safety of ramps and recent studies have focused on the alignment standard of urban interchange ramp, the design of the signs and markings and safety evaluation of ramp. Different types of restraint export of interchange were analyzed and the minimal distance between the exit of tunnel and the exit of interchange to ensure safety was proposed (Shen 2011). The superelevation-value should be determined on the basis of real vehicle speed on ramps was suggested (Zhang et al. 2007). A safety evaluation system and the alignment index of ramps, which is called RASI, were developed (Zuo 2011). And mental characteristics of drivers when driving on the interchange ramp were mainly studied (Wang et al. 2007). Fuzzy mathematics principle was applied to the optimal design of alignment of ramp and design method of ramps was put forward (Zhou 2007). Most of these related researches concentrated on the design of ramps and rarely involved the traffic safety of ramp in bad weather conditions.

In the 1980s, the problem of traffic safety under crosswind has drawn attention of vehicle aerodynamics researchers. Besides natural wind, complicated terrains such as tunnels and the bridges, also form strong wind field. The crosswind may increase the aerodynamic forces and cause vehicle collisions. Bettle investigated the aerodynamic characteristics of vehicle when it passing the bridge tower with wind-shield at a certain speed and analyzed the dynamic response of cars by simulation (Bettle et al. 2003). Snæbjörnsso evaluated the probability of traffic accidents considering the following factors: the speed of vehicle and crosswind, friction coefficient and so on (Snæbjörnsso et al. 2007). Charuvisit also found that the response of vehicle increased with increase of wind speed and vehicle speed (Charuvisit et al. 2004). Several measures to ensure traffic safety through researching the dynamics of container semi-trailer under crosswind were proposed (Jiang and Yu. 2001). However, less attention has been paid to the traffic safety for urban interchange ramp combined with crosswind, especially when more urban interchange ramps surrounded by complicated wind environment will be constructed in China. This paper studied the influence of crosswind conditions on vehicles, based on computational fluid dynamics simulation models.

WIND SPEED MEASUREMENT

Several similar traffic accidents occurred in the same spot of the interchange ramp near the Pudong airport station. In February, 2012, two taxis dropped off the ramp after breaking the outside guardrail. A month later, a car ran away from the lane and rushed into green belt. Meanwhile, vehicle collisions happened frequently in this spot. Therefore, wind speed measurement experiments have been done in order to explore the causes of these accidents.

Experiment equipment

The hot-wire anemometer is easy to use and carry, which can make fast measurement of real-time wind speed. The data can be collected by connecting the anemometer and computer and wind velocity error can be less than 0.01m/s at a sampling frequency of 1 Hz.

Experiment method

We tried to prove the relationship between crosswind and the occurrence of accidents happened on urban interchange ramp by using comparison experiments. At first, we did this experiment on September 28th, 2012, a sunny day. Secondly, we have considered the theory that generally wind speed will increase with the increase of height above ground. We chose the emergency parking area of traffic accident hot spot and the highest point of the whole ramp as experimental site which is about 200m away from the other spot. The specific location is shown in Fig. 1. Finally we measured the crosswind speed at the same time. This means can avoid the influence of the extreme weather and running cars on experimental results.

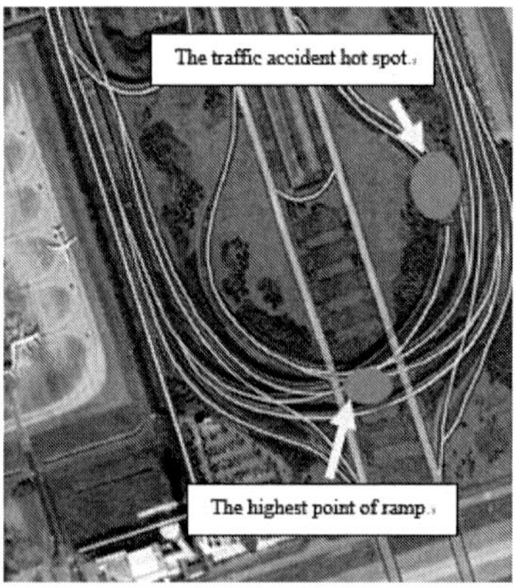

Fig. 1. Experimental site.

NUMERICAL SIMULATION SCHEME

Vehicle and ramp models

The vehicle-ramp system is composed of a typical semi-trailer and the ramp. The model shape of the ramp can be determined according to the newest standard for highway alignment design. The cross section of the prototype ramp is 8.5m wide, carrying a dual two-lane highway on its upper surface. The width of motor lane is 3.5m while the width of marginal strip is 0.25m. Moreover, the cross section of the protection rail at two sides is 0.5m wide and 1.1m high. For the vehicle model, a common and typical semi-trailer is adopted in the numerical simulation because trucks are susceptible to the influence of crosswind generally. The detailed dimension of the trailer model is shown in Fig.2. To increase efficiency, several parts of model which may have insignificant effects were not taken into account. Therefore, in this numerical simulation, the vehicle automobile rearview mirror and tire were not simulated and the cross section of vehicle and guard rail was simplified.

In order to study the effect which crosswind makes on vehicles, we took three situations into account: Two cars move side by side, a single car moves on the windward side of ramp and a single car moves on the leeward side. The changes of wind field during overtaking process could also be induced by analyzing the aerodynamic forces of vehicle in these three situations. Besides, another situation that a car moves on the ground was simulated in contrast with driving on the ramp.

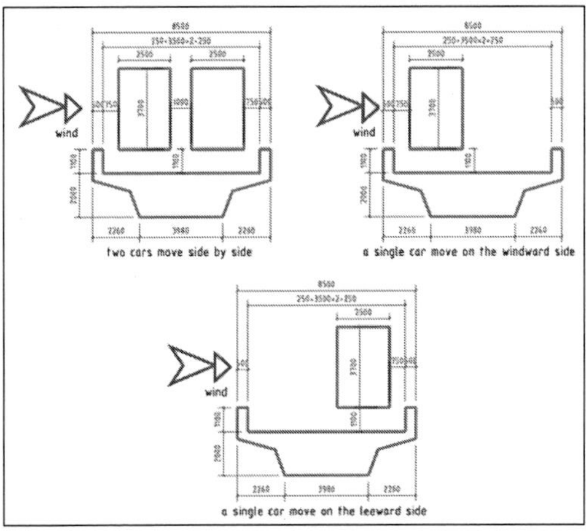

Fig. 2. Size of three simulation models in different situations.

Governing equations and numerical simulation method

The turbulent model had a decisive influence upon the accuracy of numerical model and the SST k-ω turbulence model was employed in the simulation. The basic principle behind this method is to synthesize the respective advantage of k-ε and k-ω model. After being optimized, the equations of the flow become:

$$\frac{\partial(\rho k)}{\partial t} + \Delta \bullet (\rho U k) = \frac{\partial}{\partial x_j}(\Gamma_k \frac{\partial k}{\partial x_j}) + G_k - Y_k + S_k \qquad (1)$$

$$\frac{\partial(\rho \omega)}{\partial t} + \Delta \bullet (\rho U \omega) = \frac{\partial}{\partial x_j}(\Gamma_\omega \frac{\partial \omega}{\partial x_j}) + G_\omega - Y_\omega + D\omega + S_\omega \qquad (2)$$

Where G_k = turbulent energy; ω= frequency of turbulent energy; $X_j = j^{th}$ axis in the Cartesian coordinate system; Γ_k = effective divergent terms of k; Γ_ω = effective divergent terms of ω; Y_{k} = divergent terms of k; Y_ω= divergent terms of ω; t = time; D_ω= orthogonal divergent term.

Fluent, a CFD software, was utilized here. The flow separation around the back of the car can be simulated far more accurately by SST k-ω model through comparison.

Computational domain and boundary conditions

A computational domain of rectangle shape enclosed by six outer boundaries, were formed around the vehicle-ramp system. The six outer boundaries have been named b_right, b_left, b_tail, b_head, b_down and b_up. The outer boundary b_left is the inflow face from which the wind blows in. A uniform crosswind speed of 20 m/s with a 90 degree yaw angle, dissipation ratio of 10, turbulence kinetic energy k of 0.005 and the kinematical viscosity coefficient is $1.7894*10^{-5}m^2/s$ were assigned to it. The outer boundary b_right which is parallel to b_left with an offset of X was specified for this research as a flow outlet with zero pressure. The boundaries b_tail, b_head, b_down and b_up are parallel to the direction of the crosswind. The boundaries b_up and b_down are parallel with a distance of Y between the two boundaries. The total size of the domain is shown in Fig.3. For this research, in the x direction (wind direction), an upstream length of 10m and a downstream length of 20m were assigned between the vehicle-ramp system model and the corresponding outer boundaries (a case of vehicle-road system). All the flow boundaries were enforced with mathematical boundary conditions to approximate the real situation. In addition, in this study, the flows at these boundaries were assumed to be uniform but the flow cannot penetrate the surfaces of a vehicle.

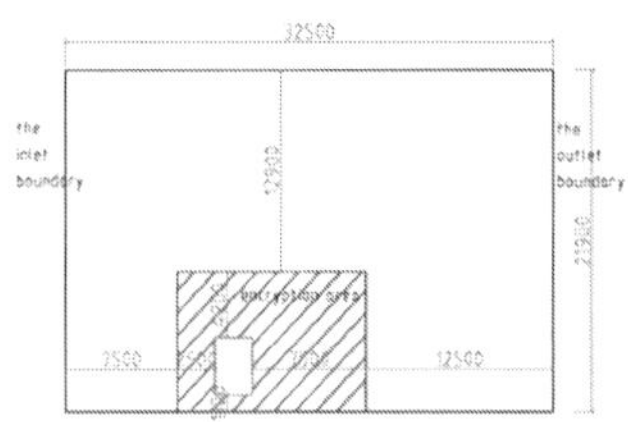

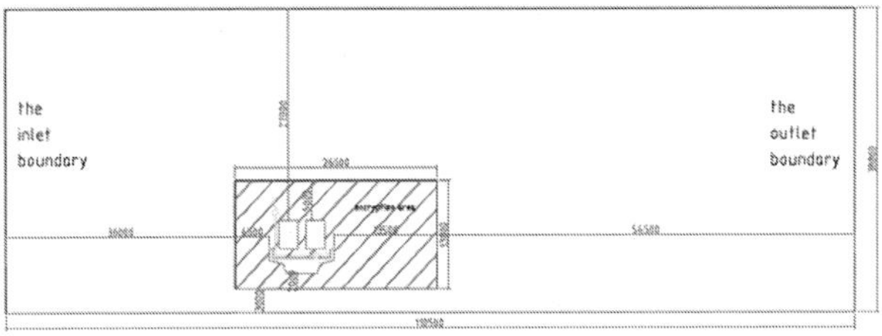

Fig. 3. Computational domain and encryption area of vehicle-ground and vehicle-ramp system.

MESHING

Four meshing schemes M1, M2, M3 and M4 with different grid sizes were generated to check the independence of the numerical results on grid sizes. Through practical operation, hybrid grids was adopted, which ensures the calculation accuracy in reasonable time. The height of the first layer grids near the surfaces of the vehicle, ground and ramp was set as 1×10^{-3}m. The mesh refinement area of different meshing schemes was determined which is shown in Fig.4.

4.1. A single car moves on the ground.

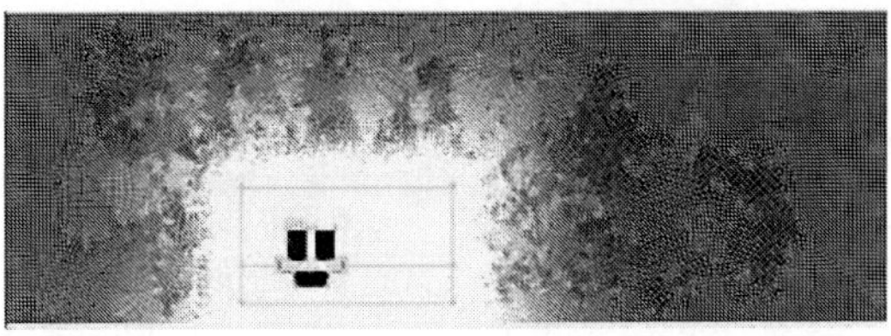

4.2. Two cars move side by side.

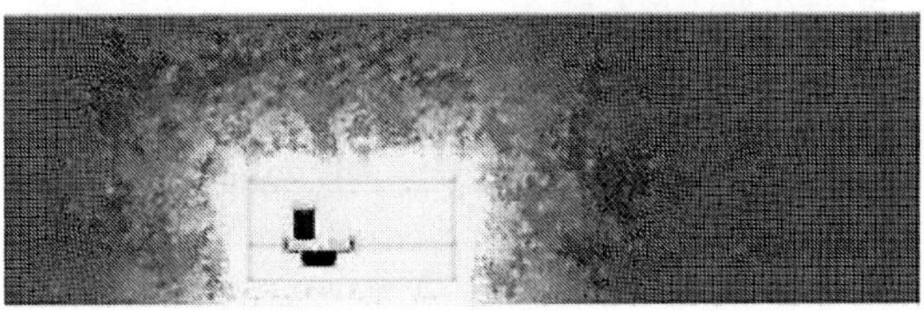

4.3. A single car moves on the windward side.

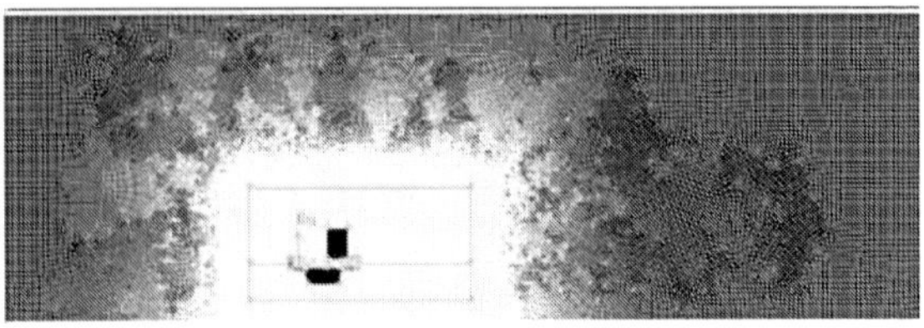

4.4. A single car moves on the leeward side.

Fig. 4. Grid distributions in these meshing schemes.

RESULTS

Measurement results

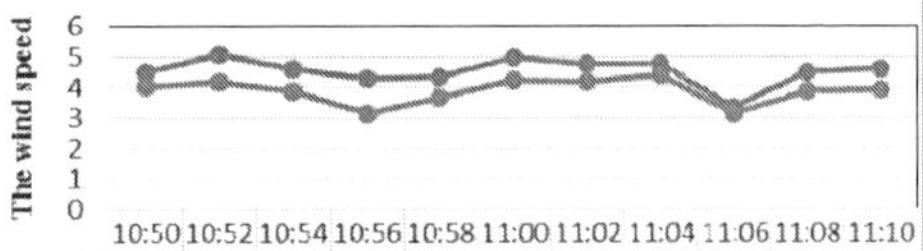

	10:50	10:52	10:54	10:56	10:58	11:00	11:02	11:04	11:06	11:08	11:10
The traffic accident hot spot	4.51	5.08	4.58	4.26	4.32	4.95	4.77	4.75	3.36	4.49	4.59
The highest point of ramp	4.01	4.17	3.85	3.14	3.63	4.24	4.16	4.4	3.13	3.86	3.93

Fig. 5. Measurement Results.

From Fig. 5, both the maximum and average value of crosswind speed at the traffic accident hot spot are bigger than the nearby the highest point of ramp, although the height of the hot spot above ground is lower. General speaking, the wind speed, which is affected by roughness length and vertical temperature, will increase with the increase of height above ground. However the results of measurement experiment fails to fit the assumption. One possibility is the influence of the narrow pipe effect. According to investigation, overturn and sideslip accidents are the most common accident types in this area. Therefore the traffic accidents seem to be associated with the crosswind. Consequently, it's necessary to study the influence of crosswind on traffic safety.

Simulation results

1) Averaged flow velocity distribution around vehicle in four situations

In Fig. 6, the wind speed around the surface of vehicle is low and the corner of vehicle and ramp is of large velocity gradient. In addition, the maximum velocity occurs on the bottom and the top of car. Therefore, the vehicle-ramp system can significantly affect the wind field.

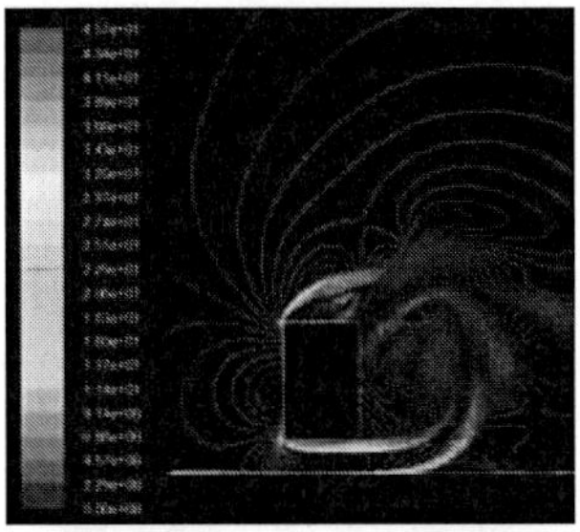

6.1. A single car moves on the ground.

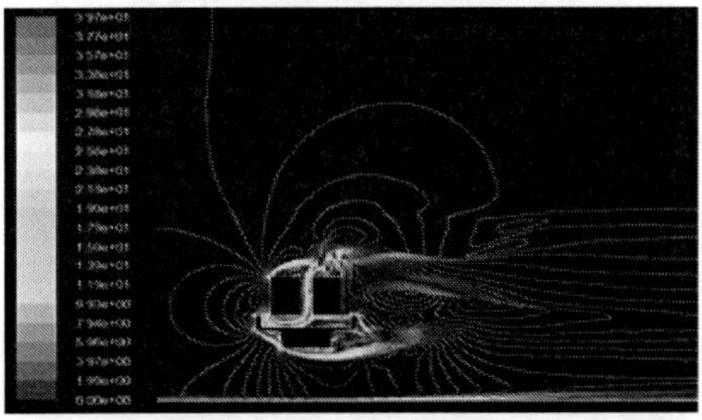

6.2. Two cars move side by side.

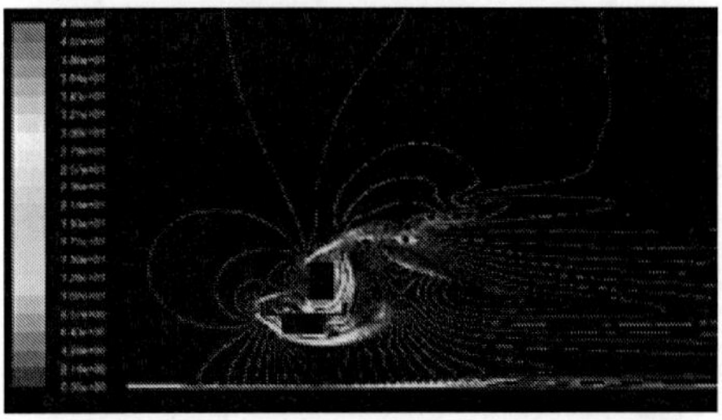

6.3. A single car moves on the windward side.

6.4. A single car moves on the leeward side.

Fig. 6. Averaged flow velocity distribution around the system.

2) Analysis of wind pressure
   The pressure contour of the four different situations in shown in Fig. 7.

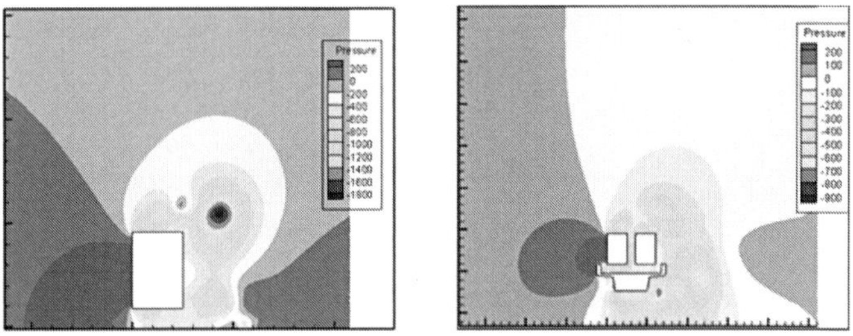

A single car moves on the ground　　Two cars move side by side

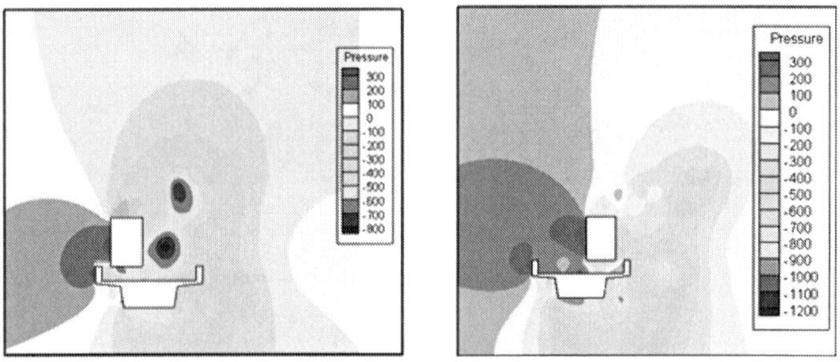

A single car moves on the windward side　　A single car moves on the leeward side

Fig. 7. Pressure contour.

3) Analysis on the aerodynamic coefficient

Through simulation, the aerodynamic coefficient of vehicles in four situations is obtained, which is shown in Table 1.

Table 1. The Aerodynamic Coefficient in the Simulation.

Situation	C_S	C_L
A single car moves on the ground	5.78	1.01
The car on the windward side(Two cars move side by side)	5.32	1.09
The car on the leeward side(Two cars move side by side)	-1.04	-0.57
A single car moves on the windward side	5.92	1.18

A single car moves on the leeward side	5.67	1.14

From these computed aerodynamic coefficient of vehicle under different situations, crosswind will generate different lift and side force then affect driving safety. These results show that C_S and C_L are biggest when a single car moves on the windward side of ramp compared to the other scenarios so this situation is most dangerous. Moreover, the side force of the car on the leeward side when two cars run side by side is in opposite direction compared with other situations in our study. Thus sudden significant change changes in the side force of the vehicle on the leeward side will occur during overtaking process, which may cause accident on the ramps.

CONCLUSIONS

In this study, we researched on driving safety of urban interchange ramp under crosswind, through wind speed measurement and numerical simulation. The research results are concluded to prove that crosswind is possible correlated with occurrence of accidents happened on urban interchange ramp and the wind field around a vehicle-ramp system is more complex than that around a single car on ground.

A numerical simulation scheme, which includes the vehicle model, the ramp model, the meshing, the boundary conditions and the SST k-ω turbulence model has been established in this study to explore the influence of crosswind on the system. Through analyzing simulation results, it has been found that the movement of a single car on the windward side of ramp is most danger and overtaking process also deserves more attention.

REFERENCES

Bettle, J., Holloway, A. G. L., Venart, J. E. S. (2003). "A computational study of the aerodynamic forces acting on a tractor-trailer vehicle on a bridge in cross-wind." *Journal of Wind Engineering and Industrial Aerodynamics*.

Charuvisit, S., Kimura, K., Fujino, Y. (2004). "Effects of Wind barrier on a vehicle passing in the wake of a bridge tower in cross wind and its response." *Journal of Wind Engineering and Industrial Aerodynamics*.

Jiang, H., Yu, Z. P. (2001). "Analysis of the driving stability of semi-trailer in the crosswind." *Journal of Tongji University*. (in Chinese)

Shen, Q. R. (2011). "Study of support technology of freeway Interchange constrained export safety." *Journal of Chang'an University*. (in Chinese)

Snæbjörnsson, J. T., Baker, C. J., Sigbjörnsson, R. (2007). "Probabilistic assessment of road vehicle safety in windy environments." *Journal of Wind Engineering and Industrial Aerodynamics*.

Wang, J., Li, S. G., Yang, H.Y., et al. (2007). "Introducing the factor-human engineering into the design of urban interchange." *Urban Bridge and Flood Protection*. (in Chinese)

Zhang, S. P., Gong, W. X., et al. (2007). "Research on the method of determining the reasonable superelevation-valu." *Journal of technology of highway*. (in Chinese)

Zhou, L. (2007). "Application of fuzzy mathematics in the optimized alignment design of interchange ramp . " *Journal of Chongqing Jiaotong University.*(in Chinese)

Zuo, W. P. (2011). "Study of interchange ramp alignment design and evaluation based on driving safety." *Journal of Chang'an University.* (in Chinese)

Research on Safety of Freeway Off-Ramp Lane Changing Behavior Based on Lateral Force Coefficient

W.J. Du[1]; G.D. Xia[2]; F. Chen[3]; and X.D. Pan[4]

[1]Key Laboratory of Road and Traffic Engineering of the Ministry of Education, Tongji University, 4800 Cao'an Road, Shanghai 201804, China.
E-mail: 414316708@qq.com
[2]Shenzhen Urban Transport Planning Center Co. Ltd., Shenzhen Guangdong 518021, China. E-mail: 379376189@qq.com
[3]Key Laboratory of Road and Traffic Engineering of the Ministry of Education, Tongji University, 4800 Cao'an Road, Shanghai 201804, China.
E-mail: chenfeng.csu@gmail.com
[4]Key Laboratory of Road and Traffic Engineering of the Ministry of Education, Tongji University, 4800 Cao'an Road, Shanghai 201804, China.
E-mail: 707670764@qq.com

Abstract

This paper evaluates the safety of lane changing behavior in freeway off-ramp section. Firstly, we discussed the relationship among lateral force coefficient, lateral stability of vehicles, and drivers' physiological status. Then, we conducted the analysis of vehicle and drivers by establishing the models of the steering angle and lateral force coefficient. Based on this, the distribution chart of lateral force coefficient in different alignment conditions was obtained. Lastly, we proposed the safe steering angles of front wheel for different alignment conditions. The results can reduce the accident rates on off-ramp sections on freeways.

INTRODUCTION

On-ramps and off-ramps are the only passageways connecting freeway to the other roads, which are under the restriction of more factors than the mainline. Especially in off-ramp sections, traffic accidents occurred frequently. According to the accident statistics, the accident rates in off-ramp region are 4~6 times bigger than the general road sections, and are 2~3 times bigger than the on-ramp sections. Therefore, traffic safety in freeway off-ramp sections has become a big concern to road safety research.

In off-ramp section, vehicles need to change lines to the outer lane. Due to the lateral force, vehicles may drive with lateral deviation along the direction of the lateral force. The lateral stability depends on the value of lateral force coefficient μ when vehicles change lines. The value of lateral force coefficient not only has a close relation with the driving stability, at the same time, an excessive lateral force coefficient also has an adverse effects on drivers' physiological and psychological status. Furthermore, it can affect driving behaviors and comfort. And an excessive physiological and psychological burden of drivers may have a negative influence on driving safety.

Several studies have analyzed the traffic safety from the view of road alignment and driver characteristics. A detailed analysis of vehicle motion parameters (driving speed and lateral acceleration) and traffic safety was made to analyze the relation between road alignment and traffic safety (Zhou 2007). The results showed that in all kinds of road alignments, curve has played an important role in traffic safety. Especially the driving speeds and lateral force coefficients on curves have a close relation with accidents. Heart rate and systolic blood pressure were selected as evaluation indexes to measure the safety and comfort of curve sections in mountainous highways (Pan et al. 2006). In the driving experiments, the quantitative relations with coefficient of transverse force were analyzed and established. And the conclusion showed that the reasonable value of the coefficient of transverse force should be less than 0.2. A driving test and data analyzed the relation among heart rate of drivers, coefficient of transverse factor, radius and driving speed on the horizontal curve of freeway without other traffic flow. Thus the model among heart rate, radius and driving speed were obtained (Zhen et al. 2003).

A statistical model for defining the relation between traffic accidents, highway geometric design elements and traffic volumes for interchange ramps and speed-change lanes was developed (Bauer and Harwood 1998). Bared et al. (1999) estimated accident frequencies for entire ramps, as a function of speed-change lane length, among some other variables. An economic analysis was then presented to evaluate the cost-effectiveness of extending speed-change lanes. Chen et al. (2009) evaluated the impacts of the number and arrangement of lanes on freeway exit ramps on the safety performance of freeway diverge areas.

The objective of this study is to explore the safe steering angle of front wheel in lane changing process. This paper takes the freeway off-ramp sections as the research object. By establishing the relation model between the steering angle of front wheel and the lateral force coefficient, the distribution chart of lateral force coefficient in different alignment conditions is formed. Then taking drivers' psychology and physiology comfort degrees as the judgment index of traffic safety, the safe steering angle of front wheel in different conditions is proposed. Thus the analysis on safe steering angle can be used as a scientific basis to ensure traffic safety in freeway off-ramp sections.

ANALYSIS METHODOLOGY

Definition of lateral force coefficient

Centrifugal force generates when vehicles drive on curves. The action point of centrifugal force is located in the center of gravity of vehicles. And it is in the opposite direction of circle center. In order to reduce the centrifugal force on curves, superelevation is formed by heightening the lateral pavement. When vehicles drive on curves with superelevation, the horizontal component of gravity of the vehicles can offset part of centrifugal force, and the remaining part is balanced by the friction between the tire and pavement. Fig. 1 shows the force analysis of vehicles on curve.

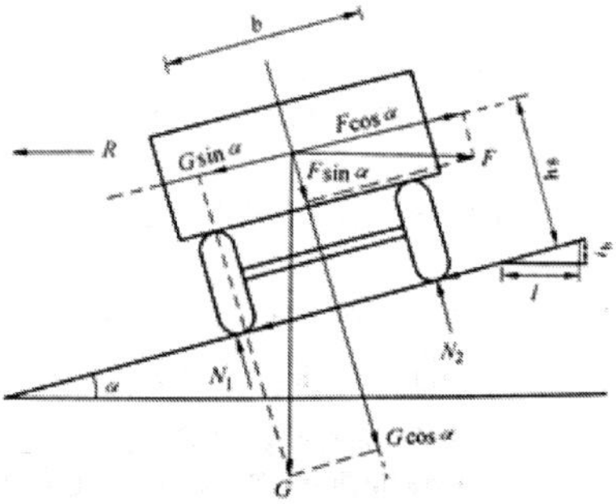

Fig. 1. The force analysis of vehicles on curve.

The lateral force can be calculated by:

$$Y = F\cos\alpha - G\sin\alpha \quad (1)$$

Where α= the lateral angle of pavement, the value of which is infinitesimal, approximately $\cos\alpha \approx 1$, $\sin\alpha \approx \tan\alpha = i_h$; F= centrifugal force(N); G= vehicle gravity(N).

Then define the lateral force coefficient as $\mu = {}^{Y}/_{G}$, which can be calculated by:

$$\mu = \frac{Y}{G} = \frac{F - G\sin\alpha}{G} = \frac{F - G\sin\alpha}{G}\frac{V^2}{gR} \pm i_h = \frac{V^2}{gR} \pm i_h \quad (2)$$

Where V= driving speed, (km/h); g= gravitational acceleration, $10m/s^2$; R= circular curve radius (m); i_h= superelevation rate.

The relation between lateral force coefficient and lateral stability of vehicles

Due to the lateral force, vehicles may drive with lateral deviation along the direction of the lateral force. In order to avoid the lateral deviation, the lateral force must be less than or equal to the lateral adhesion between tire and pavement.

$$X \le Y\varphi_h \approx G\varphi_h \quad (3)$$

$$\mu = X/G \le \varphi_h \quad (4)$$

Where G= vehicle gravity (N); φ_h= lateral adhesion coefficient, normally $\varphi_h = (0.6\sim0.7)\varphi$, φ= adhesion coefficient.

The relation between lateral force coefficient and drivers' physiological status

The value of lateral force coefficient is used to represent the intensity of the centrifugal force. If the value of lateral force coefficient μ is excessive, vehicles cannot drive continuously and steadily, even sometimes need to slow down. When vehicles drive on small radius curves, the turning radius is large, which easily makes vehicles run out of the lane and cause accidents. For drivers, if the value of μ exceeds a certain value, drivers have to take measures to increase the stability of vehicles on curves, which could cause the degree of psychological stress during driving. For drivers, as the value of μ increases, they will also feel uncomfortable. According to Technical Standard of Highway Engineer in China (JTG B01-2003), drivers' physiological reactions changing with the value of μ are as follows:

When $\mu<0.10$, drivers haven't notice the existence of curve and feel very stable;

When $\mu=0.15$, drivers start noticing the existence of curve and feel relatively stable;

When $\mu=0.20$, drivers have noticed the existence of the curve and feel slightly unstable;

When $\mu=0.35$, drivers have noticed the existence of the curve and feel a little unstable;

When $\mu\geq0.40$, drivers feel very unstable and have a sense of fear that the car will roll.

Relation model of the steering angle and lateral force coefficient

Chen and Guo (2007) investigated the steering angle of vehicles in freeways, the results show when the driving speed is constant, the steering angle of lane changing usually swings between a concrete numerical value. As the driving speed increases, the steering angle δ has a decreasing trend, but there are still few driving behaviors leading to a large the steering angle. The investigation results can be referenced as shown in Fig. 2.

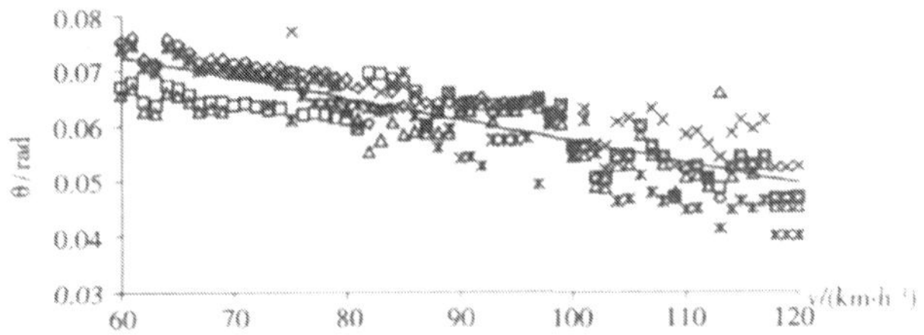

Fig. 2. The relation between steering angle and driving speed.

It can be obtained from the figure3 that the steering angle is in the range of roughly between 0.04~0.08 rad. According to data fitting of the sample data, the linear model between the steering angle and driving speed is estimated by:

$$\delta = [-0.0219V + 5.4688]/180 \tag{5}$$

Where V= driving speed. (km/h)

Combined with two-degree-of-freedom vehicle model, the model between the steering angle and lateral force coefficient is estimated by:

$$\mu = \frac{u}{g} \frac{u/L}{1+\frac{m}{L^2}\left(\frac{a}{k_2}-\frac{b}{k_1}\right)u^2} \delta + i_h \tag{6}$$

Where u= center-of-mass velocity component in x direction. (km/h); L= wheelbase.(m); a= front wheelbase.(m); b= rear wheelbase.(m); k_1= Effective cornering stiffness of front wheel.(kg· m^2); k_2= Effective cornering stiffness of rear wheel.(kg· m^2); g= gravitational acceleration, $10m/s^2$; i_h= superelevation rate.

Parameter Values of the Model

This paper uses vehicle parameters of Hongqi CA770 to solve the model and analyze the results. The parameter of Hongqi CA770 is shown in Table 1.

Table 1. Parameter List of Hongqi CA770.

Parameter	Symbol	Dimension	Parameter values
Entire vehicle mass	m	kg	3018
Front wheelbase	a	m	1.84
Rear wheelbase	b	m	1.88
Moment of Inertia about Z axis	I_Z	kg· m^2	10437
Effective cornering stiffness of front wheel	k_1	N/rad	23147
Effective cornering stiffness of rear wheel	k_2	N/rad	38318

RESULTS AND DISCUSSION

The safety analysis of off-ramp lane changing on linear alignment

When the alignment of the off-ramp is linear, the superelevation rate of highway i_h =0.02 is taken for tentative calculation. According to sample data above that the steering angle on freeway ranges between 0.04~0.08 rad, the value range of steering angle is δ=0.04~0.2 rad. And the value range of driving speed is 60~130 km/h considering of the overspeed condition. The different front wheel steering angle and the values of lateral force coefficient under different driving speeds in lane changing process are shown in Table 2. It can be seen from Fig. 3 that the lateral force coefficient increases with the increase of front wheel steering angle and driving speed. When front wheel steering angle is large, the value of lateral force coefficient increases rapidly with the increase of driving speed, and the increase rate is larger than that when the steering angle is small.

Table 2. Lateral Force Coefficients of Different Driving Speeds and Steering Angles.

Speed(km/h) / Steering angle(rad)	60	70	80	90	100	110	120	130
0.20	0.526	0.575	0.612	0.640	0.663	0.68	0.694	0.705
0.15	0.400	0.436	0.464	0.485	0.502	0.515	0.525	0.534
0.10	0.273	0.297	0.316	0.330	0.341	0.350	0.357	0.363
0.09	0.248	0.270	0.286	0.299	0.309	0.317	0.323	0.328
0.08	0.222	0.242	0.257	0.268	0.277	0.284	0.290	0.294
0.07	0.197	0.214	0.227	0.237	0.245	0.251	0.256	0.26
0.06	0.172	0.186	0.198	0.206	0.213	0.218	0.222	0.226
0.05	0.147	0.159	0.168	0.175	0.181	0.185	0.188	0.191
0.04	0.120	0.131	0.138	0.144	0.149	0.152	0.155	0.157

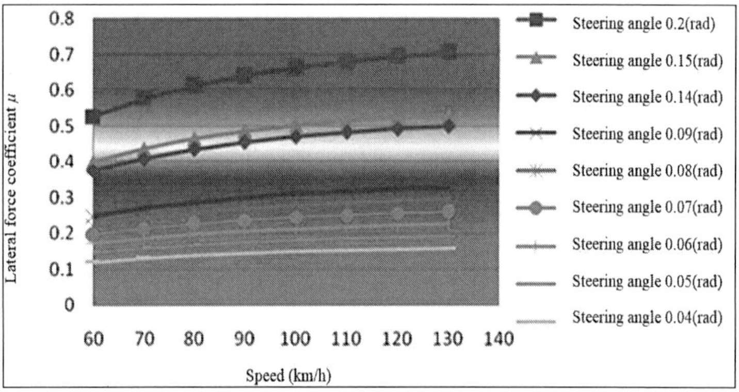

Fig. 3. The relation among driving speeds, the steering angle of front wheel and lateral force coefficients (linear alignment).

The significance of different colors in Fig. 3 is explained in Table 3.

Table 3. The Corresponding Value of μ to Different Colors.

Color	The corresponding value of μ	Drivers' feeling and the driving status
Green	<0.1	drivers haven't notice the existence of curve and feel very stable
	0.15	drivers start noticing the existence of curve and feel relatively stable
	0.2	drivers have noticed the existence of the curve and feel slightly unstable
Blue	0.35	drivers have noticed the existence of the curve

		and feel a little unstable
Yellow	>0.4	drivers feel extremely unstable and have a sense of fear that the car will roll
Red	>0.55	sideslip or rollover has occurred

In order to ensure the vehicle won't sideslip in lane changing process, lateral force coefficient must be controlled that $\mu \leq \varphi_h$. For dry asphalt pavement, $\varphi_h=0.55$, and for wet asphalt pavement, $\varphi_h=0.39$. It can be seen from the Fig. 3 that on dry asphalt pavement, when the steering angle of front wheel $\delta \leq 0.15$ rad (under dangerous situations), vehicles can keep driving stability of lane changing at high driving speed ($V \leq 130$ km/h). However, when the steering angle or the driving speed gets larger, the vehicles have a risk of sideslip. Similarly, on wet asphalt pavement, when the steering angle of front wheels $\delta \leq 0.11$ rad (under emergency situations), vehicles can keep driving stability of lane changing at high driving speed. And when the steering angle or the driving speed get larger, the vehicles also have a risk of sideslip. Two cases of the above analysis are extremely dangerous situation.

On the other hand, from the point of drivers' physiological reactions, Fig. 3 shows when $\delta \leq 0.05$ and $V \leq 130$ km/h, the increase of drivers' heart rate and blood pressure stay in a stable range, which means the physiological and psychological pressure is low. When $\delta \leq 0.06$ and $V \leq 90$ km/h, it is the same as above; when the driving speed continues to increase, the driver's physiological and psychological pressure will increase rapidly and it will affect the driving safety; when $\delta \geq 0.07$, it shouldn't be adopted on freeways, for that even if the easiest lane changing behavior may cause drives feel uncomfortable. Comparing with the samples of steering angles on freeways, when vehicles are at high driving speed, drivers seldom make the steering angle $\delta \geq 0.07$. It can explain that drivers will adjust their driving behaviors to the most comfortable state according to their feelings.

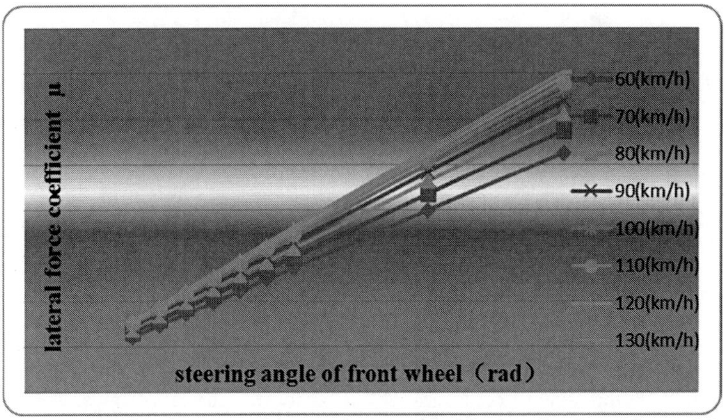

Fig. 4. The relation between the steering angle of front wheel and lateral force coefficient at different driving speeds(linear alignment).

The Fig. 4 shows the relation between the steering angle of front wheels δ and the lateral force coefficient. The latter rises linearly as the former increases. With the increase of driving speed, the increase rate of the lateral force coefficient becomes large. At the same driving speed, the lateral force coefficient on the right side section of the steering angle will increase.

The safety analysis of off-ramp lane changing on curve alignment

Compared with the liner sections of highways, curve sections are mainly different in superelevation rate, regardless of the linearity. Here the superelevation rate of horizontal curve i_h =0.08 is taken for tentative calculation.

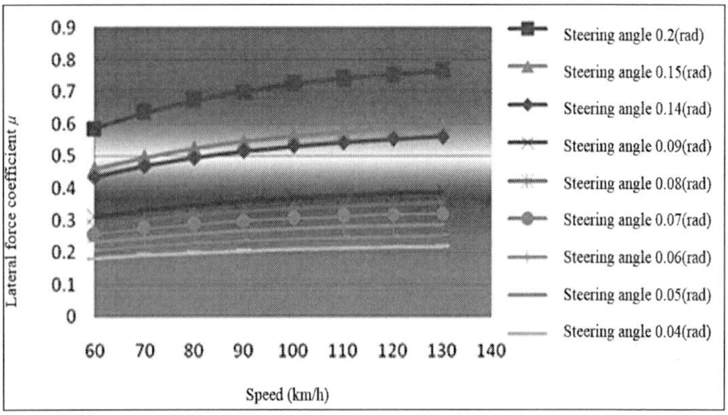

Fig. 5. The relation among driving speed, the steering angle of front wheel and lateral force coefficient (curve alignment).

It can be seen from the Fig. 5 that on dry asphalt pavement, when the superelevation rate of horizontal curve i_h =0.08 and the steering angle of front wheel δ≤0.14 rad (under dangerous situations), vehicles can keep driving stability of lane changing at high driving speed (V≤130 km/h). However, when the steering angle or the driving speed gets larger, the vehicles have a risk of sideslip. Similarly, on wet asphalt pavement, when the steering angle of front wheels δ≤0.09 rad, vehicles can keep driving stability of lane changing at high driving speed. And when the steering angle or the driving speed get larger, the vehicles also have a risk of sideslip. Two cases of the above analysis are extremely dangerous situation.

Fig. 5 shows when δ≤0.03 and V≤130 km/h, the increase of drivers' heart rate and blood pressure stay in a stable range, which means the physiological and psychological pressure is low. When δ≤0.04 and V≤90 km/h, it is the same as above; when the driving speed continues to increase, the driver's physiological and psychological pressure will increase rapidly and it will affect the driving safety; when δ≥0.05, it shouldn't be adopted on freeways, for that even if the easiest lane changing behavior may cause drives feel uncomfortable.

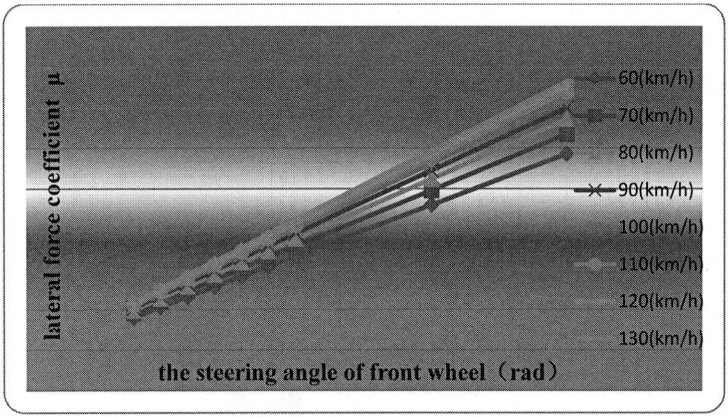

Fig. 6. The relation between the steering angle of front wheel and lateral force coefficient at different driving speeds(curve alignment).

Fig. 6 shows the relation between the steering angle of front wheels δ and the lateral force coefficient. The later rises linearly as the former increases. With the increase of driving speed, the increase rate of the lateral force coefficient becomes large. At the same driving speed, the lateral force coefficient on the right side section of the steering angle will increase larger.

CONCLUSIONS

Off-ramps are the key points of freeway and are under the restriction of more factors than the mainline, which poses a great threat to the traffic safety. This paper presents criteria of a safe steering angle of front wheel in lane changing process, in order to reduce the accident rates on off-ramp sections of freeways.

Due to the superelevation, the lateral force is formed which may lead the vehicles to sideslip. At the same time, as the lateral force coefficient increases, drivers will feel uncomfortable.

By establishing the relation model of the steering angle and lateral force coefficient, the research is conducted from two aspects, which are the vehicle driving stability and drivers' physiological status. The model indicates that no matter driving on linear or curve alignments, drivers are more sensitive to the increase of steering angle. The result of this study can be applied in off-ramp design of specific engineering projects.

ACKNOWLEDGEMENTS

This work is supported by National Natural Science Foundation of China under Grant 51278361.

REFERENCES

Bared, J., Giering, GL., and Warren, DL. (1999). "Safety Evaluation of Acceleration and Deceleration Lane Lengths" *ITE Journal*, 69(5), 50-54.

Bauer, K.M., Harwood, D.W. (1998). *Statistical Models of Accidents on Interchange Ramps and Speed-Change lanes*. Alexandria, USA.

Chen, B., Guo Y.H. (2007). "Simulation Model of Lane Changing on Freeways" *Journal of Xihua University*, 26(3), 35-40. (in Chinese)

Chen, H. Y., Liu, P., Lu, JJ., and Behzadi. B. (2009). "Evaluating the Safety Impacts of the Number and Arrangement of Lanes on Freeway Exit Ramps" *Accident Analysis and Prevention*, 41(3), 543-551.

Pan, X. D., Lin, T., and Yang, Z. (2006). "Relationship Between the Variation of Drivers' Heart Rate and Systolic Blood Pressure and the Coefficient of Transverse Force of Mountainous Highways" *Journal of Tongji University*, 34(6), 748–751. (in Chinese)

Technical Standard of Highway Engineer in China (JTG B01-2003). Ministry of Transport of the People's Republic of China. Beijing, 2003.

Zhen, Z., Rong, J., and Ren. F. T. (2003). "Analysis of the Relation Among Driving Tenseness of Drivers, Horizontal Radius and Driving Speed" *China Civil Engineering Journal*, 36(7), 57-60. (in Chinese)

Zhou, J. (2006). "Modeling and Safety Evaluation of Lateral Force Coefficient on Freeway", Master Thesis, Kunming University of Science and Technology. (in Chinese)

Unlock DDI's Capacity by Re-Routing Left-Turns at Nearby Intersections

Wei Zhang, Ph.D., P.E.[1]; and Nopadon Kronprasert, Ph.D.[2]

[1]Office of Safety R&D, Federal Highway Administration, 6300 Georgetown Pike, HRDS-10, McLean, VA 22101. E-mail: wei.zhang@dot.gov

[2]Department of Civil Eng., Chiang Mai University, 239 Huay Kaew Road, Muang District, Chiang Mai 50200, Thailand. E-mail: nopkron@gmail.com

Abstract

Since the opening of the first Diverging Diamond Interchange (DDI) in the U.S. in June 2009, 34 DDIs have been completed as of March 2014. At least 40 more DDIs are being planned/designed/constructed. The sheer number of DDI projects indicates this innovative design is gaining wide acceptance. Field results demonstrate that this design is much more cost effective than the conventional improvement designs. By converting the 3-phase signal at traditional diamonds into 2-phase, DDI provides more effective green time for moving traffic. Most DDIs constructed to date are near large business, retail, and/or medical centers; and the intersections adjacent to DDI crossovers often have high traffic demands and require 6 or more phases and longer cycle length to serve the demands from conflicting movements. Such condition limits the DDI's capacity utilization. This study explored three approaches to reduce the cycle length at nearby signals—relaxed Bowtie, Superstreet, and Quadrant Roadway intersections. They can all convert the adjacent intersections into 2-phase operation and enable the DDI to operate more efficiently. The ideas were tested in a simulated environment at a proposed DDI in Anchorage, Alaska. The simulation results indicate level-of-service at nearby intersections may be improved from D/E to A/B.

INTRODUCTION

The concept of DDI design was originated in France in the 1970s, but systematic implementation of this design started in the U.S. with the successful opening of the DDI located at I-44 and MO 13 in Springfield, MO. in June 2009. Since then, in less than 5 years, 34 DDIs have been constructed and opened to traffic in 16 states, as shown in Fig. 1. As practitioners accumulating experiences with the various aspects of DDI design, the traffic handling capacity of DDI is gaining broader recognition. The DDI population growth by year chart in Fig. 1 (the bar chart didn't include the 2

DDIs opened in first quarter of 2014) vividly depicts the upward trend of DDI deployments. The reasons behind DDI's popularity are its simplicity in operation, proven effectiveness in boosting interchange capacity, flexibility in accommodating non-motorized traffic, and most importantly, significantly lower cost than competing design options.

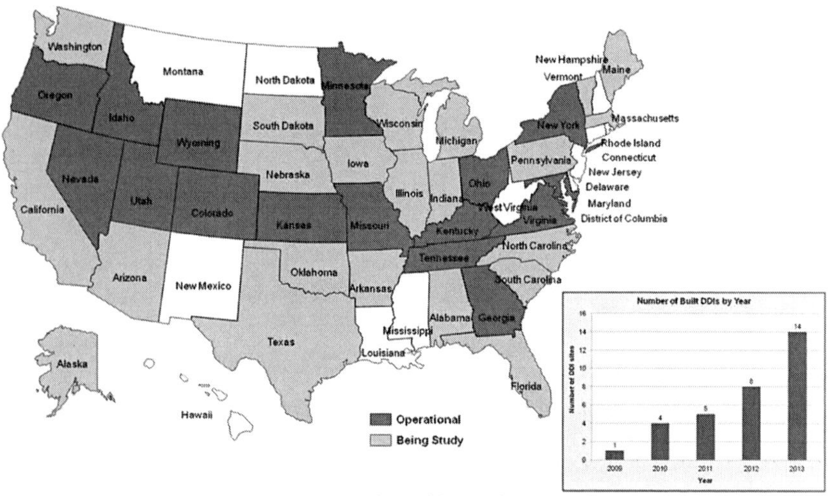

Fig. 1. Population of DDIs in the U.S.

Figure 2 shows the DDI located at KY 4 and US 68 in Lexington, KY. The DDI concept, its advantages over conventional diamond interchange, and its operational constrains imposed by the adjacent signals can all be explained by examining this DDI and the traffic scene photo taken during the morning peak period. Looking first at the WB through and left-turn traffics on the surface street, and the SB left-turn traffic from freeway off-ramp in Fig. 2, one can see that at the east crossover junction, the WB through traffic is channeled to the left side of US 68 in WB direction, and the SB left-turn traffic is channeled to the left side of road in EB direction. Their conflict location of WB through and SB Left is the east crossover junction, which is controlled by a 2-phase signal. With this configuration, the SB off-ramp left-turn traffic can move simultaneously with the WB through traffics until it reaches the east crossover. For the WB left-turn traffic (on-ramp going south onto the freeway), once it passes the east crossover, it can travel to the west crossover junction and turn left onto the freeway on-ramp without any conflict. At the west crossover junction, the WB through traffic is channeled back to the right side of the road in WB direction. At a conventional diamond interchange, the conflict between SB left-turn (off-ramp) and WB through and WB left-turn traffics occur at the west ramp junction, so as the conflict between the WB left-turn (on-ramp) and the EB through traffic. A minimum of 3 signal phases is required to resolve the given conflicts.

By changing the typical 3-phase signal operation at conventional diamond interchanges into 2-phase operation, DDI configuration gains more effective green time for moving traffic, which translates into reduced congestion, increased throughput, and better safety. The capacity is achieved without adding any traffic lanes. This presents a very cost effective way of retrofitting existing interchanges that are structurally sound but suffer capacity constrain, thereby avoiding the need of massive infrastructure re-constructions (Hughes et al, 2010). Since DDIs are often planned and constructed near large business, retail, and/or medic centers, congested signal intersections often exist next to the DDI crossover junctions. As such, the capacity utilization of these DDIs are limited by the capacity of nearby traffic signals that need to operate in 4 or more phases and require longer cycle lengths. This constrain is evident in the traffic scene photo taken during morning peak shown in Fig. 2. In order to better utilize the DDI's traffic handling capacity, improvements must be made to adjacent signal intersections to reduce their constrain to DDI's capacity utilization.

Fig. 2. Plane view of DDI at KY 4 and US 68 in Lexington, KY.

PROPOSED SOLUTIONS

As discussed before, when adjacent signal intersections are congested, vehicles leaving the DDI will queue up at adjacent intersections, back up to the DDI crossover junction, and create the traffic scene shown in Fig. 2. To mitigate this problem, both the number of phases and cycle lengths at nearby signals must be reduced so they can operate more in concert with the optimal signal cycles of the DDI. The key of reducing congestion at nearby intersections lies in the creative handling of the left-turn traffic. This paper illustrates three approaches to re-configure the adjacent intersections: relaxed bowtie, superstreet, and quadrant road intersections. All have the potential to convert a multi-phase signal operation into 2-phase operation, but each individual method is suited to a particular traffic demand pattern.

Relaxed Bowtie Intersection

Hummer and Boone are credited with conceiving the Bowtie Intersection concept, as shown in Fig. 3(a) (Hummer, 1998). This design requires two full-size roundabouts be installed at intersections on minor road next to the main intersections. All left-turn traffics are re-routed through the roundabouts and then back to the main intersection. At the main intersection, only through and right-turn movements are permitted, which makes it a 2-phase operation. To our knowledge, no bowtie intersection has ever been constructed. We hypothesize that the reason behind no implementation of this design is the cost of the two full-size roundabouts, which can easily exceed one million U.S. dollars, making it not cost effective.

In this research, we introduce the concept of relaxed Bowtie Intersection. This design requires two mini-roundabouts rather than two full-size roundabouts, and only major road left-turn traffic is re-routed through the mini-roundabouts, minor road left-turn traffic is permitted at the main intersection. The relaxed Bowtie intersection is conceived for intersections where major road /minor road traffic split is 90/10 or more. To our experience, in many places, the cost of installing two mini-roundabouts can be done at $15,000 to $30,000 each, making it a fiscally viable design. The signal operation at a relaxed bowtie intersection is 2-phase. Pedestrians use the green time assigned to vehicles to cross the roads.

Superstreet

The Superstreet layout is shown in Fig. 3(b). Many superstreet intersections have been built in North Carolina. With this design, the minor-road through and left-turn traffics must turn right at the main intersections, get into the U-turn pocket, and execute a U-turn to return to the main intersection to continue their intended trips. This design turns the original intersection into 4 T-intersections. Only 2 signal phases are needed at each T-intersection, and the traffic movements in each direction can be timed and coordinated independently. At the main intersection, one phase serves the major-road left-turn traffic and minor-road right-turn traffic; another phase serves the major-road through traffic. This design is versatile and can be used under a wide range of traffic demand patterns.

Quadrant Roadway Intersection

The Quadrant Roadway Intersection concept is shown in Fig. 3(c). This design makes use of the spare capacity of available back roads that connect the side street to the major road. The ideal application of this design is to identify the major-road approach that has the heaviest left-turn demand, turn that left-turn traffic into a through traffic at the main intersection, followed by 3 right-turns through the quadrant roads, and back to the main intersection as a minor road through traffic. Other left-turn traffics can also be re-routed through the intersections on quadrant road. In a quadrant road configuration, the main intersection operates in 2 phases, and the secondary intersections may operate in 2 or 3 phases.

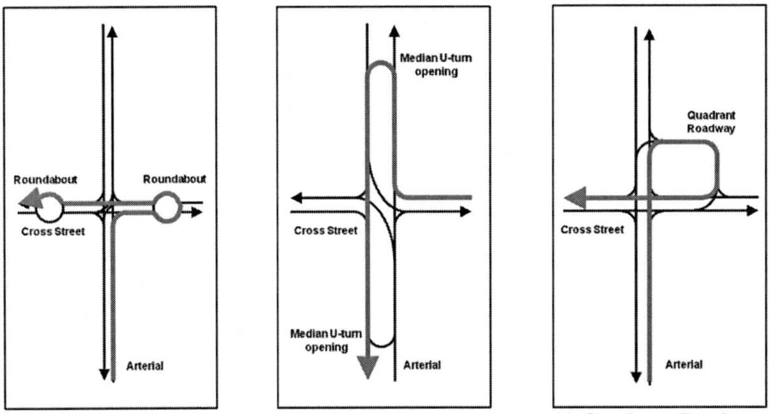

(a) Relaxed Bowtie **(b) Superstreet** **(c) Quadrant Roadway**

Fig. 3. Concept of alternative left-turn treatments

APPLY THE SOLUTIONS IN REAL PROJECT

Fig. 4 shows the Glenn-Muldoon interchange (Glenn Hwy and Muldoon Rd) in Anchorage, Alaska. This is a deficient bridge that needs improvements. (Alaska DOT, 2013), and the authors were asked to perform traffic analyses for the different design options being considered. DDI is one of the design options. The authors collaborated with customers in Alaska DOT and its consultant, to evaluate the traffic operations of different designs under existing and future traffic demands. The 2040 afternoon peak hour network traffic demand is also shown in Fig. 4. Due to page limit, only simulation results involving DDI design under the above traffic demand is presented in this paper. It should be noted that the planned construction project is limited to the interchange only, and will not include any improvements at Intersections #1, #2, and #5, which are outside the interchange. Nevertheless, the authors evaluated all five intersections as a system when exploring ways of improving DDI's capacity utilization. Based expected future traffic demand, the proposed DDI will have 3 lanes in each direction, double left-turn lanes at on-ramps and single right-turn lane at off-ramps. Regardless of the final interchange design option (partial clove-leaf or DDI) selected, the proposed improvement options for Intersections #1, #2 and #5 are expected to help improve the system performance of the 5-intersection network.

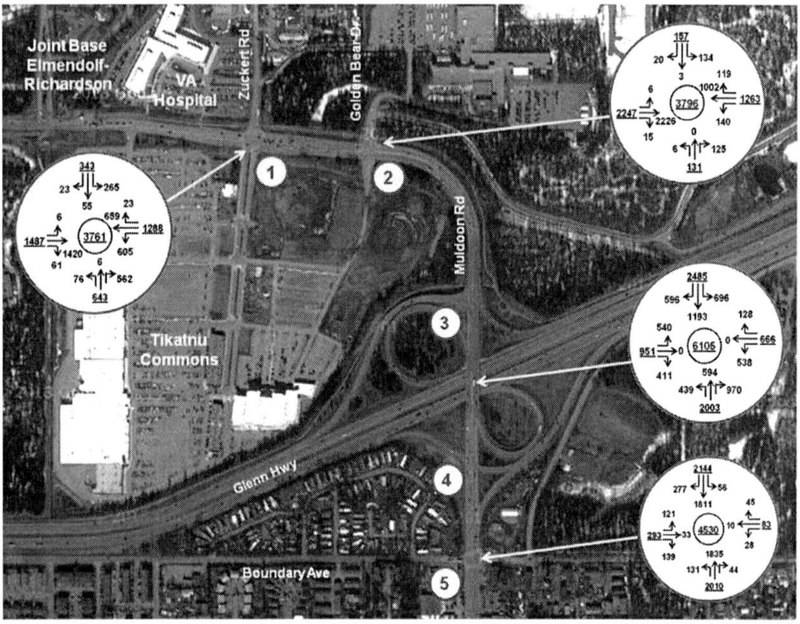

Fig. 4. Glenn-Muldoon interchange, Anchorage, Alaska

Choosing Proper Treatment for Each Intersection

Refer to the network traffic demand in Fig. 4, at Intersection #1 there are heavy demands for WB left-turn and NB right-turn. This traffic pattern is favorable for superstreet configuration, which enables the above two movements be served in one phase, and the EB through movement be served in another phase. Comparing to the original intersection which requires 6 phases, the superstreet design can cut the signal cycle length roughly in half. One can see in Fig. 4 that a parking lot exists at the NW quadrant of Intersection #1, and the road next to parking lot provides access to the major road at south end, and to the minor road at east end. Therefore, a quadrant road intersection design may also work here. This configuration turns the heavy WB left-turn traffic into a through traffic at the main intersection, followed by 3 right-turns to become minor road through traffic. At Intersection #2, the left-turn demand on any approach is relatively low, either a relaxed bowtie or a superstreet may be used here. Both designs can effectively reduce the signal cycle length at Intersection #2 and turn it into 2-phase operation. The superstreet design may offer better coordination with the DDI due to their similarity in moving traffic in one direction at a time. At Intersection #5, on major road in both directions, the through traffics are heavy, but left-turn traffics are low; on minor road, traffic demands are low in both directions. Here the major/minor traffic split is 92/8. A relaxed bowtie configuration is suitable for this traffic pattern. With this design, the major road left-turn vehicles are re-routed through the mini-roundabouts and being positioned as minor road through traffic while the major road through traffic is being served. Minor road left-turn traffic is permitted at the main intersection, or forced to go through the main intersection, and then do a U-turn followed by a right-turn to continue the intended trip route.

Three combinations of alternative intersection treatments were formulated as shown in Table 1 based on the above assessment. The microscopic traffic simulator, VISSIM, was used to build the traffic network and simulate the 3 scenarios. Refer to Table 1 scenario definitions, Case 1 simulates the network operation of constructing a 6-lane DDI only; Case 2 simulates the performance of the 6-lane DDI plus converting Intersections #1 and #2 into Superstreets, and Intersection #5 into a Relaxed Bowtie; Case 3 simulates the performance of the 6-lane DDI plus Quadrant Roadway at Intersection #1, and Relaxed Bowties at Intersections #2 and #5. One can see that each case was designed to build upon the previous case to further improve the corridor's system performance.

All cases were simulated under the same network traffic demand shown in Fig. 4. Relevant measures of effectiveness, such as queues, delays, and travel times, etc., were recorded to compare their system performances. The network models were built from the geometric layout of proposed DDI and Google street map. The simulation models used the design speeds of Muldoon Rd. and typical driver behaviors recommended in Highway Capacity Manual (TRB, 2010). Based on estimated turning movement counts and design configurations, the signal operations for each design alternative were evaluated using the HCM procedure. Optimization software SYNCHRO was used to find the optimal signal cycle length, the green time required for each approach, and the signal cycle offsets. The signal coordination plans used for each case are summarized in Table 1.

Table 1: Combinations of Alternative Designs at Different Intersections

Intersections	Design Treatment Scenarios		
	Case 1	Case 2	Case 3
Intersection#1 (Muldoon Rd and Zuckert Rd)	**Signal** 6-phase, 160-sec cycle	**Superstreet** 2-phase, 110-sec cycle	**Quadrant** 2-phase, 120-sec cycle
Intersection#2 (Muldoon Rd and Golden Bear Dr)	**Signal** 6-phase, 160-sec cycle	**Superstreet** 2-phase, 110-sec cycle	**Relaxed Bowtie** 2-phase, 120-sec cycle
Intersection#3 (North ramp)	**DDI** 2-phase, 160-sec cycle	**DDI** 2-phase, 110-sec cycle	**DDI** 2-phase, 120-sec cycle
Intersection#4 (South ramp)	**DDI** 2-phase, 160-sec cycle	**DDI** 2-phase, 110-sec cycle	**DDI** 2-phase, 120-sec cycle
Intersection #5 (Muldoon Rd and Boundary Ave)	**Signal** 6-phase, 160-sec cycle	**Relaxed Bowtie** 2-phase, 110-sec cycle	**Relaxed Bowtie** 2-phase, 120-sec cycle

RESULTS AND DISCUSSION

Fig. 5 summarizes the MOEs of Cases 1, 2, and 3, representing three different combinations of alternative treatments at nearby intersections under 2040 PM traffic. The MOEs include the average and maximum queue lengths, and delays and level-of-service (LOS) by approach and by intersection. Refer to the legend located at the lower left corner of each case for meanings of the numbers and letters in Fig. 5.

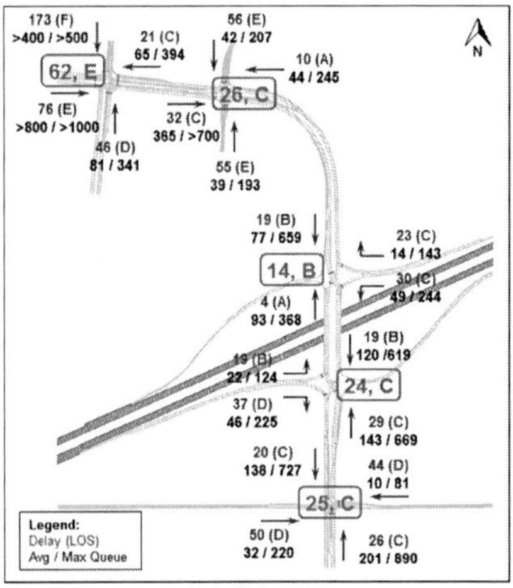

(A) CASE 1

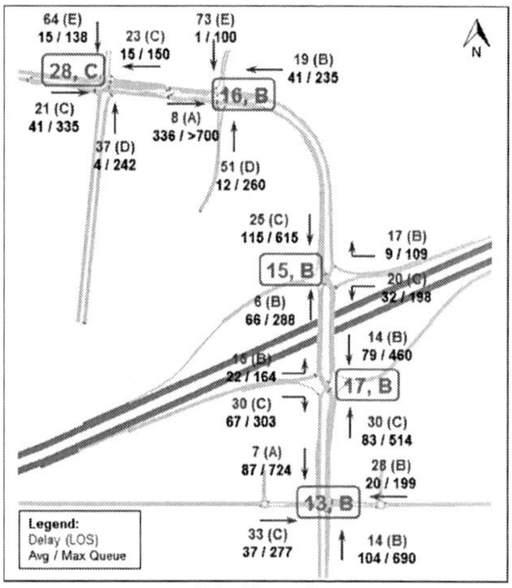

(b) Case 2

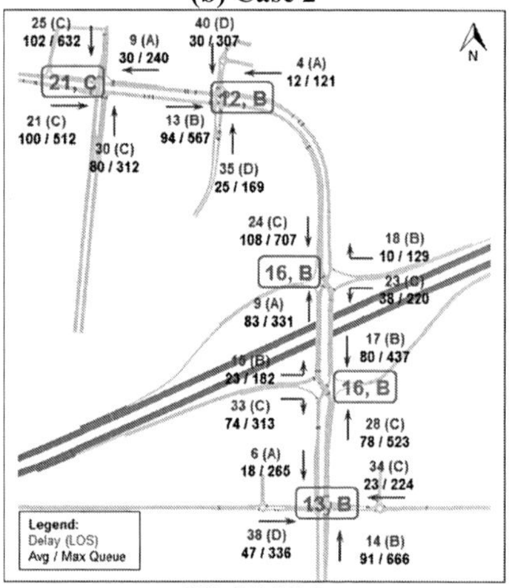

(C) CASE 3

Fig. 5. Comparison of MOEs among different design alternatives

The results from Case 1 (DDI alone) showed that the proposed DDI design will perform well at both ramp intersections, but the queue lengths and LOS at nearby intersections are at the unacceptable states. Results of Case 2 show that by introducing superstreets at Intersections #1 and #2, and Relaxed Bowtie at Intersection #5, intersection delays at the DDI crossovers will be reduced (although not enough to change LOS ratings), and LOS at the three intersections outside the DDI will be significantly improved. Results of Case 3 showed similar improvements. Although the quadrant roadway design introduces some extra travel distance to the WB left-turn traffic and adds more traffic to the minor-road SB approach, the overall delays were improved at the two north-side intersections and their LOS would be improved from E and C to C and B, respectively.

Next, the operation benefits of all cases are compared using Case 1 as the base. Fig. 6 shows the network throughputs achieved under different combinations of alternative treatments. Case 1(DDI only) can achieve 6,326 VPH network throughput, Case 2 (DDI plus Superstreets at #1 and #2, and Relaxed Bowtie at #5) and Case 3 (DDI plus Quadrant Road at #1, and Relaxed Bowties at #2 and #5) can deliver 6,648 VPH and 6,736 VPH network throughputs, respectively. Table 2 compares the network-wide total vehicle throughputs, travel distance, travel time, and annual travel cost savings under 2040 traffic demand. The results show that by strategically re-routing some left-turn traffic at nearby signal intersections, additional vehicle-miles of travel can be served, and sizable user cost savings can be achieved. Improvements associated with Case 2 and case 3 reduce the network travel time by 21%, and 18%, respectively, during the peak period, and lead to the total cost savings of $287K, and 235K.

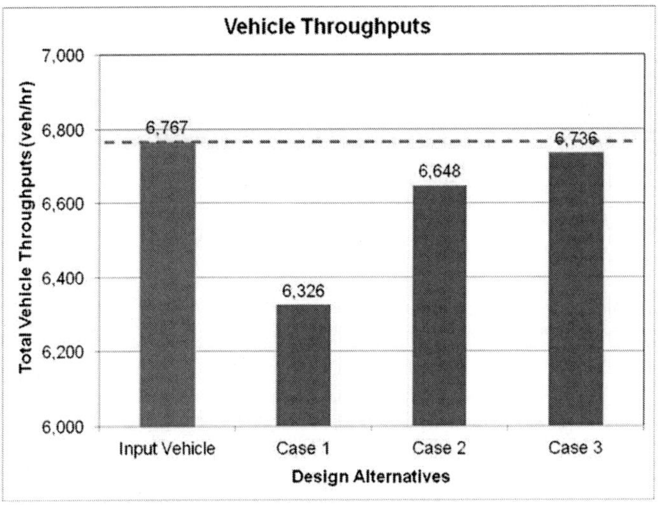

Fig. 6. Comparison of network throughputs

Table 2: Comparison of Performances among Different Design Scenarios

Performances	Design Scenarios		
	Case 1	Case 3	Case 4
Total input traffic demand (veh/hr)	6,767	6,767	6,767
Total vehicle throughputs (veh/hr)	6,326	6,648	6,736
No. of vehicles re-routed (veh/hr)	—	717	1,278
Percent of vehicles re-routed	—	(11%)	(19%)
Veh-mile of travel (1,000 veh-mi) served	4.64	5.00	5.14
Extra veh-mile of travel (1,000 veh-mi) achieved	—	0.35	0.50
Vehicle-hour of travel (veh-hr)	357.9	282.6	291.9
Travel time savings (veh-hr)	—	75.3	66.0
Annual cost saving ($US)	—	$287,600	$234,600

* The $13.67/hr user cost is the adjusted 2013 value of time ($/hr) recommended by Office of Secretary of Transportation (2011). The cost savings calculation is for week days (52 week/yr x 5 days/week = 260 days/yr), and during PM peak hour only.

CONCLUSIONS

This study explored 3 alternative treatments for improving capacity at DDI crossover adjacent signals. The following conclusions were derived:

1. DDI's capacity utilization is often limited by adjacent signal intersections.
2. Broader implementation of DDI design is dependent on finding cost effective ways of reducing cycle length and signal phasing at nearby intersections.
3. Relaxed Bowtie, Superstreet, and Quadrant Road intersections are suitable and cost effective treatments for boosting capacity at nearby signals.
4. Specific treatment for re-routing the left-turn traffic should be decided based on the given traffic demands patterns at each adjacent intersection.

REFERENCES

Alaska DOT (2013). "Glenn highway and Muldoon road interchange improvements." http://www.dowlhkm.com/projects/GlennMuldoon/meetings.html (January 2014).

Chlewicki, G. (2003). "New interchange and intersection designs: The synchronized split-phasing intersection and the diverging diamond interchange." *2nd Urban Street Symposium*, Anaheim, CA.

Hughes, W., Jagannathan, R., Sengupta, D., and Hummer, J. E. (2010). *Alternative intersections/interchanges: informational report*, FHWA, McLean, VA.

Hummer, J. E. (1998). Unconventional left-turn alternatives for urban and suburban arterials—Part one. *ITE Journal*, 68(9), pp.26-29.

Hummer, J. E. and Reid, J. D. (1999). "Unconventional left-turn alternatives for urban and suburban arterials: An update." *TRB Circular E-C019: Urban Street Symposium Conference Proceedings*, Dallas, TX.

Reid, J. D. (2004). *Unconventional arterial intersection design, management and operations strategies*. Parsons Brinckerhoff Inc., New York, NY

TRB (2010). *Highway Capacity Manual 2010, 5th ed.*, Transportation Research Board, Washington, D.C.

A Macro-Meso-Micro Approach for Safety Evaluation of Highway Access Management

Liping Cheng[1]; Weijie Lu[2] ; Linjun Lu[3]; and Jian Lu[4]

[1] School of Transportation, South East University, Si Pai Lou, Nanjing 210096, China.
[2]Transportation Research Center School of Naval, Architecture, Ocean and Civil Engineering, Shanghai Jiao Tong University, 800 Dongchuan Road, Mulan Building A906, Shanghai 200240, China. E-mail: linjun.lu@yahoo.com
[3]Transportation Research Center School of Naval, Architecture, Ocean and Civil Engineering, Shanghai Jiao Tong University, 800 Dongchuan Road, Mulan Building A906, Shanghai 200240, China. E-mail: linjun.lu@yahoo.com
[4]Transportation Research Center School of Naval, Architecture, Ocean and Civil Engineering, Shanghai Jiao Tong University, 800 Dongchuan Road, Mulan Building A906, Shanghai 200240, China. E-mail: jianjohnlu@sina.com

Abstract

In all kinds of safety measures, access management techniques have been proved the most successful in improving the safety and efficiency of arterial roads. The purposes of this research, therefore, were to objectively and comprehensively evaluate the highway safety from the aspect of access management. In this paper, we propose a three-layered (micro-meso-macro) framework that provides safety evaluation at different levels of details. At the macro level, system-wide goals are achieved, we considered highway safety with road functions and land-use; on meso level, safety is concerned with access distance and access types; and at macro-level, we made the consideration of safety protection designs at access points. We believe that the combination of access management at each of the levels may lead to a more efficient and sustainable use of the highway infrastructure.

INTRODUCTION

With the development social and economic, road traffic volume grows quickly, China National Highway accesses become accident-prone areas. Linear regression showed positive correlation between crash severity score and access management at an 81% confidence level. Therefore it is necessary to develop a systematic access safety evaluation and improve the road safety level.

The main concerns of access safety research over the past were the physical characteristics of the main roads, such as the spacing of signal control sections, no signal control spacing, median opening and other factors; many scholars evaluated road safety from the perspective such as accident rates, type of conflict, conflict severity (Lu 2010; Lu 2013). In fact, factors that affecting the safety entrance were in a combination way. Research conducted by Butorac and Wen (2007) provided a synthesis on the effect of crossroads in the vicinity of interchanges and the impact these crossroads have on capacity and safety. In the previous UDOT research, relationships between crash rate and the characteristics of the roadway segment was identified. Grant (2007) divided the categories of AADT into three number groups and categorized the road segments using the selected values for volume and signals per mile .Furthermore, land use was identified to play a significant role in the safety of arterial roads as those arterial road segments with adjacent commercial land use tended to have higher crash rates and severity. Strasburg and Crawley (2005) found that 90.0 percent of urban median crossover crashes occurred within 1.0 mile of an interchange, while 90 percent of rural median crossover crashes occur within 3.0 miles of an interchange. It shows that in safety evaluation, we need to differentiate road safety between city environments and rural environments.

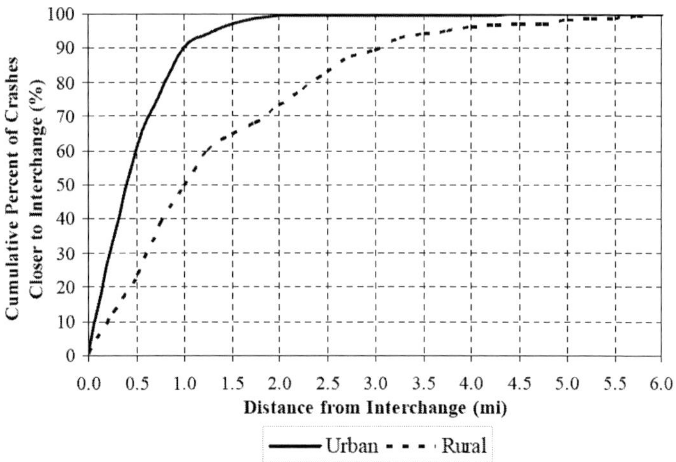

Fig. 1. Crash Distribution for Urban and Rural Median Crossover Crashes.

The results of regression coefficient, standard error, t-statistic, and p-value for each independent variable for right-angle collisions indicated that right-angle collisions were positively correlated with commercial adjacent land use; volume, and speed limit, while not directly related to access management, were also correlated to the severity score. Many facts showed that recurring congestion or accident-prone were not necessarily the problem of road itself, but had relationship with the environment and access conditions. Therefore, the establishment of a "macro-meso-micro" evaluation index system was necessary, only in this way could we make a systematic and rational evaluation of the access safety.

MICRO-MESO-MACRO APPROACH

Our macro-meso-micro approach aims to evaluate road safety at different levels, tackling the particular problems of each of them. The end goal of our approach is that the combination of the access management in each of the levels leads to a more sustainable and efficient use of the road infrastructure.

The macro level deals with infrastructure or system-wide goals, such as providing appropriate transitions from one classification of roadway to another and making an efficient use of the road network. The decision making at this level is based on the road function analysis, land use and road surrounding environment.

Macro indicator mainly reflects the China National Highway entrances operating conditions, including the China National Highway surrounding land use and the surrounding environment for socio-economic development level, micro indicators include access horizontal and vertical curves, sight distance, access point, access channelization compounds rational, signs and markings, signal control, traffic safety facilities.

The meso level is concerned with access spacing, providing circulation system and preserving functional area of intersections and interchanges. Well-planned communities provide a supporting network of local and collector streets to accommodate development. Alternatively, commercial strip development with separate driveways for each business forces even short trips onto arterial roadways, thereby reducing safety and impeding mobility.

Finally, at the micro level, we deal with intersection safety facilities such as left /right turning lanes setting, clearance distance, intersection conflicting points. Drivers make more mistakes and are more likely to have collisions when they are presented with the complex driving situations created by numerous conflict points. Conversely, simplifying the driving task contributes to improved traffic operations and fewer collisions. A less complex driving environment is accomplished by limiting the number and type of conflicts between vehicles, vehicles and pedestrians, and vehicles and bicyclists.

This micro-meso-macro approach has already been studied in other fields, such as evolutionary economics, complex systems, or organic computing. All these approaches identify the meso level as the level that acts as a bridge between the other two levels, accounting for systems dynamics and being responsible of changes in agents behaviours and the relationships among them. Although the approach is presented as having three levels, there could be multiple nested meso levels, as shown in Figure 2, depending on the level of detail at which the system is inspected. Moreover, an individual could belong to more than one meso level system at the same time (e.g. one for negotiating the cruise speed in a platoon with neighbouring vehicles, and another one for deciding what route to take, which may involve a different set of vehicles, depending on its destination).

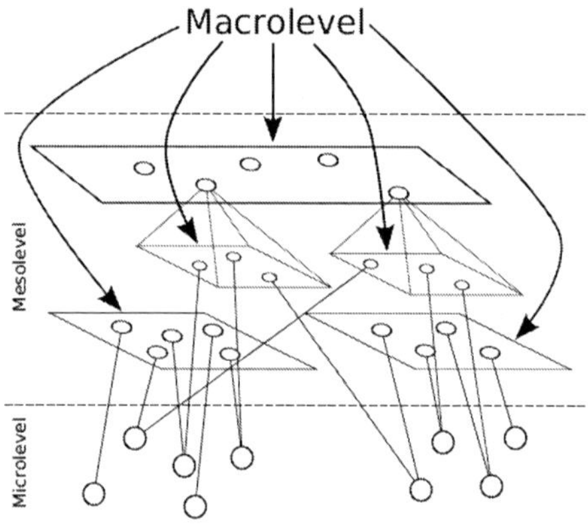

Fig. 2. Micro-Meso-Macro Approach.

In the following sections we provide some implementations of each of the levels in our approach. We should note that each level may implement several, possibly complementary, systems, and so those presented here are only examples to show what kind of intelligence can be achieved with our approach.

MACRO LEVEL: ROAD FUNCTION AND ROAD ENVIRONMENT

Macro-scale urban road network indicators as independent variables and passenger transport fatalities as the dependent variable are analyzed in this research using GLM analysis. From this study, it can be concluded that the length of road per million inhabitants, blocks per area, nodes per selected areas and length of motorway per ten thousand inhabitants are the urban road network factors that are associated with passenger transport fatalities. The macro factors for highway access management included:

- Arterial functional classification (e.g., principal arterial, minor arterial);

- Area type of urban, suburban, or urbanizing;
- Land use of residential, commercial, or mixed-use;
- Intersection Hierarchy.

Arterial Functional Classification

An efficient transportation network provides appropriate transitions from one classification of roadway to another. For example, freeways connect to arterials through an interchange that is designed for the transition.

Macro indicator mainly reflects the China National Highway entrances operating conditions, including the China National Highway surrounding land use and the surrounding environment for socio-economic development level, average running speed, average road service levels, traffic accidents and other indicators; meso index mainly reflects entrance security situation, including the access road type, access intersection types, access density, entrances average delay; micro indicators include access horizontal and vertical curves, sight distance, access point, access channelization compounds rational, signs and markings, signal control, traffic safety facilities.

Land Use

Land use means different building environments or land development states. Depending on the classification of the arterial road segment according to its volume and signals per mile, road segments having either adjacent commercial or residential land use greater than at rural sections, and to increase sharply as interchange spacing decreases in urban areas. Many research results show that building environment and road infrastructure play a substantial role in the occurrence of road accidents. At national highways, different building environments that beside sides of national highway may trigger different traversing traffic demands, pedestrians and non-motor vehicle that had the activities of crossing the trunk road at intersections produced most of traffic accidents, as traverse activity has lots of conflicts with the through traffic. The three land use types included residential, commercial, and mixed use. The three area types included urban (metropolitan area with population of at least 250,000), suburban (nearby areas with population of 50,000 to 250,000), and urbanizing (areas with build out plans to reach or exceed population of 50,000).

Access Intersection Hierarchy

Traffic flow produced in the highway should be guided into the main arterial in certain hierarchy, only in this way could avoid traffic conflicts of different grade roads. In the typical case, traffic flow guiding grade is introduced shown as below, some lower grade roads are forbidden accessing to the main trunk roads. the synthetic of access road hierarchy could be defined as table 1.

Table 1. Access Hierarchy Standard.

Access Permitting \ Type	Main trunk roads	Secondary trunk roads	Main collector roads	Secondary collector roads	Local highways
Main trunk roads	Permitted	Permitted	Partly limited	Limited	Forbidden
Secondary trunk roads		Permitted	Permitted	Partly limited	Limited
Main collector roads			Permitted	Permitted	Permitted
Secondary collector roads				Permitted	Permitted
Local highways					Permitted

MESO LEVEL: ACCESS DISTANCE AND ACCESS TYPE

In 2009, the FHWA initiated a study to investigate the safety impacts of access management policies and techniques. A panel of state and local representatives subsequently identified priority access management principles and design factors that should be included in a corridor-level crash prediction model. Based on this input and availability of data, the following access management strategies were selected.

Unsignalized Access Spacing (Intersections and Driveways)

Commonly referred to as driveways or intersections, access points introduce conflicts and friction into the flow of traffic along a roadway (Lu 2014). Increasing the spacing between access points improves traffic flow and safety along the roadway by:

• Reducing the number of conflicts per mile.

• Providing a greater distance for motorists to anticipate and recover from turning maneuvers.

• Providing opportunities for the construction of acceleration lanes, deceleration lanes, or exclusive left-turn or right-turn lanes.

Traffic Signal Spacing

The proper spacing of traffic signals in terms of frequency and uniformity has an important effect on arterial safety and traffic flow. Frequency refers to the number of traffic signals for a given length of roadway and is sometimes referred to as "signal density." It is typically expressed as the number of signals per mile. Uniformity refers to the variation in the distances between individual traffic signals along a given length of roadway. It is desirable to minimize this variation and to space traffic signals at uniform distances. Closely-spaced or improperly-spaced traffic signals can result in increased crash rates, frequent stops and unnecessary delays for motorists and pedestrians, as well as increased fuel consumption, and excessive vehicular emissions. Properly-spaced traffic signals allow for the efficient progression of motor vehicle and pedestrian traffic, as well as provide an agency with greater flexibility in developing signal timing plans to reduce traffic conflicts.

Median Opening Spacing

A median opening is an opening in a non-traversable median that provides for crossing and turning traffic. A "full" median opening allows all turning movements, whereas a "partial" median opening allows only specific movements and physically prohibits all other movements. To realize the safety benefits, median openings should not encroach on the functional area of another median opening or intersection.

Interconnected Street and Circulation System

Well-planned communities provide a supporting network of local and collector streets to accommodate development, as well as unified property access and circulation systems (Strasburg and Crawley 2005). Interconnected street and circulation systems support alternative modes of transportation and provide alternative routes for bicyclists, pedestrians, and drivers. Alternatively, commercial strip development with separate driveways for each business forces even short trips onto arterial roadways, thereby reducing safety and impeding mobility.

Functional Area of Intersections and Interchanges

The functional area of an intersection or interchange is the area that is critical to its safe and efficient operation. This is the area where motorists are responding to the intersection or interchange, decelerating, and maneuvering into the appropriate lane to stop or complete a turn. Access connections too close to intersections or interchange ramps can cause serious traffic conflicts that result in crashes and congestion.

MICRO LEVEL: CONFLICT POINTS AND LANE SETTING

In this level we focus on the intersection safety design level considering the conflicting points, channel lane setting, and corner clearance.

In total, the study corridors represented more than 600 miles and contained more than 1,500 signalized intersections, 3,500 unsignalized intersections, and 15,000 driveways. The following list identifies the variables collected for each corridor.

Presence of Left and Right Turn Lane(s) on Major and Minor Road

Remove Turning Vehicles from Through Traffic Lanes: Turning lanes allow drivers to decelerate gradually out of the through lane and wait in a protected area for an opportunity to complete a turn (Zeitz 2003). This reduces the severity and duration of conflict between turning vehicles and through traffic and improves the safety and efficiency of roadway intersections.

Turning Restrictions on Major and Minor Road

Roadways that serve higher volumes of regional through traffic need more access control to preserve their traffic function. Frequent and direct property access is more compatible with the function of local and collector roadways.

Limit the Number of Conflict Points

Drivers make more mistakes and are more likely to have collisions when they are presented with the complex driving situations created by numerous conflict points. Conversely, simplifying the driving task contributes to improved traffic operations and fewer collisions (Twomey 1993). A less complex driving environment is accomplished by limiting the number and type of conflicts between vehicles, vehicles and pedestrians, and vehicles and bicyclists.

CONCLUSIONS AND RECOMMENDATIONS

In this paper we have proposed a three-layered framework to deal with highway access management system, notably regarding safety and sustainability issues. Each level of the framework addresses particular problems at different scales of detail, starting with a global view of road network (macro level), going through access spacing controlling (meso level), and ending with intersection safety setting (micro level).

While each of the levels can be seen as a standalone system, we believe that the combination of all of them may lead to a much more efficient and sustainable use of the highway access system. This would be an incredibly useful tool for the authorities and transportation planners, with which they could implement certain access management policies and control how the road infrastructure is used.

Research showed that macro level access management would be the main contents of the future access management research, we could achieved by making laws to prevent disordering and over-exploitation economic activities on both sides of the main roads and limit lower hierarchy roads accessing to the highways; On meso levels, we needs to improve the rational setting of intersection functional areas, access spacing and other issues; on micro-level we should improve intersection channelization to protect pedestrians and non-motorized.

REFERENCES

American Association of State Highway and Transportation Officials (2004). *A Policy on Geometric Design of Highways and Streets*, Washington, DC.

Brase, C. H., and C. P. Brase.(1995). "Understandable Statistics: Concepts and Methods". D.C. *Heath and Company,* Lexington, MA.

Huang, B., Lu, J.J. (2014). "A new access density definition and its correlation with crash rates by microscopic traffic simulation method", *Accident Analysis Prevention*, Volume 64, March 2014, Pages 111-122

Kurtz, M. (1991). "Section 7: Transformation of space in computer graphics." *Handbook of applied mathematics for engineers and scientists,* McGraw-Hill, New York, 7.1-7.42.

Lu, J.J., Chen, S. (2013). "A programmable calculation procedure for number of traffic conflict points at highway intersections", *Journal of Advanced Transportation.;* 47(8).

Lu, J.J., Pan, F.J. (2010). "Safety Evaluation of Signalized Highway Intersections Based on the Concept of Level of Safety Service" *Journal of Transportation Safety & Security 01*; 2(3):279-29

Roess, R. P., E. S. Prassas, and W. R. McShane.(2004). *Traffic Engineering*, 3rd Ed., Pearson Education, Upper Saddle River, NJ.

Schultz, G. G., and K. T. Braley. (2007). "A Prioritization Process for Access Management Implementation in Utah". Report UT-07.05, Utah Department of Transportation Research and Development Division, Salt Lake City, UT,

Strasburg, G., and L. C. Crawley. (2005). "Keeping Traffic on the Right Side of the Road." *Public Roads,* Vol. 68, No. 4 , pp. 34-37.

Transportation Research Board of the National Academies (2003). *Access Management Manual*, Washington, DC,

Twomey, J. M., M. L. Heckmand, J. C. Hayward, and R. J. Zuk. (1993). "Accidents and SafetyAssociated with Interchanges". In Transportation Research Record: Journal of the Transportation Research Board, No. 1385, *Transportation Research Board of the National Academies,* Washington, D.C., pp. 100-105.

Zeitz, R. (2003). "Low-Cost Solutions Yield Big Savings". *Public Roads,* Vol. 67, No. 3, pp.48-50.

Comprehensive Access Planning Is Key to Economic Sustainability

Chris Huffman[1] and Michelle Winkelmann[2]

[1]Huffman Corridor Consulting, LLC, 534 Kansas Avenue, Suite 1108, Topeka, KS.
E-mail: chris@huffmancorridorconsulting.com
[2]Huffman Corridor Consulting, LLC, 534 Kansas Avenue, Suite 1108, Topeka, KS.
E-mail: winkelmann@pbworld.com

Abstract

The purpose of this paper is to relate a planning procedure for the systematic retrofit of a four lane suburban expressway to a six lane urban freeway. The retrofit is likely years away and will require a number of projects over multiple funding programs. To develop an implementation-ready corridor access management plan, it is necessary to balance priorities for land-use (intensity, livability, walkability, and sustainability) and desires for public safety and accessibility on the public roadway system (safety and capacity constraint). The desired result is a plan that will systematically retrofit a suburban expressway into an urban freeway through a combination of access management, corridor preservation, and development coordination. Land-uses and transportation networks exist in a demand/supply relationship just as any economic system, and a general equilibrium is sought throughout the years required for full retrofit. Negative economic and environmental impacts will compound over time, and the public safety will be compromised, if inappropriate access connections and intersection controls are allowed. An area-wide access management plan was developed beginning with an understanding of future traffic generation/distribution characteristics and the capacity constraint requirements of the various regulatory partners. A VISSIM model was developed with the capacity constraints and future traffic demands were added. Points of interchange with the freeway system were identified, and then "nodes" of connection for the supporting street network were added. With a transportation system framework in place, a land-use plan was developed. Development characteristics were identified that will fully leverage the capacity of the planned transportation network. The foundation of an implementation plan was developed to allow for cooperation and coordination between multiple levels of government. Execution of this plan will provide for improved economic and environmental sustainability and public safety.

INTRODUCTION

Retrofits of partially access controlled expressway facilities into fully access controlled freeways are fraught with challenges. The corridor that is the subject of this paper is approximately 5 kilometers in length and exists in the center of the United States on the eastern side of the Wichita, Kansas metropolitan planning area.

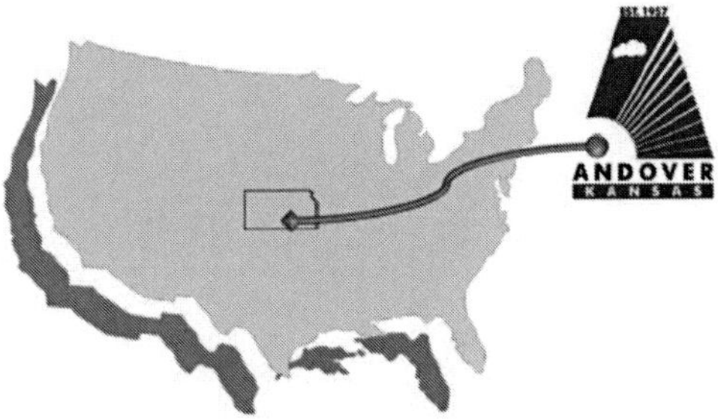

Fig. 1. Study corridor location.

The US-54/400 corridor in this area is currently a four-lane, divided expressway with partial control of access, and at-grade intersections with minor roadways. To the west, the City of Wichita, Kansas has been systematically retrofitting this corridor into a six-lane urban freeway with one-way pair collector/distributor roads to provide for circulation capacity. Due in part to lack of advance corridor planning, and to poor land use development coordination, this retrofit process has become prohibitively expensive. Costs of tens of millions of dollars per kilometer have been seen, and a very recent project saw right-of-way and relocation expenses alone in excess of $150,000,000.

The City of Andover is utilizing a corridor planning process to manage these costs more effectively, and to minimize impacts to future developments and economic activities. The purpose of this paper is to relate this corridor planning process, and to discuss the results of its implementation, to date.

Transportation and Market Area Characteristics

The existing corridor is a four-lane divided expressway with partial control of access and at-grade intersections with minor roads. Average Annual Daily Traffic (AADT) within this corridor currently vary from approximately 33,000 vehicles per day (vpd) in the western extreme of the corridor, to approximately 18,000 vpd in the eastern extreme of the corridor. These volumes are expected to grow aggressively over the planning horizon, varying from 1.3% to 2.6% annually. Growth rates on some of the intersecting roadways are expected to be much higher on routes, as many of these routes have significant growth potential.

Land uses within the project area include retail uses such as fuel stations, restaurants, grocery and other merchant shops. There are also some light industrial uses including warehouses and light manufacturing facilities. Finally, there are civic uses such as a public golf course, and a community recreation center. This commercial frontage rapidly transitions to residential uses both north and south of the corridor. The land uses within the project footprint (5 kilometers east-west and 2 kilometers north-south) have developed over approximately 30 years with little consideration to comprehensive planning. As a result, there are widely varying lot sizes and on-site traffic circulation patterns.

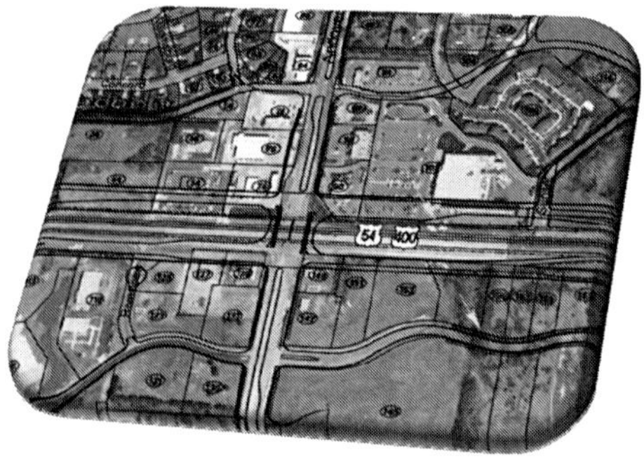

Fig. 2. Varying existing lot sizes.

These variances in land development patterns make transportation planning especially challenging. Since land-use and transportation exist in a supply/demand relationship (just like any economic system), it is important to balance the supply and demand to prevent catastrophic collapse of the area served by the corridor.

Since access is the gateway through which people and goods flow, appropriate location and design of access is critical to supporting and sustaining economic activity.

Social Priorities and Goals

An extensive public involvement effort established the priorities of the community and of the regulatory partners responsible for the corridor. The community is largely a "bedroom" town (people live here, but commute to work), and is proud of its small town environment and quality of life. The community understands that the freeway retrofit is eventually necessary, and that the continuing development of the corridor is inevitable. However, as redevelopment occurs, stakeholders are determined to accomplish the following:

- To increase, and sustain a diverse tax base.
- To protect its "home-town" image.
- The freeway corridor will not constitute a "river of concrete" dividing the community.

Simultaneously, the need for the regional freeway corridor to move large volumes of traffic at higher speeds is a priority of state and regional authorities.

Future Transportation System

One of the principle challenges with long range transportation planning in the central portion of the United States is that population densities are generally low. This has resulted in arterial and expressway class corridors being forced to serve functions of local access as well as through movement. One of the principle purposes of this project (1) was to identify a complete system of freeway/arterial/collector/local class facilities that can, as a system, serve the needs of through movement and local access safely and efficiently. After determining the need for the various roadway functional classifications, the next step is to determine cross-sections (number of lanes) for each functional classification.

The social priorities and transportation demands for a six-lane urban freeway with a depressed grade in order to take it below existing ground level in order to preserve the line of sight across the freeway are depicted in Fig. 3. Additionally, three or four land reverse access roads are needed to provide access, at-grade, to the various land uses in the development zone. The public participants, government officials, business owners, and other key stakeholders expressed strong resistance to massive, multi-lane pavement cross-sections dominating the community. The social priorities of lifestyle (pedestrian amenities, green-space, etc.) meant that maximum cross-sections of the various roadway classifications had to be established before network modelling could begin.

Fig. 3. Depressed freeway system cross-section.

With pavement cross-sections and future traffic demands in hand, a VISSIM model was developed. It is not common in the U.S. for simulation modelling to take place from a capacity constrained persepective, but approaching the model from this perspective allowed the Team to focus upon the balance of the supply and demand. Grade separated interchange nodes were analysed, and analysis revealed that a single interchange in the middle of the corridor did not provide acceptable levels of service. Consideration was also given to an extended diamond design with ramps and the eastern and western ends of the corridor only, but that also yielded unacceptable levels of service. A system of three interchange nodes, one on the east end, one in the center, and one on the west end was selected, as shown in Fig. 4. Intermediate north/south roadways between the interchanging roadways will overpass, but will not interchange, the freeway.

The first, and perhaps most surprising, result of this modelling was that the transportation network selected for analysis had more than enough capacity for the future traffic demands. Therefore, it was possible to generate a land use plan, as shown in Fig. 5,that allowed for much higher land use densities, while preserving acceptable levels of service.

The floor area ratios that are supported by this network, with acceptable levels of service, are more than double what is typically seen in this part of the U.S. This means additional rentable floor area generating economic activity, yielding increased property tax and sales tax revenues. In order to support this land use density, however, it is vital that the access to the system is managed appropriately. Utilizing Synchro, queuing analysis is performed at each intersection (node) in order to establish the limits of required access control, and identifying the access window within which access points may be located.

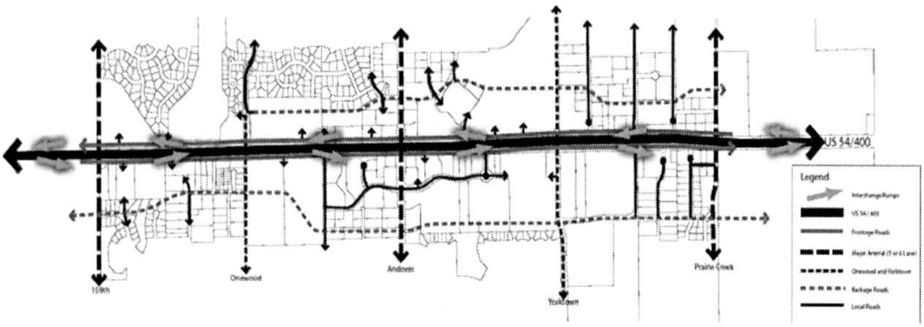

Fig. 4. Selected roadway network.

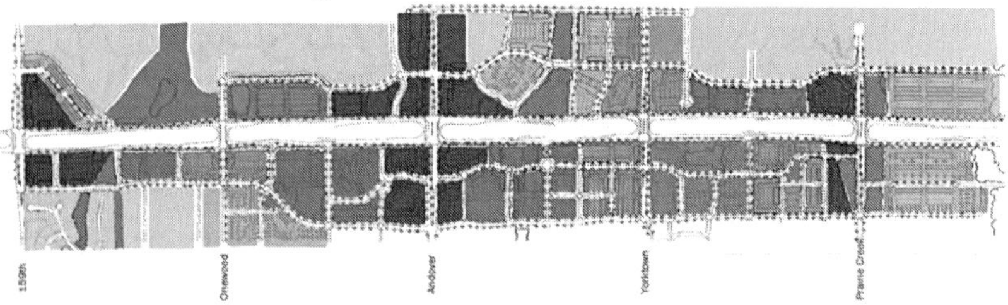

Fig. 5. New future land use plan.

Particularly exciting is the fact that areas of exceptionally high value development can now be identified, and positive guidance (regarding site design requirements and access) and development incentive can now be offered. Such nodes of mixed use development have the potential to offer mixtures of higher density residential, retail, and office, as well as lifestyle amenities that make them especially valuable. One particular example is an area that lies on the north side of the freeway corridor, and will have multi-modal access (vehicular, pedestrian, bicycle) across the freeway to recreational and educational amenities on the south side of the freeway.

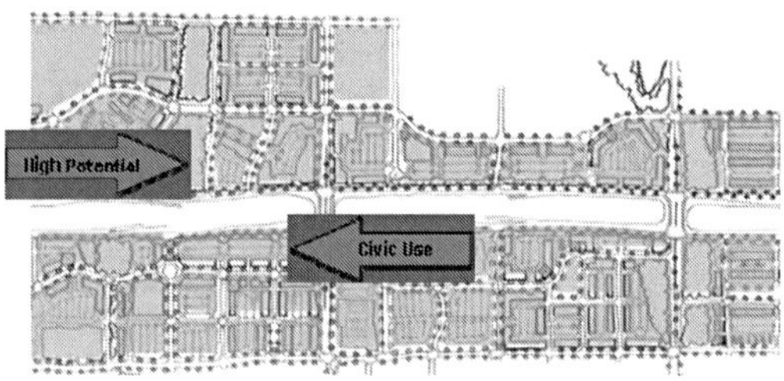

Fig. 6. Area of high economic potential.

Integration of the City's bicycle and walking trail plan was also necessary in order to ensure the goals of liveability identified by the stakeholders, and in oder to support the lifestyle goals of the community.

CONCLUSIONS

In conclusion, this corridor plan begins with clear statements of social policy and measures of effectiveness that identify and prioritize the goals of the key stakeholders, and result in a clear vision of a successful corridor. This qualitative effort then supports the quantitative analyses that result in a transportation plan that will support and sustain a land use plan. That transportation systems respond much more slowly to changing conditions than do the economic forces that drive land uses means that transportation professionals find themselves in a losing game of attempting to respond to changing demands. Thus, it is advantageous to all stakeholders if the transportation elements can be fixed in plan, then the land uses guided to take advantage of the transportation network. This is due to the fact that, while transportation and land-use exists in a supply and demand relationship just as any economic system, changes in land use (demand) happen much more rapidly than corresponding changes in transportation (supply). This results in transportation networks becoming overburdened, congested, and unsafe and leads to significant challenges of implementation. This also results in significant future transportation improvements becoming much more difficult and expensive. The next step of this effort is to integrate the specific recommendations of the study into the various regulatory frameworks of the responsible parties. This increases the probability of consistent implementation of the study recommendations, and helps to ensure eligibility for a variety of funding sources. The City of Andover has, as of the date of this paper, already begun the process of advance acquisition of needed rights of way as opportunities present themselves. Compared to a recent project approximately 6 kilometers to the west on US-54/400, where right-of-way and relocation expenses along exceeded $150 million, this approach of advance acquisition within the framework of a corridor plan provides remarkable advantages.

REFERENCES

Parsons Brinckerhoff, Inc., Huffman Corridor Consulting, Poe & Associates, Inc., and TranSystems. 2011 City of Andover, US-54/400 Corridor Study

Transportation Research Board, 2003, Access Management Manual

Gluck, J., H.S. Levinson, and V. Stover. Impacts of Access Management Techniques. NCHRP Report 420, Transportation Research Board, National Research Council, Washington, DC (1999)

National Highway Institute, 2007 NHI 133078: Access Management, Location, and Design

National Highway Institute, 2010 NHI 134073: Leap not Creep, Accelerating Innovation

http://www.andoverks.com/DocumentCenter/Index/65 US-54/400 Corridor Study website by the City of Andover, Kansas

http://www.ksdot.org/burProgProjMgmt/Local/sepl.asp US-54/400 Retrofit on Kansas Department of Transportation's long range S.T.I.P.

Application of Access Management Techniques to Reduce Wrong-Way Driving Near Interchange Areas

Mahdi Pour-Rouholamin[1]; Huaguo Zhou[2]; Mohammad Jalayer[3]; and Michael Williamson[4]

[1]Ph.D. Graduate Student, Department of Civil Engineering, Auburn University, Auburn, AL 36849-5337. E-mail: mahdipn@auburn.edu
[2]Associate Professor, Department of Civil Engineering, Auburn University, Auburn, AL 36849-5337. E-mail: zhouhugo@auburn.edu
[3]Ph.D. Graduate Student, Department of Civil Engineering, Auburn University, Auburn, AL 36849-5337. E-mail: jalayer@auburn.edu
[4]Instructor, Department of Civil Engineering, Indiana State University, Terre Haute, IN 47809-1902. E-mail: Michael.williamson@indstate.edu

Abstract
Past studies indicated that interchange configurations, access control, and geometric design are related to Wrong-Way Driving (WWD) and minor ramp geometric changes can be effective in reducing the number of wrong-way entries onto freeways. In this paper, access management techniques and geometric elements, which are capable of discouraging wrong-way maneuvers, are identified and discussed. Additionally, every aspect of these elements and their relationship to WWD is investigated. These geometric elements include interchange types, exit ramp terminals, frontage roads, raised medians, channelizing islands, and control radius. The aforementioned elements should be given a special consideration during the design stage of interchanges and intersections.

INTRODUCTION

Wrong-Way Driving (WWD) is defined as the driving movement against the main direction of traffic flow along high-speed, physically-divided highways (i.e., freeways, expressways, and interstate highways) and their access ramps (NTSB 2012; Zhou et al. 2014; Zhou and Pour Rouholamin 2014a). An analysis on eight years of crash data (2004-2011) extracted from the Fatality Analysis Reporting System (FARS) database, revealed that an average of 359 victims result from 269 fatal WWD crashes annually (ATSSA 2014). To reduce the possibility and severity of these types of crashes, many efforts have been made and various countermeasures have been implemented by state and local agencies (Zhou and Pour Rouholamin 2014b). In addition to traffic signs, pavement markings, traffic signals, and intelligent transportation systems (ITS) technologies, various geometric design and access management techniques have also been found to yield promising results.

Access management techniques that can be applied as WWD countermeasures are typically used to achieve a balance between accessibility and mobility. One of the primary goals of access management strategies is to provide a safer movement of vehicles while maintaining vehicular throughput and access to adjoining facilities and lands. WWD calls for further studies on its various techniques and geometric elements, such as exit and entrance ramp characteristics, frontage roads, control radius, and sight distance. In this paper, these objects of study are identified to help reduce driver confusion near interchange areas, which is where a high proportion of wrong-way entries occur. Further studies are also conducted to determine their characteristics and to judge their effectiveness in reducing the possibility of wrong-way entries.

RESEARCH BACKGROUND

Several previous studies have focused on the application of access management to reduce WWD at intersections and interchange terminals. They have focused predominately on the effect roadway geometric characteristics and modifications have on the safety of the studied locations.

A recent study by the North Texas Tollway Authority (NTTA) investigated median modifications to reduce WWD incidents at a location in Dallas, Texas. The study location was identified to be the originating point of several WWD incidents, which were caused by the existence of a side street close to an exit ramp. This situation confused drivers turning left from the crossroad toward the side street, causing some of the drivers to enter the nearby exit ramp mistakenly. To mitigate WWD activities, the NTTA proposed closing the median opening to prohibit left turns into the side street from the crossroad. After the median closure, no WWD incidents were recorded at this location (Ouyang 2013; ATSSA 2014).

In their research, Chassande-Mottin and Ganneau (2008) proposed to reduce the complexity of intersections connecting to interchanges by using various methods. The presence of more than one island at the exit ramp terminal was identified as an undesirable choice that may cause driver confusion; however, constructing roundabouts became a way to control access while reducing the possibility of wrong-way entries on interstates.

Zhou et al. (2012; 2014) collected and analyzed a six-year period of crash data from 2004 to 2009 in an attempt to identify contributing factors regarding WWD crashes on Illinois freeways. In their study, a total of 217 WWD crashes were used to identify contributing factors, as well as pertinent countermeasures for each specific location. A guideline for reducing wrong-way movements on freeways was developed as the second phase of this project. This paper summarizes some key findings from the guideline on the application of access management techniques to reduce wrong-way entries near interchanges. These techniques include raised medians, channelizing islands, turning radius, and exit ramp placement.

ACCESS MANAGEMENT TECHNIQUES

Exit/Entrance Ramps

There are various possibilities to enter a freeway system in the wrong direction (i.e., executing a U-turn on the freeway mainline, crossing the median through an emergency turnaround, or entering from an exit ramp). Therefore, when it comes to WWD, exit ramps are critical points to consider for the application of access management techniques. Geometric attributes of exit ramps, such as layout, connecting angles with crossroad, and cross section, can reduce WWD activities at interchange terminals, if designed properly. For instance, adjacent exit and entrance ramps (parallel, side by side) may be more prone to wrong-way maneuvers. A side street or access driveway close to exit ramps may also cause driver confusion. Therefore, special consideration should be given to the geometric characteristics that make up exit ramps, for the purpose of making them less susceptible to WWD.

The angle at which an exit ramp connects to a roadway (crossroad or frontage) depends on the functionality of the roadway. If left turns from exit ramps are prohibited because of a connecting one-way roadway or the presence of a raised median on the roadway, an acute angle should be used to connect exit ramps to the roadway. Sweeping connections of exit ramps to crossroads, such as outer connections, loops, and some diamond ramps, are less susceptible to WWD. This is primarily due to the inherent capability of the formed acute angles with the crossroads, which causes turning movements in either direction difficult (AASHTO 2011). On the other hand, when exit ramps cross two-way roadways where left turns are allowed, a right angle should be used to connect exit ramps to crossroads.

Reducing ramp throat width is another access management technique that can be effective in decreasing wrong-way entries. Raised channelizing islands are often used to fulfill this need. This change will make exit ramps uninviting to drivers, especially at multilane exit ramps. Additionally, right-way movements are encouraged by providing a wider entrance ramp throat using either flat radii or removing raised islands that separate adjacent entrance and exit ramps at partial cloverleaf (parclo) interchanges (Cooner and Ranft 2008; USDOT 1996).

In summary, several types of exit ramps, their configuration, and nearby access points should be avoided as they are more susceptible to WWD, including:

- Adjacent entrance and exit ramps intersecting a crossroad (e.g., parclo interchanges);
- Isolated exit ramps (NJDOT 2011; Caltrans 2012);
- Left-side exit ramps (drivers usually expect to make right turns to enter freeways) (Cooner et al. 2004);
- One-way exit ramps connected as unchannelized T-intersections (AASHTO 2011);
- Exit ramps intersecting two-way frontage roads (IDOT 2010);
- Less common arrangements of exit ramps (e.g., button-hook or J-shaped ramp connected to a parallel or diagonal street or frontage road) (Fig. 1) (AASHTO 2011);

- Temporary ramp terminals at work zones;
- Freeway feeders (where exit ramps transition into local roads) (IDOT 2010); and
- Side streets adjacent to exit ramps (AASHTO 2011).

Fig. 1. Button-hook ramp connected to parallel frontage road with proper (green) and wrong (red) movements.

Frontage Roads

Frontage roads in the National Cooperative Highway Research Program (NCHRP) Report 420 are defined as one of the access management techniques to reduce the frequency and severity of crashes on roadways (Gluck et al. 1999). However, improper design and connections of frontage roads with freeway exit ramps may cause driver confusion and increase WWD incidents.

The connection of an exit ramp and a two-way frontage road are more vulnerable to WWD than to a one-way frontage road. A study by Schrocks et al. (2005) revealed that in cases of having both one-way and two-way frontage roads in a certain region, two-way frontage roads have a higher potential for making drivers confused and leading to wrong-way maneuvers. This result can be traced to the essential complexity of making turning movements at the intersections of exit ramps and two-way frontage roads versus the simplicity at one-way frontage roads.

A study by the Minnesota Department of Transportation (MnDOT 2009) found that despite providing more favorable access from controlled-access highways to local streets, continuous multilane frontage roads with numerous intersections may be undesirable and may lead to a high potential for WWD crashes, especially on two-way frontage roads. The study also found that the slip ramps connected to two-way frontage roads at an acute angle offered a higher potential for WWD caused by the exit ramps resembling the extension of a frontage road, thus confusing drivers.

Raised Median

According to the AASHTO Green Book (2011), a median is an elongated divisional island built as a portion of a highway, which serves primarily to separate opposing directions of traffic on the same roadway. These geometric elements have a fundamental capability for managing access to freeways and deterring wrong-way movements. For instance, the presence of raised medians or median barriers between two abutting exit and entrance ramps (i.e., in trumpet interchanges) can help avoid wrong-way entries (Moler 2002).

A non-traversable median on a crossroad is an effective treatment to discourage wrong-way left-turn entry onto diamond, parclo, and full cloverleaf interchanges (AASHTO 2011). This modification is sometimes implemented by narrowing median openings on arterial highways, making left-turn movements onto exit ramps extremely difficult.

Longitudinal channelization devices can also be used as low-cost countermeasures by transportation agencies to fulfill various permanent and temporary channelization needs. Fig. 2 depicts a plan view of a location before treatment with longitudinal channelizer as a wrong-way entry countermeasure. This treatment was implemented by the Michigan Department of Transportation (MDOT) after relating approximately one-third of WWD crashes in 2010 to one specific location. Preliminary results after this improvement (Fig. 3) demonstrated that there were no WWD incidents recorded at this site after implementation of longitudinal channelization devices (Morena and Ault 2013).

Fig. 2. Plan view of a treated intersection before treatment with proper (green) and wrong (red) movements.

Despite being beneficial in removing undesirable turning movements, elongated raised medians should not be used to divide the same direction of traffic (AASHTO 2011). This style may introduce potential WWD incidents because drivers tend to expect the raised median to separate two different directions of travel. Fig. 4 shows a raised median installed to separate dual left-turns from through movements onto an exit ramp. The crash data and field study indicated that this type of design can be misleading to drivers and has caused a high number of WWD incidents at this specific location (Zhou et al. 2012).

Fig. 3. Application of longitudinal channelization in restriction of left-turn access (Morena and Ault 2013).

Fig. 4. Raised median for separating same direction of traffic on an exit ramp.

Channelizing Islands

Channelizing islands define the desirable path, separate conflict points, and enhance safety near an intersection (Wolshon 2004; FHWA 2013). These elements can also be used to block the prohibited turns at intersections wherever necessary and practical, including wrong-way turns, or at least discourage their completion.

Raised channelizing islands that are adequately reflectorized can impede wrong-way movements effectively. These elements exclusively target older drivers' poor contrast visibility by providing greater contrast and, therefore, making geometric characteristics of the downstream intersection more visible. Conversely, those channelizing islands which are not properly reflectorized can adversely affect older drivers' vision and when struck, these islands might be a source of serious injuries and even fatalities (WSDOT 2013).

When using channelizing islands, a height of at least four inches needs to be considered where it is intended to prohibit or prevent traffic movements such as WWD. Lower height islands may prove unsatisfactory for their intended purpose and can be easily traversed (IDOT 2010).

"Scissors channelization" occurs when there is a two-way frontage road adjacent to the freeway and exit and entrance ramps are connected to the frontage road. This arrangement has proven to mislead drivers and has frequently resulted in wrong-way maneuvers (AASHTO 2011). For these reasons, it is seldom used by highway designers.

Control/Corner Radius

The control radius of an intersection refers to the minimum left-turn path for a design vehicle that affects the radius of the intersection corner as well as the location and opening length of the median (Harwood and Glauz 2000). The control radius can be used to prevent WWD at exit ramp terminals.

At the intersection of the left edge of exit ramps and the right edge of crossroads, a short-radius curve or angular break discourages wrong-way right turns from the crossroads. While circular curves with larger radii may encourage a wrong-way right turn onto the exit ramp, the angular corner or tight radii make this movement difficult (WSDOT 2013).

When considering this element, ensuring that the control radius is tangent to the crossroad centerline and not tangent to the edge is critical. Having the control radius tangent to the centerline (and not the edge) makes wrong-way right-turning movements less likely (AASHTO 2011). This design element can define the raised median opening and position it to extend far enough to make the wrong-way left turn an awkward move. The red curved line in Fig. 5 is tangent to the centerline of the roadway.

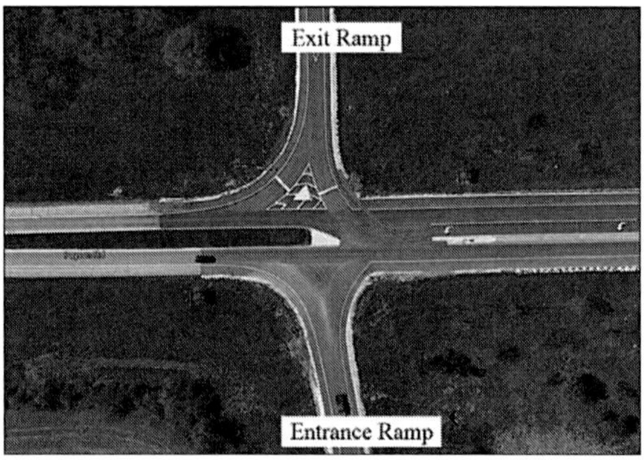

Fig. 5. Control radius (red curve) tangent to the centerline at a ramp-crossroad intersection (Map data: Google).

Sight Distance

Providing drivers with an open sight distance of entrance ramps can help reduce WWD. An adequate sight distance not only provides drivers on crossroads with a better view of ramp terminals, but also helps them distinguish between entrance and exit ramps (i.e., when they are closely spaced) by the approaching right-way drivers' headlamps.

Moreover, uniform lighting levels for both entrance ramps and exit ramps facilitate drivers' vision of intersection, improve their perception of intersection configuration, and lessen, if not eliminate, the possibility of wrong-way movements (Zhou et al. 2012).

Another way to improve sight distance is to move stop lines for left turns at the intersection of two-way ramps and crossroads (i.e., parclo interchanges) from crossroads forward so that motorists have a better view of the entrance ramp and improve their turning radius. A good practice is to locate the stop line between 50% and 60% of the way through the intersection (Fig. 6) to provide an appropriate intersection balance (WSDOT 2013).

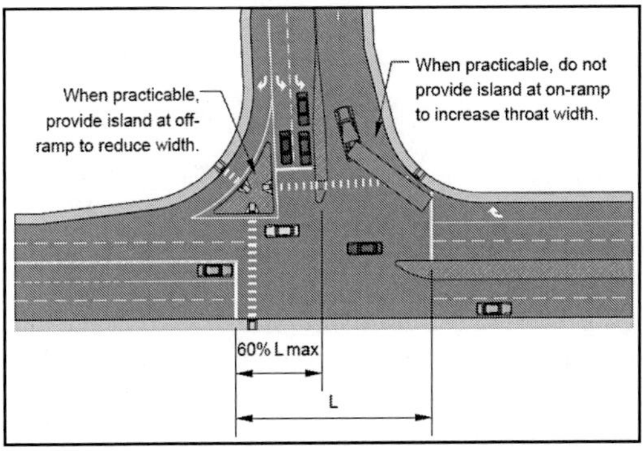

Fig. 6. Intersection balance for two-way ramps to address wrong-way issues (WSDOT 2013).

Excessive grade differentials between ramps (i.e., exit ramp or two-way ramp) and crossroads should be avoided. A large difference between grades of ramps and crossroads can lead to a sight distance problem and increase the likelihood of WWD.

Median barriers should not be extended all the way to the stop line on two-way ramps because they might block the drivers' view of the entrance ramp. A recent study (Morena and Leix 2012) showed that a guardrail installed between two adjacent exit and entrance ramps, as a median barrier, blocked left-turn drivers' view of the entrance ramp terminal and increased the possibility of making a wrong turn onto the exit ramps, resulting in an increase in WWD crashes.

CONCLUSION

This study identified and summarized various access management techniques and geometric elements which are often used to reduce WWD incidents near interchange areas. Each of the identified elements plays its own role in the context of WWD as follows:

- Exit/entrance ramps types, connecting angles with crossroads, and nearby access control are related to WWD activities.
- Two-way frontage roads are identified to be more prone to wrong-way entries compared to one-way frontage roads.
- Raised medians are effectively capable of controlling/restricting wrong-way left-turn access into exit ramps.
- Channelizing islands can narrow multilane exit ramp throats to make them less inviting.
- A short-radius curve or angular break as a control/corner radius not only makes the wrong-way right turn less likely, but also defines the width of the ramp throat, reducing the appeal.
- Providing an open sight distance at the ramp-crossroad intersection helps drivers to distinguish exit ramps from entrance ramps, especially when they are close in proximity. Avoiding excessive grade differentials and proper stop line placement can be helpful for improving sight distance.

ACKNOWLEDGMENTS

This study was based on a guidebook for reducing WWD on freeways developed for the Illinois Department of Transportation (IDOT) and the Illinois Center for Transportation (ICT). The authors would also like to extend their sincere thanks to the Technical Review Panel (TRP) members for all their feedback and input.

REFERENCES

American Association of State Highway and Transportation Officials (AASHTO). (2011). "A Policy on Geometric Design of Highways and Streets." Washington, DC.

American Traffic Safety Services Association (ATSSA). (2014). "Emerging Safety Countermeasures for Wrong-way Driving." Fredericksburg, VA.

California Department of Transportation (Caltrans). (2012). "Highway Design Manual (HDM)." Los Angeles, CA.

Chassande-Mottin, R., and Ganneau, F. (2008). "How to Fight Wrong Way Driving Accidents." *Proc., Transport Research Arena Europe (TRA),* Ljubljana, Slovenia.

Cooner, S.A., and Ranft S.E. (2008). "Wrong-way driving on freeways: problems, issues and countermeasures." *Proc., 87th Annual Meeting of the Transportation Research Board*, TRB, Washington, DC.

Cooner, S.A., Cothron, A.S., and Ranft, S.E. (2004). "Countermeasures for Wrong-Way Movement on Freeways: Guidelines and Recommended Practices." Texas Transportation Institute (TTI), College Station, TX.

Federal Highway Administration (FHWA). (2013). "Signalized Intersections Informational Guide, 2nd Edition." <http://safety.fhwa.dot.gov/intersection/signalized/13027/index.cfm> (Feb. 13, 2014).

Gluck, J.S., Levinson, H.S., and Stover, V.G. (1999). "Impacts of access management techniques: NCHRP Report No. 420." Transportation Research Board, National Research Council, Washington, DC.

Harwood, D.W., and Glauz, W.D. (2000). "Synthesis of Highway Practice 281: Operational Impacts of Median Width on Larger Vehicles." National Cooperative Highway Research Program (NCHRP), Transportation Research Board, Washington, DC., 26.

Illinois Department of Transportation (IDOT). (2010). "Bureau of Design and Environment Manual (BDE Manual)." Springfield, IL.

Minnesota Department of Transportation (MnDOT). (2009). "Design-Build Modifications to the Road Design Manual." St. Paul, MN.

Moler, S. (2002). "Stop. You're Going the Wrong Way!" *Public Roads*, 66(2), <http://www.fhwa.dot.gov/publications/publicroads/02sep/06.cfm> (March 07, 2014).

Morena, D.A., and Ault, K. (2013). "Michigan Wrong-Way Freeway Crashes." *Proc. 2013 National Wrong-Way Driving Summit*, Edwardsville, IL.

Morena, D.A., and Leix, T.J. (2012). "Where These Drivers Went Wrong." *Public Roads*, 75(6), 33–41.

National Transportation Safety Board (NTSB). (2012). "Highway Special Investigation Report: Wrong-Way Driving." Washington, DC.

New Jersey Department of Transportation (NJDOT). (2011). "Roadway Design Manual." <http://www.state.nj.us/transportation/eng/documents/RDM/sec7.shtm> (Mar. 14, 2014).

North Texas Tollway Authority (NTTA). (2009). "Keeping NTTA Roadways Safe: Wrong-Way Driver Task Force Staff Analysis." <https://www.ntta.org/newsresources/safeinfo/wrongway/Documents/WWDAnalysisAUG2011.pdf> (Oct. 13, 2013).

Ouyang, Y. (2013). "North Texas Tollway Authority's (NTTA) Wrong Way Driving Program: from a Traffic Engineer's Perspective." *Proc., 2013 National Wrong-Way Driving Summit*, Edwardsville, IL.

Schrock, S.D., Hawkins, H.G., and Chrysler, S.T. (2005). "Effectiveness of lane direction arrows as pavement markings in reducing wrong-way movements on two-way frontage roads." *Transportation Research Record: Journal of the Transportation Research Board*, 1918, 63-67.

U.S. Department of Transportation (USDOT), National Highway Traffic Safety Administration (NHTSA). (1996). "The Highway Safety Desk Book: NHTSA, People Saving People." Advisory Committee on Highway Safety of the International Association of Chiefs of Police, Washington, DC.

Washington State Department of Transportation (WSDOT). (2013). "Design Manual." Olympia, WA.

Wolshon, B. (2004). "Toolbox on Intersection Safety and Design." Louisiana State University, Baton Rouge, LA.

Zhou, H., and Pour Rouholamin, M. (2014a). "Guidelines for Reducing Wrong-Way Crashes on Freeways." Illinois Center for Transportation (ICT), Rantoul, IL.

Zhou, H., and Pour Rouholamin, M. (2014b). "Proceedings of the 2013 National Wrong-Way Driving Summit." Illinois Center for Transportation (ICT), Rantoul, IL.

Zhou, H., Zhao, J., Fries, R., and Pour Rouholamin, M. (2014). "Statistical Characteristics of Wrong-Way Driving Crashes on Illinois Freeways." *Proc., 93rd Annual Meeting of Transportation Research Board*, TRB, Washington, DC.

Zhou, H., Zhao, J., Fries, R., Gahrooei, M.R., Wang, L., Vaughn, B., Bahaaldin, K., and Ayyalasomayajula, B. (2012). "Investigation of Contributing Factors Regarding Wrong-Way Driving on Freeways." Illinois Center for Transportation (ICT), Rantoul, IL.

Estimation of Vehicle Queue Lengths Based on Driveway Access Design

J. S. Choi[1]; S. K. Kim[1]; M. S. Jin[1]; D. C. Min[1]; S. Y. Kim[2]; and Y. S. Jang[3]

[1]Department of Transportation Engineering, University of Seoul, 163 Siripdae-ro, Dongdaemun-gu, Seoul 130-743, Republic of Korea. E-mail: traffic@uos.ac.kr; ic@uos.ac.kr; latentkid@uos.ac.kr; roadmin@uos.ac.kr
[2]Department of Saemangeum and Regional Development, Jeonbuk Development Institute, 1696 Kongjwi-patjwi-ro, Wansan-gu, Jeonju-si, Jeollabuk-do 560-860, Republic of Korea. E-mail: road@uos.ac.kr
[3]Wirye Project Division, Korea Land and Housing Corporation, 879 Heolleung-ro, Sujeong-gu, Seongnam-si, Gyeonggi-do 461-200, Republic of Korea. E-mail: ysjang@lh.or.kr

Abstract
This paper presents the result of a study carried out to identify vehicle queue effects of driveway geometry based on statistical analysis of field survey data. The study is necessary to help transportation engineers apply highway access control more efficiently to achieve its full benefits such as diminished queue lengths in main roads as well as in driveways. We applied the multiple regression analysis to establish evidence based on vehicle queues and driveway geometric features. Our research showed that vehicle queues actually vary closely with driveway geometric conditions and that the presence of the speed bump and the crosswalk are the most important factors for increasing vehicle queue lengths. The research findings will contribute to our increased understanding of proper driveway designs when highway access control is the main research interest.

INTRODUCTION

Transportation engineers work to provide safe and efficient vehicle flows based on a high level of highway mobility and building accessibility. In performing their work, transportation engineers usually apply a design approach that draws upon the hierarchical classification of highway network, which is called the highway functional classification.

The essence of the functional classification is the grouping of highways by the character of service they provide. The function of a highway link is determined by predetermining the amount of expected accessibility and movement for each highway link. Once determined, it is important to preserve the determined function whatever occurs to the surrounding condition of the highway link. This is a design criterion to be ensured by transportation engineers.

In practice, however, transportation engineers are challenged by continuous requests from the public wanting to modify highway functions, e.g., a request from a property owner to reflect the increased accessibility of the area adjacent to roadway due to strip developments along major arterials. Consequently, this will result in locations where highways with different classification of function are intersecting, and a significant amount of conflicts will occur. To minimize this condition and preserve initial highway functions, transportation engineers are using the access control (AASHTO, 2011; MLTMA, 2012; TRB, 2003). Access control refers to the design, implementation and management of entry and exit points (i.e., driveways, entrances or exits) between roadways and adjacent properties (TRB, 2003). This technique has been used quite extensively in many nations, whereas South Korea is very slow to use this technique. It is desired to apply the access control and realize its benefits.

One of the challenging issues in applying the access control is that South Korean transportation engineers do not have the empirical evidence of how vehicle flow improves or deteriorates by using different geometric features of the entry and exit points, i.e., at present, they are unable to answer what are the best combination of throat width, throat length, curve return radius, and number of driveway lanes for a certain amount of vehicle volume on main roads.

The objective of the research is to identify the effect of drive geometric features on vehicle delays within driveway adjacent areas, so that transportation engineers may use the findings to develop the best driveway access designs.

ACCESS DESIGN AND RELATED ISSUES

Regulating access is called access control (AASHTO, 2011), and access control is achieved through the regulation of public access rights to and from properties abutting the highway facilities (AASHTO, 2011). In the US, the regulations within the context of access design generally are categorized as follows (AASHTO, 2011).

- Full control of access
- Partial control of access
- Access management
- Driveway/entrance

Among these categories, this paper is centered on driveway/entrance because they represent the areas in South Korea where the geometric standards are lacking and contributions are possible with academic research.

In the US, driveway refers to a type of private road for local access to one or a small group of structures, and is owned and maintained by an individual or group (AASHTO, 2011; Gattis, 2010; Gluck, 1999; ITE, 2008; Koepke, 1992; TRB, 2003). In contrast, In South Korea, we do not use 'driveway' but uses the expression of 'entry and exit point'. Besides, the entry and exit point actually covers a relatively wide range of access types, e.g., the driveway that can be seen in the US, exit/entrance areas between the local street and the collector, the collector and the minor arterial, and even between the local and the major arterial. This paper mostly discusses the collector and the minor arterial.

Based on the finding from existing research, we come to understand that the proper geometric design for the entry and exit point is essential to the successful operation of traffic flow on urban streets. So we are determined to carry out an empirical analysis that can explain how vehicle delays on the arterial or the driveway vary depending upon the geometric design features of the entry and exit point.

RESEARCH METHODOLOGY

Data Collection

We carried out a field study to investigate major traffic operation and geometric factors that were expected to influence operational conditions of the driveway such as vehicle queue length. The survey team composed of five small teams, and two survey crews constitute one small team. Field data were collected on weekdays from 7 AM to 7 PM during a two-week period of February 2014, and a 15-minute interval was applied. The survey sites were located in Seoul and a total of 30 sites were randomly selected. Vehicle queue length was the representative traffic flow data, while other related information including ingress and egress volume within the driveway, pedestrian volume on the driveway, and average vehicle speed of the arterial were also recorded for a supplementary purpose. In addition to traffic flow data, the survey crews also collected geometric information, e.g., width of driveway, radius of driveway, grade of driveway, number of driveway lanes, the presence of the speed bump, the presence of the crosswalk, channelization geometry, length of deceleration lane in the arterial, driveway spacing, corner clearance to adjacent signalized intersections, spacing between the driveway and bus-stops, and number of lanes for the arterial. We collected geometric conditions of the driveway based on field surveys, but also used roadway drawings when available.

We also applied the camera recorder to better capture the real world traffic information for survey sites. Figure 1 shows the configuration of the driveway connecting to the arterial (or other classification of highway). Attention was given to cover the entire roadway space where vehicle queues may occur. Vehicle queues occur in the driveway or in the arterial, so each queue was recorded separately in the field survey.

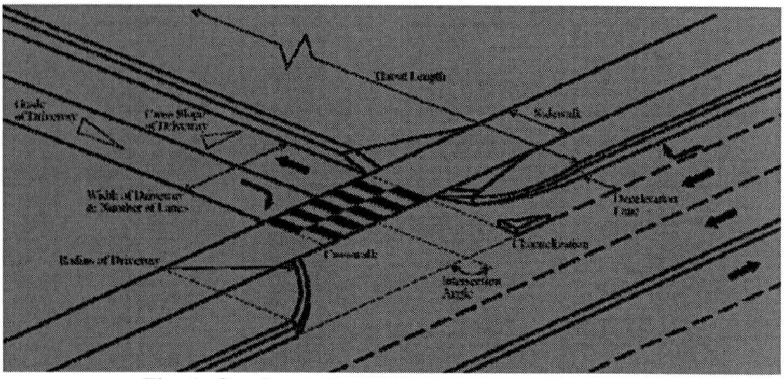

Fig. 1. Configuration of the driveway, the arterial.

Data Reduction and Analysis

It was found based on the field survey result that there were three vehicle queue types depending upon their location in the driveway or the main road surrounding the site. Figure 2 illustrates the types. Case A refers to a queue of the driveway ingress movement waiting on the main road, Case B is for a queue of the driveway egress movement waiting on the driveway, and Case C is a queue waiting on the main road that is one of the approaches of a signalized intersection located downstream of the driveway. Examining the data collected in the survey, we found that Case C was a rare event requiring a huge effort to successfully detail its flow characteristics. So we decided to preclude Case C from our subsequent analyses. As a result, our research effort became focused on Case A and Case B.

With this result of vehicle queue patterns and the result of our review for existing research, we assumed that vehicle queues for Case A would grow if the vehicle volume of egress movement in the driveway and the pedestrian volume crossing the driveway increase. We also assumed that the length of vehicle queues for Case B would generally increase with the vehicle volume on the main road. As a follow-up stage, we proceeded to identify major influencing factors for vehicle queue lengths of the driveway and the contributory effect of each factor by investigating vehicle queue lengths in conjunction with the driveway geometric features.

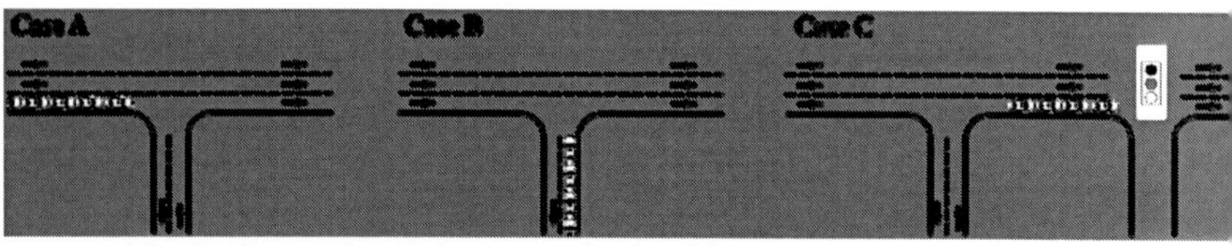

Fig. 2. Cases of vehicle queues in the driveway and the main road.

We first examine how vehicle queues grow or diminish based on the curb return radius. Figure 3 shows the result. For both Case A and Case B, it is observed that the queue length diminishes with longer radii, indicating potential effects of driveway geometric features on vehicle queues generated on driveway adjacent areas.

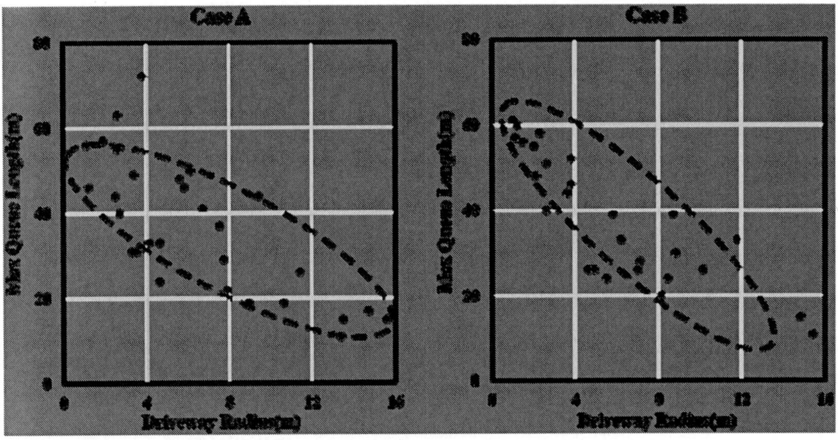

Fig. 3. Vehicle queue length with driveway curb return radius.

We then examine how vehicle queues in the driveway vary for different volume levels in the main road adjacent to driveways. Figure 4 shows the result.

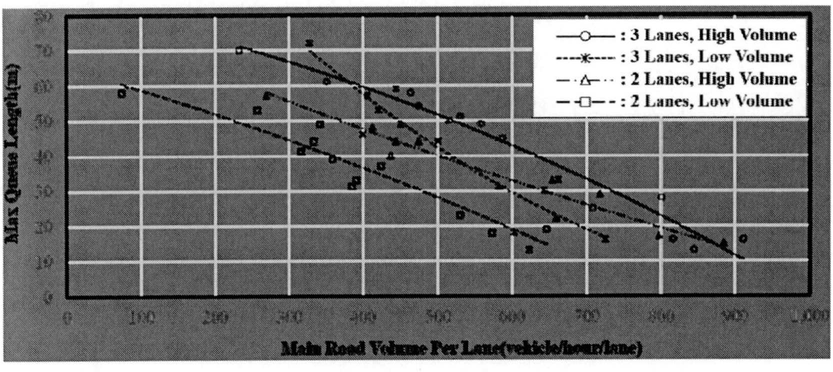

Fig. 4. Vehicle queue length with vehicle volumes in main roads.

We select the vehicle queue lengths for Case A and Case B as the dependent variable. For the independent variables, we want to group them according to their characteristics of generating vehicle queues in the driveway area, which are traffic operational and driveway geometric variables. Figure 5 illustrates the driveway geometric design elements investigated in the research and the independent variables in conjunction with their location in the driveway area. Table 1 lists the variable definitions and their brief summary.

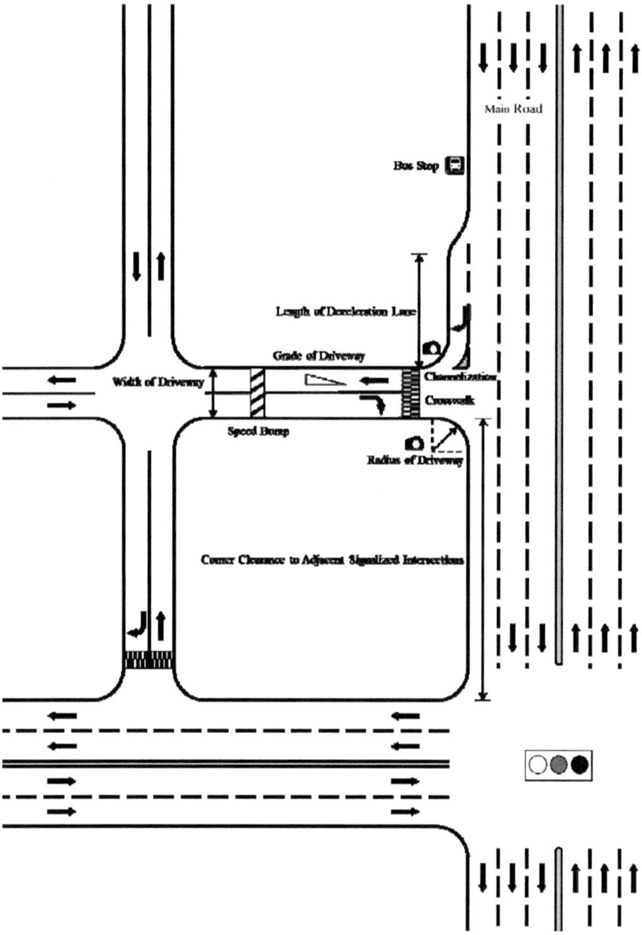

Fig. 5. Driveway geometric design elements, and the placement of camera recorders.

Table 1. Variable definition and brief summary.

Variable Name				Definition and Unit	Mean
Dependent	Y_A	Max. Queue Length $_{\text{Case A}}$		Meter	37.53
	Y_B	Max. Queue Length $_{\text{Case B}}$		Meter	38.95
Independent	X_1	Case A	Driveway Ingress Volume	Vehicle/hour	378
		Case B			197
	X_2	Case A	Driveway Egress Volume	Vehicle/hour	174
		Case B			329
	X_3	Case A	Pedestrian Volume Crossing the Driveway	Pedestrian/hour	955
		Case B			1195
	X_4	Case A	Average Main Road Vehicle Speed	Km/hour	28.46
		Case B			28.42
	X_5	Case A	Vehicle Volume for Lane 1 of the Main Road	Vehicle/hour	723
		Case B			713
	X_6	Width of Driveway		Meter	6.63
	X_7	Radius of Driveway		Meter	4.07
	X_8	Grade of Driveway		Percent	2.76
	X_9	Number of Driveway Lanes		Lanes	1.60
	X_{10}	Presence of the Speed Bump		1=Yes, 0=No	0.33
	X_{11}	Presence of the Crosswalk		1=Yes, 0=No	0.53
	X_{12}	Channelization Geometry		1=Yes, 0=No	0.17
	X_{13}	Length of the Deceleration Lane in the Main Road		Meter	12.37
	X_{14}	Driveway Spacing		Meter	70.20
	X_{15}	Corner Clearance to Adjacent Signalized Intersections		Meter	102.93
	X_{16}	Presence of Driveway Adjacent Bus Stop		1=Yes, 0=No	0.40
	X_{17}	Number of Main Road Lanes		Lanes	3.07

3.3 Statistical Analysis

We use the length of vehicle queues in the main road adjacent to the driveway as well as the queue length in the driveway itself to assess the effects from driveway geometric features. Queue lengths are continuous variables, so we have no other choice but to apply the regression analysis. There are studies in other research areas based on the same approach (Lee 2008; Kim 2011; Qu 2008). We apply the multiple regression analysis because we use many explanatory variables, so that potential effects of driveway geometric features on traffic flow in the main road may be detailed successfully.

We also apply the stepwise regression analysis to detail the characteristics of vehicle queue lengths and their contributory variables. In statistics, the stepwise regression includes regression models in which the choice of predictive variables is carried out by an automatic procedure, e.g., forward selection, backward elimination, and bidirectional elimination (Kim, 2013). We apply forward selection and first select the candidate variables from traffic operational condition, e.g., driveway ingress volume, driveway egress volume, pedestrian volume crossing the driveway, average speed of the main road adjacent to driveways, and vehicle volume of lane 1 of the main road. We then select the candidate variables from driveway geometric features. The variable list includes the width of driveway, the radius of driveway, the grade of driveway, the number of driveway lanes, the presence of the speed bump, the presence of the crosswalk, the presence of channelization, the length of the deceleration lane in the main road, driveway spacing, the distance from the driveway to adjacent signalized intersections, the distance from the driveway to bus-stops, and the number of lanes in the main road.

The expected model of vehicle queues in driveway adjacent areas is given in Eqn. (1).

$$Y = \beta_0 + \beta_1 f(Operation\ factors) + \beta_2 f(Geometric\ factors) + C \quad (1)$$

ANALYSIS AND RESULT

Multiple Regression Model

Applying the regression analysis of assessing explanatory variables, we need to apply a normality test to determine if a data set is modeled well by a normal distribution (Kim, 2013). We apply Shapiro-Wilk test. The Shapiro-Wilk test utilizes the null hypothesis principle to check whether a sample x_1, ..., xn came from a normally distributed population. The test statistic is:

$$W = \frac{(\sum_{i=1}^{n} a_i x_{(i)})^2}{\sum_{i=1}^{n} (x_i - \overline{x})^2} \quad (1)$$

The test result shows as in Table 2 that the probability of the null hypothesis being less than a 0.05 significance level is greater than 0.05, so the null hypothesis that the data came from a normally distributed population cannot be rejected.

Table 2. Result of the normality test.

Variable Name	Average	Min	Max	Standard Deviation	N	P-value
Max. Queue Length $_{\text{Case A}}$	36.90	11	72	16.816	30	0.245
Max. Queue Length $_{\text{Case B}}$	38.40	11	61	14.864	30	0.121

Carrying out the stepwise regression analysis, we come up with a set of models describing how vehicle queues in driveway adjacent areas vary based on explanatory variables, which will include traffic- and geometric-related characteristics. Table 3 and Table 4 show the models from this analysis for Case A and Case B, respectively.

Table 3. Summary of developed model, Case A.

Classification			Model 1	Model 2	Model 3	Model 4	Model 5
Constant			**-17.375***	**-324.256***	**-46.364***	**-32.226***	**-26.252***
Operation Related	X_i	X_{A1}	**0.060***		**0.052***	**0.058***	**0.062***
		X_{A2}	0.018		0.036	0.025	
		X_{A3}	**0.027***		**0.043***	**0.036***	**0.019***
		X_{A4}	0.021		0.010	0.018	
		X_{A5}	-0.002		-0.005	-0.004	
	$Ln(X_i)$	$Ln(X_{A1})$		**38.092***			
		$Ln(X_{A2})$		10.043			
		$Ln(X_{A3})$		8.180			
		$Ln(X_{A4})$		1.069			
		$Ln(X_{A5})$		-2.873			
Geometric Related	X_i	X_6			**-1.413***		**-1.367***
		X_7			**-1.634***		**-1.169***
		X_8			-0.075		
		X_9			2.238		
		X_{10}			0.783		
		X_{11}			-1.168		
		X_{12}			1.174		
		X_{13}			-0.127		
		X_{14}			-0.080		
		X_{15}			0.034		
		X_{16}			**6.686***		**6.671***
		X_{17}			-0.588		
	$Ln(X_i)$	$Ln(X_6)$				-6.202	
		$Ln(X_7)$				-0.471	
		$Ln(X_8)$				-0.597	
		$Ln(X_9)$				2.567	
		$Ln(X_{10})$				0.862	
		$Ln(X_{11})$				-1.057	
		$Ln(X_{12})$				0.949	
		$Ln(X_{13})$				-0.046	
		$Ln(X_{14})$				-4.149	
		$Ln(X_{15})$				2.828	
		$Ln(X_{16})$				2.611	
		$Ln(X_{17})$				-1.845	
R_a^2			0.988	0.974	0.947	0.896	0.990

* variable is significant at the 0.05 level

Table 4. Summary of developed model, Case B.

Classification	Model 1	Model 2	Model 3	Model 4	Model 5
Constant	**-19.416***	**-381.804***	**-17.699***	**-19.465***	**-20.355***

Operation Related	X_i	X_{B1}	-0.023		-0.020	-0.038	
		X_{B2}	**0.061***		**0.067***	**0.103***	**0.082***
		X_{B3}	**0.044***		**0.018***	**0.014***	**0.026***
		X_{B4}	-0.079		-0.035	-0.056	
		X_{B5}	-0.004		0.001	0.002	
	$Ln(X_i)$	$Ln(X_{B1})$		-30.178			
		$Ln(X_{B2})$		**37.414***			
		$Ln(X_{B3})$		**55.972***			
		$Ln(X_{B4})$		-0.946			
		$Ln(X_{B5})$		-1.006			
Geometric Related	X_i	X_6			**-0.296***		**-0.132***
		X_7			**-1.848***		**-0.982***
		X_8			0.224		
		X_9			-3.252		
		X_{10}			**8.669***		**7.100***
		X_{11}			**6.269***		**4.997***
		X_{12}			0.078		
		X_{13}			**-0.063***		**-0.069***
		X_{14}			-0.006		
		X_{15}			-0025		
		X_{16}			**-3.568***		**-4.991***
		X_{17}			0.784		
	$Ln(X_i)$	$Ln(X_6)$				**-3.781***	
		$Ln(X_7)$				**-5.086***	
		$Ln(X_8)$				0.509	
		$Ln(X_9)$				2.053	
		$Ln(X_{10})$				**4.148***	
		$Ln(X_{11})$				**4.303***	
		$Ln(X_{12})$				1.369	
		$Ln(X_{13})$				**-1.126***	
		$Ln(X_{14})$				-1.392	
		$Ln(X_{15})$				0.426	
		$Ln(X_{16})$				-4.223	
		$Ln(X_{17})$				1.347	
R_a^2			0.985	0.976	0.992	0.989	0.989

* variable is significant at the 0.05 level

Based on the stepwise regression analysis, we finally develop models that can be used to assess the maximum length of queue in driveway adjacent areas. Traffic operational variables in the final models include driveway ingress volume, driveway egress volume, and pedestrian volume crossing the driveway. Geometric condition variables include the width of driveway, the radius of driveway, the presence of the speed bump, the presence of the crosswalk, the presence of the driveway adjacent bus stop, and the length of the deceleration lane in the main road. Eqn. (3) and Eqn. (4) are the developed models.

$$Max.Queue\,Length_{Case\,A} = -26.252 + 0.062X_1 + 0.019X_3 - 1.367X_6 - 1.169X_7 + 6.671X_{16}$$

(3)

$$Max.Queue\,Length_{Case\,B} = -20.355 + 0.082X_2 + 0.026X_3 - 0.132X_6 - 0.982X_7 + 7.100X_{10} + 4.997X_{11} - 0.069X_{13} - 4.991X_{16}$$

(4)

Where, X_1 : Driveway Ingress Volume (Vehicle/hour)
X_2 : Driveway Egress Volume (Vehicle/hour)
X_3 : Pedestrian Volume Crossing the Driveway (Pedestrian/hour)
X_6 : Width of Driveway (m)
X_7 : Radius of Driveway (m)
X_{10} : Presence of the Speed Bump (0 or 1)
X_{11} : Presence of the Crosswalk (0 or 1)
X_{13} : Length of Deceleration Lane in the Main Road (m)
X_{16} : Presence of the Driveway Adjacent Bus Stop (0 or 1)

As a follow-up stage, we wonder how each independent variable influences the queue length, so we apply the elasticity analysis to our models. The elasticity refers to the ratio of the percentage change in one variable to the percentage change in another variable, when the latter variable has a causal influence on the former (Kim, 2011). Table 5 provides the analysis result and shows that traffic operation variables involve higher values of elasticity than geometric condition variables, implying that the length of vehicle queue in driveway adjacent areas is more sensitive to how engineers run traffic operations in such areas.

Table 5. Result of the elasticity analysis.

Classification	Case A	Case B
Driveway Ingress Volume	0.624	-
Driveway Egress Volume	-	0.693
Pedestrian Volume Crossing the Driveway	0.483	0.798
Width of Driveway	0.242	0.022
Radius of Driveway	0.127	0.103
Presence of the Speed Bump	-	0.060
Presence of the Crosswalk	-	0.068
Length of the Deceleration Lane in Main Road	-	0.022
Presence of the Driveway Adjacent Bus Stop	0.071	0.051

Validation

To assess the statistical validity of the result from our research effort, we apply the F-test and examine the adjusted coefficient of determination, R_a^2, because the coefficient of determination is automatically increasing when extra explanatory variables are added to the model in the multiple regression analysis (Kim, 2011). The F-test in one-way analysis of variance is used to assess whether the expected values of a quantitative variable within several pre-defined groups differ from each other (Kim, 2011). Table 6 provides the result of the F-test.

Table 6. Result of the F-test.

Variable Name	Classification	Sum of Square	N	Mean Square	F-value	p-value
Max. Queue Length $_{\text{Case A}}$	Model	8120.057	5	1624.011	483.322	0.000
	Residual	80.643	24	3.360		
	Sum	8200.700	29			
Max. Queue Length $_{\text{Case B}}$	Model	6333.604	8	791.700	225.905	0.000
	Residual	73.596	21	3.505		
	Sum	6407.200	29			

Table 7 shows the result of the adjusted coefficients of determination, which are both very close to 1.0. A statistical model explains sample data points well if its adjusted coefficient of determination shows a value that is greater than 0.7 (Kim, 2013). Hence, we find that the model developed in this analysis is statistically significant.

Table 7. Result of adjusted coefficients of determination, Case A and Case B.

Variable Name	R_a^2
Max. Queue Length $_{\text{Case A}}$	0.990
Max. Queue Length $_{\text{Case B}}$	0.989

FINDINGS AND DISCUSSION

It is intended to investigate the operational effect of driveway geometric conditions such as driveway configuration, the acceleration lane (or the deceleration lane), the vertical grade of driveway. The result shows that vehicle queue lengths actually vary with driveway geometric conditions. Based on vehicle queue lengths according to their relative location to a driveway, we identify the following major geometric factors influencing the queue length: For case A driveway, the pedestrian volume crossing the driveway, the presence of the crosswalk, the presence of bus stops adjacent to the driveway. For case B driveway, all factors for Case A plus additional factors of the presence of the speed bump and the length of the deceleration lane in main roads.

As clearly stated by Stover and Koepke in 'Transportation and land development,' driveway geometric conditions will influence driveway traffic operation in many ways through driveway entry and exit maneuvers, intersection sight distance, and driveway entry speed (ITE, 2002). We come to find the empirical evidence of this influence. Moreover, our model successfully captures contributory effect from each driveway geometric element.

Interestingly, the presence of bus stops adjacent to the driveway provides contrasting effects depending upon where vehicle queues are located, i.e., it has different queue effects in main roads and driveways. For example, when there is a bus stop within the intersection sight distance of 105m for 60km/h design speed, it will provide a positive effect to driveway egress movement, because the bus stop will block driveway ingress movement from the main road. Obviously, vehicle queues in the main road tend to grow with the presence of bus stops in driveway adjacent areas.

It is noted that, while our model explains how vehicle queues will grow or diminish based on Driveway geometric conditions such as the width of the driveway, the radius of the driveway, and the radius of the curb return, etc., it also explains how vehicle queues change with pedestrian movement in driveway adjacent areas. This will contribute to our increased understanding of proper driveway designs based on vehicle queues as well as pedestrian volume.

CONCLUSIONS

We aimed to establish evidence that vehicle queues in driveway adjacent areas grow or diminish based on driveway geometric conditions, and we conducted field surveys in Seoul to compare how vehicle queue lengths vary with driveway geometry in driveway adjacent areas. Our research showed that vehicle queues actually vary closely with driveway geometric conditions and the length of the queue can be explained by applying empirical models developed in the research.

The following points summarize our additional findings:

- Vehicle queue lengths in driveway adjacent areas were modeled based on driveway geometry.
- The multiple regression analysis was applied to determine how vehicle queues occur depending upon different driveway geometric conditions. This analysis demonstrated that transportation engineers can produce real-world results using traffic and pedestrian flow surveys in driveway adjacent areas. The model we developed shows that the presence of the speed bump and the crosswalk are the most important factors for increasing vehicle queue lengths.

We expect that our research will be helpful for understanding vehicle queue changes for different driveway designs as well as for the operational benefits gained from applying optimal driveway designs.

REFERENCES

AASHTO (2011). *A Policy on Geometric Design of Highways and Streets,* Washington, DC.

Fitspatrick. K. and Potts, I. (2012). "Turn Lane Warrants." *4th Urban Street Symposium,* Transportation Research Board, 1-14.

Gan, C.T. and Long, G. (1997). "Effects of Inadequate Driveway Corner Clearances on Traffic Operations, Safety, and Capacity." *Transportation Research Record 1579,* Transportation Research Board, 35-42.

Gard, J.T. (2001). "Estimation of Maximum Queue Lengths at Unsignalized Intersections." *ITE Journal,* Institute of Transportation Engineers, 71(11), 26-34.

Gattis, J.L. and Duncan, L.K. (2009). "Effect of Grade on the Speed and Travel Time of Vehicles Entering Driveways." *Transportation Research Record 2092,* Transportation Research Board, 66-74.

Gattis, J.L., Gluck, J.S., Barlow, J.M., Eck, R.W., Hecker, W.F., and Levinson, H.S. (2010). "Guide for the Geometric Design of Driveways." *NCHRP Report 659,* Transportation Research Board, Washington, DC.

Gluck, J.S., Levinson, H.R., and Stover, V.G. (1999). "Impacts of Access Management Techniques." *NCHRP Report 420,* Transportation Research Board, Washington, DC.

ITE (2002). *Transportation and Land Development 2nd edition,* Institute of Transportation Engineers, Washington, DC.

ITE (2008). *Urban Street Geometric Design Handbook,* Institute of Transportation Engineers, Washington, DC.

Kim, S.Y. (2011). "The Development of Safety Evaluation Criteria Based on Vehicle Speeds and Crash Data for Multi-lane Highways." Ph.D. Thesis, University of Seoul, Korea.

Kim, S.Y. and Choi, J.S. (2013). "Effects of Preceding Geometric Conditions on Operating Speed Consistency of Multilane Highways." *Canadian Journal of Civil Engineering,* Canadian Society for Civil Engineering, 40(6), 528-536.

Koepke, F.J. and Levinson, H.S. (1992). "Access Management Guidelines for Activity Centers." *NCHRP Report 348,* Transportation Research Board, Washington, DC.

Lee, H.Y. (2008). *Research Methodology,* Cheongram, Seoul.

MLTMA (2012). *Highway Geometric Design Standard and Regulation,* Ministry of Land and Transportation and Maritime Affairs, Seoul.

MLTMA (2012). *Traffic Impact Analysis,* Ministry of Land and Transportation and Maritime Affairs, Seoul.

Prassas, E.S. and Chang, J.I. (2000). "Effects of Access Features and Interaction among Driveways as Investigated by Simulation." *Transportation Research Record 1706,* Transportation Research Board, 17-28.

Rawlings, J. and Gattis, J.L. (2008). "Detailed Study of Driveway Collision Patterns in an Urban Area." *Compendium of Paper, 87th Annual Meeting*, Transportation Research Board, Washington, DC.

Stover, V.G. (1981). *Guidelines for Spacing of Unsignaled Access to Urban Arterial Streets,* Technical Bulletin 81-1, Texas Engineering Experiment Station, Texas A&M University, College Station.

TRB (2003). *Access Management Manual,* Transportation Research Board, Washington, DC.

Qu, Z., Schultz, G. and Almaik, B. (2008), "Multiple Regression Analysis on the Safety of Transportation and Social Economic Development." *Transportation and Development Innovative Best Practices,* 210-215.

Experimental Analysis of a Direct Access Driveway at a Roundabout: Performance with One or More Slip Lanes

Majed Al-Ghandour, Ph.D., P.E.[1]

[1]Engineering Manager, Program Development Branch, North Carolina Department of Transportation, 1534 Mail Service Center, Raleigh, NC 27699-1534. E-mail: malghandour@ncdot.gov

Abstract

This paper studies the performance of single-lane roundabouts with a direct driveway access road, one lane driveway right-out movements only, to allow facility (land parcel) driveway volumes to access directly at a roundabout. VISSIM micro- simulation assessment considers experimental origin-destination (O-D) balanced flow scenarios with three different driveway volumes (low, moderate, high, and compared to no driveway) and with and without slip lane scenarios (no slip lane, one slip lane, and two slip lanes). Simulated right-turning traffic volumes range from 50 to 500 vehicles per hour. Experiment results indicate that average delay of a roundabout is sensitive to changing driveway's volumes. As expected, results indicate a direct access driveway increases the roundabout vehicle average traffic delay. At high driveway's volumes level, the total average roundabout delay was increased by 50% at a moderate vehicle traffic level and by 71% at a high vehicle traffic level, before oversaturation. The total average driveway traffic delay also was increased by 8% at a moderate traffic level and by 12% at a high traffic level. Finally, this study also suggests that having more than one slip lane, at different roundabout approaches, reliefs total roundabout average delay, provides an efficient direct roadway driveway access, and enhances access management.

Key words: Roundabout, Driveway, Average delay, VISSIM, Slip lanes.

INTRODUCTION

Access Management enhances vehicular access points to land parcels adjacent to roadways including a direct access to roundabouts. KDOT (2014) restricts accesses directly at a roundabout, but allows accesses near a roundabout that meet certain conditions. The KDOT Access Management Policy provides guidance on access management at roundabouts, including access spacing criteria.

Malik et al. (2011) studied the driveways with direct access, in commercial areas, and found that the driveways can create operational as well as safety concerns for both mainline and driveway traffic. They focused on the left turns in and out of developments by using microscopic simulation models. They recommended ranges for driveway left turns, as results, in terms of delay for the mainline and driveway traffic, and queues for the mainline traffic.

Roundabouts can be used as an alternate intersection design to facilitate major traffic turning movements and to enhance operational and safety performance. A slip lane, a separate lane that facilitates right-turning traffic flow, reduces approach delay by allowing right-turning movements to bypass the roundabout, thereby reducing vehicle conflicts. Though roundabouts are an increasingly common form of intersection control in the U.S., research has yet to quantify slip lane contributions to Access Management operational and safety improvements when slip lanes are installed.

NCHRP Reports 572 (2007) and 672 (2010) define two types of slip lane: a non-yield slip lane, merging with the roundabout exit leg and forming a new acceleration (free-flow) lane adjacent to exiting traffic; and a yield slip lane, terminating at a sharp angle with the roundabout exit approach so that right-turning traffic is yielding.

Al-Ghandour et al. (2011) experimentally studied roundabout with a and without slip lane performance by simulating both balanced (total traffic flow into and out of each roundabout approach is the same) and more realistic unbalanced flow scenarios (traffic flow into and out of different roundabout approaches is different) for a range of volume levels, likewise using a microscopic simulation package (VISSIM) (PTV2007). They found that a roundabout with a free-flow or with a yield slip lane exit types significantly reduce total roundabout average traffic delay, compared to no slip lane.

So far, the research has rarely developed "direct driveway access" at roundabouts based on vehicle delay. Therefore, it would be useful to explore an experimental procedure that considers some driveway traffic volumes patterns. This paper explores overall roundabout and driveways' operational performance in a simulated environment by quantifying average delay with and without slip lanes options.

APPROACH

In this study, single-lane roundabout-based experimental simulation was used to determine the average delay of roundabout and driveway with different driveway's volumes and with slip lanes configurations. Roundabout entry and exit flow volumes for each approach are assumed the same (symmetric), although the dominant right-turning traffic percentages are same (75%), Al- Ghandour et al. (2011). Six experimental scenarios with equal traffic percentage turning volume distributions were assumed for balanced scenarios (traffic flow into and out of each roundabout approach is the same) and with different slip lanes configurations

(Figure 1). The base scenario, a single-lane roundabout with no slip lane, is shown in S1. In S2, similar as S1, with direct access driveway at a roundabout circle and is placed at the northbound entry approach; in S3, similar to S1, is a single-lane roundabout with a slip lane placed at the northbound entry approach, in S4, similar to S2 and S3, is a single-lane roundabout with a slip lane and a driveway is placed at the northbound entry approach; in S5, is a single-lane roundabout with two slip lanes are placed at the northbound and the southbound entry approaches; finally, in S6, similar to S5 with two slip lanes and with a driveway, (Figure 1).

The scenarios (S1 to S6) were initialized, analyzed, and then controlled through several iterations. Several variables were tested across the traffic percentage distribution scenarios: 1) right-turning traffic volume as the dominant turn (from 50 vehicles per hour to 500 vehicles per hour, in increments of 50—representing low, moderate, and high volumes), 2) traffic percentage distribution flow patterns (75%), and 3) three different driveway volumes vehicles per hour (DRW50: low: 50 vehicles per hour, DRW100: moderate: 100 vehicles per hour, and DRW300: high: 300 vehicles per hour).

These six traffic volumes distribution scenarios were coded into VISSIM to evaluate the performance of roundabout. The total minimum number of VISSIM runs was 20 (1 hour each) based on statistical equation to be met with a 95% confidence interval and acceptable error. In total, 60 scenarios were modeled: 1 traffic O-D (75%) x 10 traffic volumes (50 to 500 vehicles per hour) x 6 scenarios (S1-S6). For each simulation scenario, 20 VISSIM runs, executed using different random number seeds, resulted in a total of 1,200 simulations. The driveway were modeled, as VISSIM allows, for the driveway, the vehicle type and vehicle model were both selected as cars (no trucks or buses). The driveway geometry was constructed using links modeled as (one lane driveway right-out movements only - 12 feet.) Drivers on the driveway must yield the right-of-way to roundabout circle or slip lane traffic crossing (priority rule).

Average roundabout delay (in seconds) for all vehicles entering the roundabout and average of delay for the driveway (vehicles in seconds) is used in this study as the Measure of Effectiveness (MOE) of the roundabout and the driveway.

ANALYSIS AND RESULTS

A sample of total approach (V_a) volume flows for scenarios S1-S6 is summarized in Table 1. At high traffic volumes (V_{rt} = 500 vehicles per hour), northbound approach volume (V_a) is 677 vehicles per hour. Scenario (S2), which has driveway, sustains more vehicles circulating than the other scenarios S3 to S6.

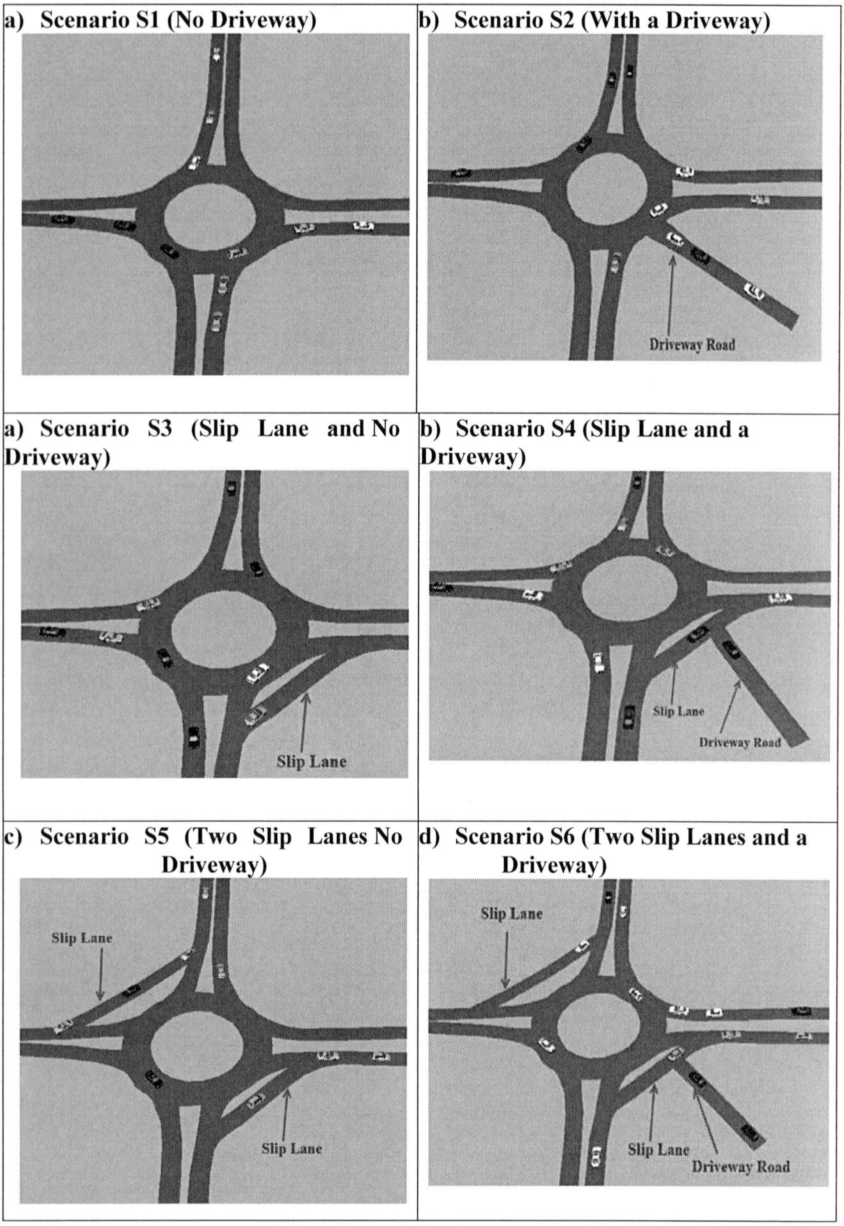

Figure 1. VISSIM Snapshots with a Driveway and Slip Lanes Scenarios (S1-S6).

Table 1. Sample of Total Approach Volumes, Vehicles per Hour.

V_{rt}: Right-Turn Volume at Northbound (NB) Approach (Vehicle/hour)	V(a):Volumes (Vehicle/hour)	S1-S6 (75%)
V_{rt} = 50 (Low)	V_a	166
V_{rt} = 250 (Moderate)	V_a	333
V_{rt} = 500 (High)	V_a	677

V_a : Approach volumes. V_{rt}: A dominant right turns volumes, vehicles per hour.

Average Roundabout Delay with a Driveway Access

VISSIM results with driveway volumes for balanced scenarios were analyzed. Results showing average vehicle delay in seconds for all vehicles in the roundabout with three driveway volumes levels range from 50 to 300 vehicles per hour. Figures 2a to 2c show the roundabout average delay time per vehicle with driveway.

a) Low Volume Driveway (DRW50) 50 vehicles per hour

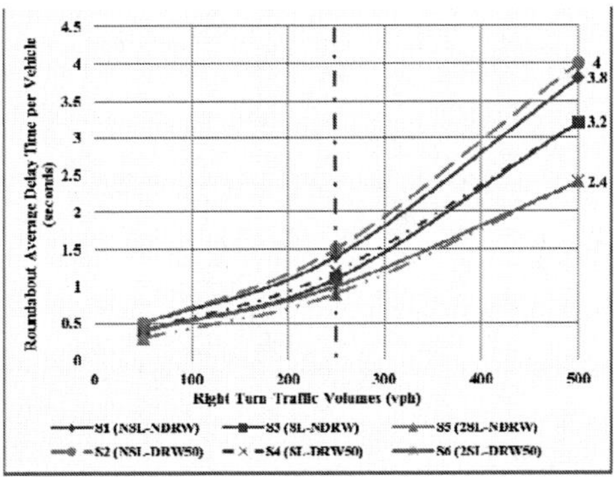

b) Moderate Volume Driveway (DRW100) 100 vehicles per hour

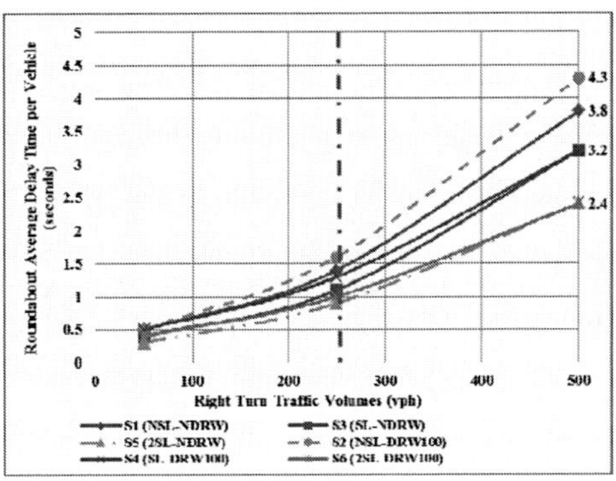

c) High Volume Driveway (DRW300) 300 vehicles per hour

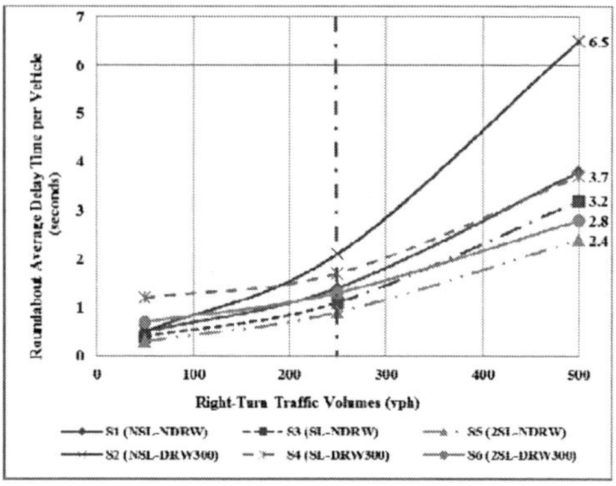

Figure 2. Sample of Outputs from VISSIM: Comparison between Roundabout Average Delay, Scenarios S1-S6.

For example, for low driveway volumes DRW50 (50 vehicles per hour), roundabout average vehicle delay (in seconds) delay decreases from 3.8 seconds per vehicle (Scenario S1 – no slip lane no driveway) to 3.2 seconds per vehicle (Scenario S3 - no slip lane and no driveway) at right-turn volumes V_{rt}= 500 vehicles per hour, a (-15.79 %) change, (Figure 2a).

At moderate driveway volumes, volumes DRW100 (100 vehicles per hour), Scenario S2 (no slip lane and driveway) shows the highest delay for the roundabout with driveway (4.3 seconds per vehicle) compared to S6 (two slip lanes and driveway) (2.4 seconds per vehicle, a (-44.19 %) change Figure 2b).

Finally, results shows at highest driveway volumes, volumes DRW300 (300 vehicles per hour), Scenario S2 (no slip lane and driveway) shows the highest delay for the roundabout with driveway (6.5 seconds per vehicle) compared to S4 (one slip lane and driveway) (3.7 seconds per vehicle, a (-43.08 %) change (Figure 2c) and compared to S6 (two slip lanes and driveway) (2.8 seconds per vehicle, a (-56.92 %) change (Figure 2c).

Figure 3 shows the "delay-induced" value which is defined as the difference between roundabout average delay with a driveway and slip lane, at specific driveway traffic volume level. Figure 3 shows the delay-induced is more significant when a roundabout include driveway and two slip lanes S6 (-1.6) compared to S2 (no slip lane and driveway), at any driveway volumes levels. When a roundabout traffic becomes oversaturated, a theoretical capacity thresholds values (limits) for right-turning volumes (V_{rt}) are estimated to be within a range of 200 to 300 vehicles per hour for various balanced traffic distribution volumes (vertical dashed line on Figure 2).

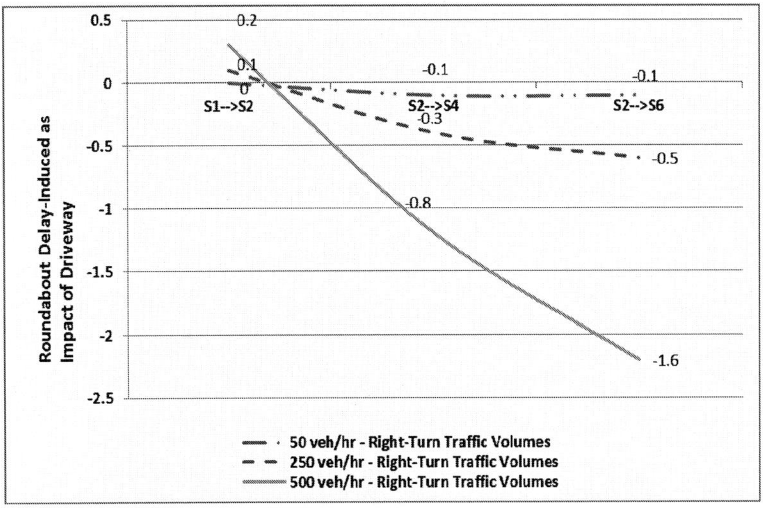

Figure 3. Sample VISSIM: Roundabout Delay-Induced as Impact of Driveway and Slip Lanes in Scenarios (S1-S6).

Average Driveway Delay

VISSIM results show average delay (sec) per vehicle on the driveway for balanced scenarios were analyzed for three driveway volume levels from 50 to 300 vehicles per hour (Figures 4a to 4c). For low driveway volumes, DRW50 (50 vehicles per hour), vehicles average delay decreases from 11.6 seconds (Scenario S2 - no slip lane) to 3.70 seconds (Scenario S4 – one slip lane) at right-turn volumes V_{rt}= 500 vehicles per hour, (Figure 4a). Scenario S6 (two slip lanes) shows the lowest delay (best performance) for the driveway with all volumes levels, Figures 4a, 4b, and 4c.

Scenario S2 shows no slip lane, creates the highest increase of driveway traffic delay, compared to S4 and S6 scenarios: from 11.6 to 2.7 seconds per vehicles (50 vehicles per hour level), from 13 to 3.2 seconds per vehicle (DRW100: 100 vehicles per hour level), and from 27.6 to 5.9 seconds per vehicle (DRW300: 300 vehicles per hour level). It is expected that driveway vehicle delay increases with increases in vehicles traffic volumes and without slip lanes. Similar results reduction on the average delay of the driveways, a (-71.37 %) change S4 (one slip lane and driveway) and compared to S6 (two slip lanes and driveway) a (-78.60 %) change.

Validation

The base model of a single roundabout, scenario S1 was validated by field data from the City of Carmel, Indiana (2009). For approximately ten single-lane roundabouts, these data provided an analysis of traffic those documents turning movement delay. The data were coded and simulated in VISSIM for validation (Al-Ghandour et al. (2011)). The same experiment was evaluated using a case study field data for S2 (no slip lane with a driveway) off Hillsborough Street and W. Morgan Street, Raleigh, NC, as shown in Figure 5. The driveway volumes were collected (18 vehicles per hour AM peak time and 36 vehicles per hour PM peak time) and matched the low volumes levels (DRW 50: 50 vehicles per hour).

Finally, a statistical validation also was tested from VISSIM results for the six experiments, based on the standard error for the percentage change of the driveway's average delay, Table 2, using a driveway is 0.15. The 95% confidence interval is ±1.96 standard errors from the average delay percentage of reduction. Therefore, the VISSIM 95% confidence interval for implementing a driveway type is estimated between 37.33 and 17.87 with high driveway's volumes for S2 Scenario (no slip lane and driveway) and is estimated between 11.01 and 4.79 with high driveway's volumes for S4 Scenario (one slip lane).

a) Low Volume Driveway (DRW50) 50 vehicles per hour

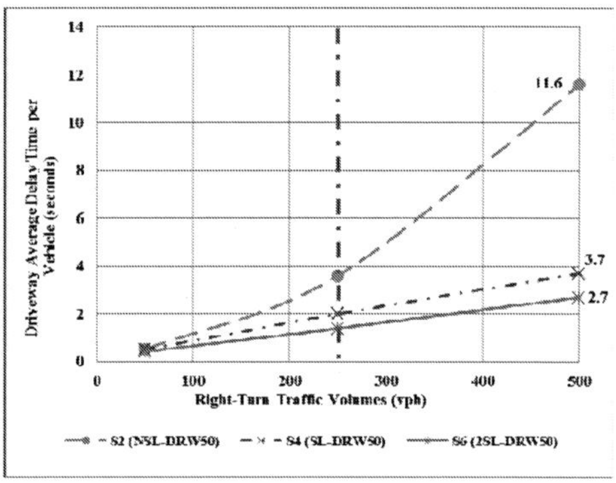

b) Moderate Volume Driveway (DRW100) 100 vehicles per hour

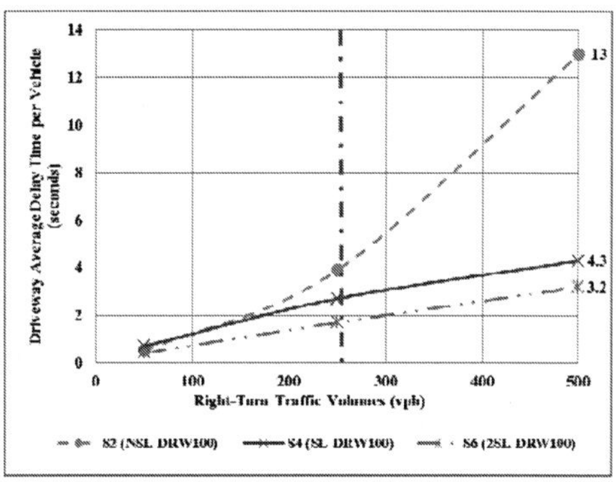

c) High Volume Driveway (DRW300) 300 vehicles per hour

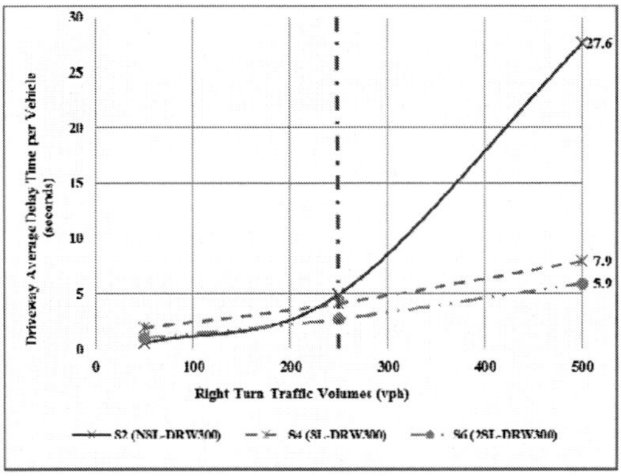

Figure 4. Sample of Outputs from VISSIM: Comparison between Driveway Average Delay, Scenarios S1-S6.

Figure 5. A Case Study: Roundabout with No Slip Lane and a Driveway off Hillsborough Street and W. Morgan Street, Raleigh, NC – Scenario S2 (with a Driveway).

Table 2. Sample of VISSIM Driveway Average Delay–Balanced Scenarios (S2 and S4)

Scenarios	V_{rt}: Right-Turn Volume at Northbound (NB) Approach (Vehicle/hour)	Average Driveway Delay (seconds) (Standard Deviation(s) Errors for 20 Runs)		
		Driveway Volumes (DRW50)	Driveway Volumes (DRW100)	Driveway Volumes (DRW300)
S2	50 (Low)	0.5 (1.2)	0.5 (0.9)	0.5 (1.4)
	250 (Med)	3.6 (4.7)	3.9 (5.1)	4.9 (6.0)
	500 (High)	11.6 (10.7)	13 (13.0)	27.6 (22.2)
S4	50 (Low)	0.5 (1.7)	0.7 (1.80)	1.9 (2.9)
	250 (Med)	2.0 (3.2)	2.7 (3.8)	4.1 (4.5)
	500 (High)	3.7 (4.8)	4.3 (5.1)	7.9 (7.1)

V_{rt}: A dominant right turns volumes, vehicles per hour. S2: Scenario S2 (with a driveway). S4: Scenario S4 (slip lane and a driveway). DRW50: Low driveway's volumes (50 vehicles/hour). DRW100: Moderate driveway's volumes (100 vehicles/hour). DRW300: High driveway's volumes (300 vehicles/hour).

CONCLUSIONS

The driveway traffic with a direct access at a roundabout may cause longer waiting time for right out traffic movements, crossing the circulating roundabout traffic and cause more vehicle conflicts. As expected, results indicate a direct access driveway increases the roundabout vehicle average delay. At high driveway's volumes level, the total average roundabout traffic delay was increased by 50% at a moderate vehicle traffic level and by 71% at a high vehicle traffic level, before oversaturation. The total average driveway traffic delay also was increased by 8% at a moderate traffic level and by 12% at a high traffic level, before oversaturation.

A hypothesis, having more than one slip lane, at different roundabout approaches, help to reduce total roundabout average traffic delay and provides efficient roadway driveway access. Finally this study suggests that incorporating slip lanes are significant sensitive per driveway which impact on the driveway access and operation for a roundabout.

RECOMMENDATION

Future research should include investigating the performance of access point driveways with full movement (right-in and right-out) access near a roundabout, including the capacity of the access point, left-turn storage needs, including access spacing criteria and sight-distance. Also futures studying the impact of pedestrian's crosswalks are crossing access point driveways.

Additional analysis should be conducted for other variables such as roundabout geometric (spacing distance from the end of the splitter island leaving the roundabout). More field collected data for different driveways (public and private) traffic volumes that affect roundabout operational performance. Finally, future analysis should be conducted to consider more variations in percentages of trucks in and out from the driveways.

ACKNOWLEDGEMENTS

The author wants to acknowledge previous work done under the direction of Dr. William Rasdorf; Dr. Billy Williams, and Dr. Bastian Schroeder of the Department of Civil, Construction, and Environmental Engineering, North Carolina State University and Institute for Transportation Research and Education (ITRE), Raleigh, NC, USA.

REFERENCES

Al-Ghandour, M. N., Williams, B. M., Rasdorf, W. J., and Schroeder, B. J., (2011). "Single- Lane Roundabout Performance with a Slip Lane under Varying Traffic Volumes and Exit Types in VISSIM." Proceedings of the 3rd International Roundabout Conference, Transportation Research Board: Roundabout Task Force, Carmel, Indiana, May 18-20, 2011.

City of Carmel, Indiana (2009). "Traffic Congestion and Safety Study Final Report, DLZ Consultants."

KDOT (2014). Kansas Department of Transportation, "Kansas Roundabout Guide, 2nd edition." Kittelson and Associates Consultants. Access Web Site: http://www.ksdot.org/burtrafficeng/Roundabouts/Roundabout_Guide/RoundaboutGuide.asp

Malik, B. Z., Siddiqui, M. A., and Lyles, R. W. (2011). "An Evaluation of Driveway Turning Restrictions in Access Management Using Microsimulation." Transportation Research Board Annual Meeting, Washington, D.C., Transportation Research Board Annual Meeting 2011 Paper#11-2589.

NCHRP Report 572 (2007). National Cooperative Highway Research Program, "Roundabouts in the United State." National Research Council, Transportation Research Board, National Cooperative Highway Research Program; Washington, D.C.

NCHRP Report 672 (2010). National Cooperative Highway Research Program, "Roundabouts: An Informational Guide." Second Edition. Transportation Research Board, National Cooperative Highway Research Program; Washington, D.C.

PTV (2007). VISSIM User's Manual 5.0, PTV AG, Karlsruhe, Germany.

Perspectives on Urban Form, Road Classification and Access Management

H. J. Stander, Pr. Eng., F.SAICE, M.ITE[1] and A. D. Abrahamson, Pr. Eng., F.SAICE, F.ITE[2]

[1]Aecom SA (Pty) Ltd, P.O. Box 112, Bellville 7535, South Africa.
E-mail: hein.stander@aecom.com
[2]A3 Transportation Engineers, P.O. Box 771, Rondebosch, Cape Town 7701, South Africa. E-mail: tony@a3.co.za

Abstract
Integral to the application of Access Management is the road classification system. Road classification in urban areas boils down to defining the higher order mobility routes (Classes 1 to 3) and then managing access provision to be commensurate with the classification. These higher order routes tend to define the urban form, as they have larger road reserves, more lanes and can end up as some form of barrier. They are also crucial for the functioning of cities (provision of mobility) as they are carrying the highest numbers of vehicles and consequently the highest vehicle-kilometer of travel. One objective of this paper is to evaluate the spacing of these higher order roads in some real world cities, in view of the ideal spacing guidelines. The existing density of higher order routes in Cape Town (South Africa) has been determined and compared (on strategic level) with four major cities in other parts of the world (Berlin, Melbourne, Dallas, and Shanghai). A further objective is to investigate the consequences of access spacing guidelines on real world situations. Issues with respect to real situations experienced in Cape Town are illustrated and compared with local and international standards. An approach to implementing the ideal standards is suggested, also in view of the guidance provided in the access management manual.

INTRODUCTION

Integral to the application of Access Management is the road classification system. In fact the Access Management Manual refers to it as the "foundation of any access management program". Road classification in urban areas boils down to defining the higher order mobility routes (Classes 1 to 3) and then managing access provision to be commensurate with the classification. These higher order routes tend to define the urban form, as they have larger road reserves, more lanes and can end up as some form of "barrier". They are also crucial for the functioning of cities (provision of mobility) as they are carrying the highest numbers of vehicles and consequently the highest vehicle-kilometre of travel.

One objective of this paper is to provide some perspective on the spacing of these higher order roads in some real world cities, in view of the ideal spacing guidelines. The existing density of higher order routes in Cape Town (South Africa) has been determined and compared (on strategic level) with four major cities in other parts of the world (Berlin, Melbourne, Dallas and Shanghai).

The road classification system, as well as the actual spacing of higher order routes is considered of high importance in access management, as the spacing of accesses is typically linked to the class of road. Understanding and acknowledging the linkage between the provision of high order roads and access standards are required. Too few or too many high order roads can impact on the application of access standards.

A further objective is to consider the consequences of spacing guidelines on real world situations. Is it possible to always comply with the relatively high standards that have been set for access spacing? Issues with respect to real situations experienced in Cape Town are illustrated and compared with local access spacing standards. How should situations that can clearly not comply with set standards be treated? An approach to implementing the ideal standards will be suggested, also in view of the guidance provided in the TRB's Access Management Manual (2003) - "AM involves trade-offs between competing objectives, so the appropriate amount and type of access have ultimately to be decided on the basis of policy".

SPACING HIGHER ORDER ROADS

Spacing in Theory

For centuries the form and shape of cities was determined by the topography and by measures to defend the city against enemies (including the need for walls around cities). Today, the former remains an important factor, but it is considered that the major road network, together with surface rail lines, has become a very important form giving element in cities. As liveable cities are desired by all, and if road classification is the foundation of access management, how do the major mobility routes influence urban form? There are at least two questions, namely (i) the ideal form of the major road network, and (ii) the spacing of the major roads.

A number of road classification systems have been developed worldwide, and although different names have been used, they are reasonably similar. Some of the typical USA road classification systems are shown in Table 1 below, together with one recently proposed for South Africa. Freeways (Principal Arterials - no at grade access) are generally considered to be Class 1 roads, arterials (major and minor) are mostly considered Class 2 and/or 3 roads, Collectors as Class 4 and Local Streets as Class 5 routes.

Form of network

Most cities have a well-defined central area (Central Business District), possibly other secondary business nodes, industrial areas, sprawling residential areas and green areas (parks, etc). There are in principle two forms for the major road network, namely radial and rectangular grid. The ideal (grid) network can be as shown in Figure 1. The boxes formed by the major roads can be served by open or closed networks (see Figure 2 below), or a combination of these. General access management principles support closed networks rather than open networks (latter with frequent accesses to the arterials).

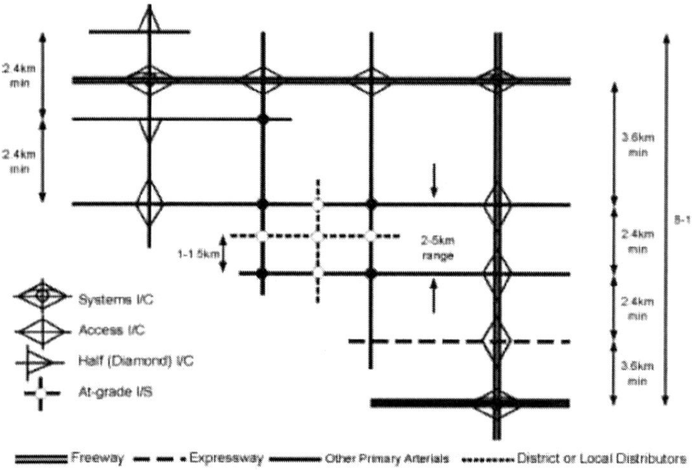

Figure 1: Typical Ideal Grid Network of Major Roads

Table 1: Typical Road Classification Systems

Class of Route	S AFRICA: Road Classification and Access Management COTO, 2012	USA: A Policy on Geometric Design of Highways and Streets (Urban) AASHTO, 2004	USA: Transportation and Land Development ITE, 2006	USA: Access Management Manual (Urban Areas) TRB, 2003
1	Principal Arterial	Principal Arterial	Freeway	Freeway
2	Major Arterial	Minor Arterial	Strategic/Principal/Major Arterial	Major Arterial
3	Minor Arterial	Collector	Minor Arterial	Minor Arterial
4	Collector	Local Street	Major and Minor Collector	Major Collector
5	Local Street		Local Street	Minor Collector
6	Ped Walkway			Local

Spacing of major roads – It is generally agreed that the spacing of major roads is influenced by development density. Guidelines for the ideal spacing of major roads have been developed and locally (COTO, 2012) have been suggested as shown in Table 2. The values in brackets are small modifications to fit into a grid network.

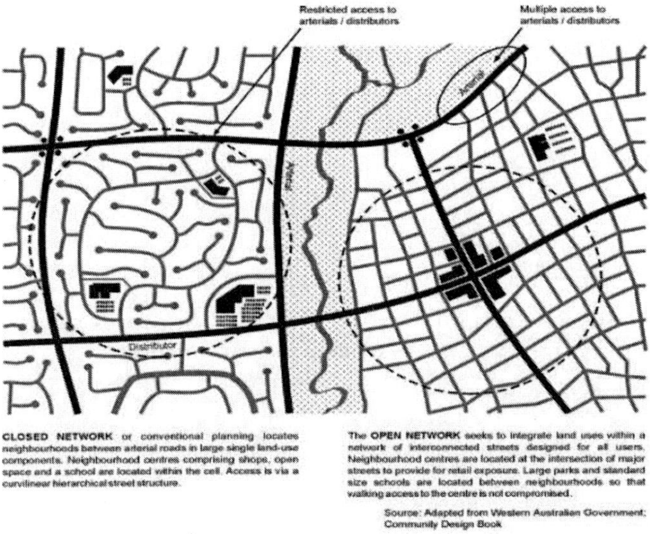

Figure 2: Illustration of Open and Closed Local Road Networks

Table 2: Guidelines for Ideal Major Road Spacing (COTO, 2012)

Road Class	High density Urban	Medium density Suburban	Low density Urban fringe
Class 1 - Principal Arterials	5 km (5 km)	8 km (6 km)	10 km (10 km)
Class 2 - Major Arterials	1.5 km (1.7 km)	3 km (3 km)	5 km (5 km)
Class 3 - Minor Arterials	1.2 km (0.85 km)	2 km (1.5 km)	As required (2.5 km)

From these modified ratios, a theoretical provision of major roads per square kilometre can be calculated. This yields the results shown in Table 3. These ratios can be used to test to what extent a road network complies to the ideal spacing. An analysis of five medium density areas in Cape Town, covering 688 km^2, or approximately half of the city, yielded the results shown in Table 4.

The result of this calculation is that for the five selected medium density areas, the provision of Class 1 and 2 routes is in fact very close to the "ideal" provision. The Class 3 routes in these areas are under provided by approximately 10%. It would have been interesting to complete the exercise for the total city, to get an overall picture, but this has not been done at this stage. It can be concluded that for at least

50% of Cape Town, the amount of major routes are complying with the national guidelines. A previous investigation (Stander, 2011) into the spacing of the major roads revealed that the Class 2 and 3 roads are severely under provided in one east/west corridor, and the opportunities for further major roads in the corridor are in effect non-existent.

Table 3: Major Road Density Required Based on Ideal Spacing

	Road length - km/square km		
Road Class	**High density Urban**	**Medium density Suburban**	**Low density Urban fringe**
Class 1	0.4	0.33	0.2
Class 2	0.8	0.33	0.2
Class 3	1.2	0.67	0.4
Total for Class 1, 2 and 3	2.4	1.33	0.8

Table 4: Supply of Major Roads in Cape Town (5 Selected Areas) Compared With Ideal Provision

City of Cape Town built-up areas
Analysis of supply of km of Class 1, 2 and 3 arterials

Zone	Area sq km	Class km			
		1	2	3	Total
Southern Suburbs	64	24	14	46	84
South S Suburbs	49	9	4	28	41
NI corridor	192	67	67	100	234
West Coast	143	48	65	75	188
South East Sector	240	81	78	127	286
Total	688	229	228	376	833
Ratio km/sqkm		0,33	0,33	0,55	1,21
Ideal ratio km/sqkm		0,33	0,33	0,67	1,33
Ideal km		227	227	461	915
Shortfall		-2	-1	85	82

COMPARISON INTERNATIONAL CITIES

In an effort to do some comparison with other world cities, four other cities have been selected for a strategic (high level) assessment. The cities are on four continents and the assessment is based on their major road networks as is shown at the back of this paper (information of cities obtained from internet and contact persons in these cities – assessment of road networks done by authors). The cities are (Cape Town included in this assessment):

Berlin: Population 3.4 million, capital of Germany, located on the river Spree and the largest city in Germany. City had relatively low car ownership of 360 cars/1000 people, versus 570/1000 people for Germany in 2008.

Dallas: Population of city 1.25 million in 2012, but part of Dallas/Fort Worth/Arlington metropolitan area with 6.7 million people – as such, fourth largest metropolitan area in USA. Vehicle ownership, as for USA, is the highest in the world – in vicinity of 800 cars/1000 population.

Melbourne: Population 4.25 million in 2012, second largest city in Australia, located on estuary of Yarra river. City has very high car ownership of 850 vehicles/1000 persons.

Shanghai: Population wise (17.8 million) the largest city in the world. Located at the mouth of the Yangtze river where it flows into the East China Sea.

Cape Town: Population around 4 million, second largest city in South Africa, at the southern tip of Africa. Vehicle ownership relatively low at 180 cars/1000 persons.

Berlin

Salient characteristics of the major road network are:

- There are at least six Class 1 routes radiating from the centre of Berlin. In a north/south direction there are two Class 1 routes varying between 3 and 8 kilometres apart, whilst in an east/west direction there are also two Class 1 routes approximately the same distance apart.
- A Class 2 ring road of approximately 5 kilometres in diameter has been completed, whilst a partial Class 1 ring road exist (western and southern part of city) – diameter of 12 to 16 kilometres from the centre of town.
- The Class 2 routes form blocks of varying size – generally between 0.8 and 2 kilometres apart.
- The blocks formed by the Class 2 routes have in many cases one or more Class 3 routes traversing the block.
- A fourth Road Class is indicated as roads of special importance.

When compared with the "ideal" spacings, it can be concluded that Berlin has a relatively dense network of higher order roads, which in many cases are even closer than the "ideal" spacing. This could be attributed to the historic nature of the city, the topography and also the affordability of high order (mobility) routes.

Dallas

The Dallas major road map shows the following characteristics:

- A radial network of at least eight Class 1 roads (freeways), focussed on the CBD, and stretching in all directions;
- Class 1 routes spaced at varying distances of between 3 and 10 kilometres apart. Close to CBD the spacing even closer.
- A completed Class 1 (freeway) ring road around the CBD forming a block of approximately 2 by 2.5 kilometres, and a second completed Class 1 ring of 28 to 30 kilometres in diameter;

- Other major routes mostly conforms to a grid network forming many blocks of 1.6 by 1.6 kilometres (one mile by one mile) and some spaced closer at 800 metres or even closer.

When compared with the "ideal" spacings, it can be concluded that Dallas also has a relatively dense network of higher order roads, which in many cases are even closer than the "ideal" spacing. This could be attributed to the topography, the high vehicle ownership and the affordability of these high order (mobility) routes.

Melbourne

The Melbourne major road map shows the following characteristics:

- Five Class 1 routes radiating out from the central area;
- Class 1 routes spaced between 10 and 12 kilometres apart;
- Class 2 and 3 routes largely following a grid pattern with many blocks of approximately 1.6 by 1.6 kilometres.
- A Class 1 road ring around the city has not been completed, but there is in effect only a 16 kilometre gap on the north-eastern side. It is not known whether it is planned to eventually close the gap. This (incomplete) ring has a diameter of 36 to 38 kilometres.

When compared with the "ideal" spacings, it can be concluded that Melbourne's freeway (Class 1) network is less dense, but still is close to the "ideal" spacing. The Class 2 and 3 routes are pretty much complying with the "ideal" standards. This could also be attributed to the high car ownership in the city and the affordability of high order routes.

Shanghai

The road classification map for Shanghai is difficult to interpret for non-Chinese speakers, but the available map appears to indicate:

- At least three ring roads exist – the smallest one has a diameter of approximately 12 kilometres, there is a second one at about 30 kilometres diameter and the third one is approximately 70 kilometres in diameter. They appear to be at least Class 2 routes, with the outer one likely to be of freeway standard (Class 1) for the full length.
- The spacing of Class 2 and 3 routes is difficult to ascertain, but from the available information it is concluded that it varies between 1 and 3 kilometres in the central area (within second ring road), to quite far apart – 4 to 8 kilometres in some outer areas.

When compared with the "ideal" spacings, it can be concluded that Shanghai has a relatively dense network of higher order roads within a 30 kilometer circle, which approximately complies with the "ideal" spacing. This could be attributed to the high population and density of the city, as well as recent growth and development in the city.

Cape Town

The Road Network Hierarchical Classification for Cape Town shows the following characteristics:

- The general form of the major road network is a radial network focussed on the CBD of Cape Town, with Class 1 routes stretching towards the north, east and south (the Atlantic Ocean being to the west).
- The Class 1 routes are spaced between 5 and 12 kilometres apart;
- The Class 2 and 3 routes by and large can be considered a grid system, and are spaced 1.5 to 3 kilometres apart in most areas;
- A previous investigation (Stander, 2011) into the spacing of the major roads in the N1 corridor revealed that the Class 2 and 3 roads are severely under provided, and the opportunities for further major roads in the corridor are in effect non-existent.

Whilst the Class 1 network (freeways) is approximately complying with the "ideal" spacing, it is concluded that the other higher order routes are generally under provided (if compared with international norms) and the importance of protecting the mobility function of the existing higher order routes, is clear.

Conclusions

This high level (cursory) evaluation of the major road networks in five selected cities, indicated that the actual spacing of higher order roads in most of them is close to the "ideal" spacings as suggested in the literature. It is concluded that the standards for the ideal spacing of higher order routes are realistic. Further research on the relationship between the presence/spacing of higher order routes and access management is required.

ACCESS SPACING AND REAL WORLD SITUATION

Development of access spacing standards

Should one accept that the real world road networks are by and large complying with the standards developed through the years, the next question is whether the suggested **access spacings** for different class routes, are practical for real world application. International literature has suggested a number of criteria that could be used for determining access spacing. A list is provided below. The dilemma is how and where to apply these criteria.

Weaving distance (WD) – mostly applicable to freeways;

Signal progression (SIG) – ideally two way progression of traffic flow should be aimed for;

Communication criteria (CC) – adequate space for signing should be available;

Stopping sight distance (SSD) – ideally this should be available at all places on a road;

Functional boundary distance (FBD) – no access should be allowed within the FBD, not clear whether FBD's could overlap;

Left turn conflict (LTC) or Right turn conflict (RTC) in countries driving on right hand side.

In South Africa, two approaches have been followed. Firstly, the Western Cape Province (PAWC, 2001) has selected these criteria to be applicable under specific circumstances (selection based on engineering judgement). The circumstances contain four variables, namely (i) development environment, (ii) class of road, (iii) type of driveway/intersection and (iv) specific operating speeds have been selected for each situation.

Secondly, the national access management guidelines (COTO, 2012) have followed the following approach:

- For mobility routes (Classes 1 to 3) – two way progression of signals, which results in intersection spacings of 600 to 800 metres. Access to property is generally not allowed.
- For access streets (Classes 4 and 5) – intersections at 150 to 250 metres with access to property at 15 to 50 metres.

Application of standards

Experience in the Western Cape Province of South Africa is that the road authorities are applying the access management standards strictly. There have been court cases with respect to the closing of not complying accesses. A question can be asked regarding the implications of these standards in real world situations. One of them is the servicing of regional shopping centres, which is discussed briefly below to illustrate the point. A regional shopping centre has around 100 000m^2 of Gross Leasable Area (GLA) (or around 1 million ft^2). At a bulk factor of around 0.3, and in order to have adequate space for parking, a site of around 600 metres by 600 metres is required. These blocks are typically formed as shown below between Class 2 and 3 routes which are in effect spaced at the "ideal" distances.

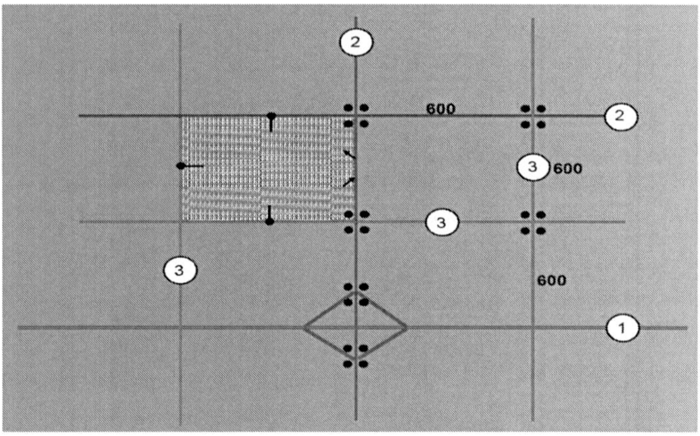

Figure 3: Typical Location for Regional Shopping Centre Within City Blocks

These regional shopping centres are open seven days a week and the peak hour trip generation has in fact reduced slightly from the days when it was not open on Sundays. Even so, empirical evidence shows that the afternoon peak hour trip generation of such a shopping centre is around 4000 to 4500 vehicles/hour. This has to be served by at least three signalised entrances. The dilemma is that the access standards do not allow for signals in the 600 metre block in which the centre is situated. A practical example of this in Cape Town is illustrated below. In this case, three signalised intersections, as well as a left-in/left-out access, had to be provided at less than half the "ideal" and prescribed spacing.

Figure 4: Servicing of Regional Shopping Centre in Cape Town

This point is also clear from the proposals for servicing shopping centres made by Stover and Koepke (ITE, 2006). In their illustration of the typical features of a large commercial site (see below), the roads to the north and east of the site are indicated as major arterials. The block required for the centre has to be around 600 metres by 600 metres, which implies that the midblock accesses should not be signalised. It is considered that they have to be signalised under the illustrated conditions.

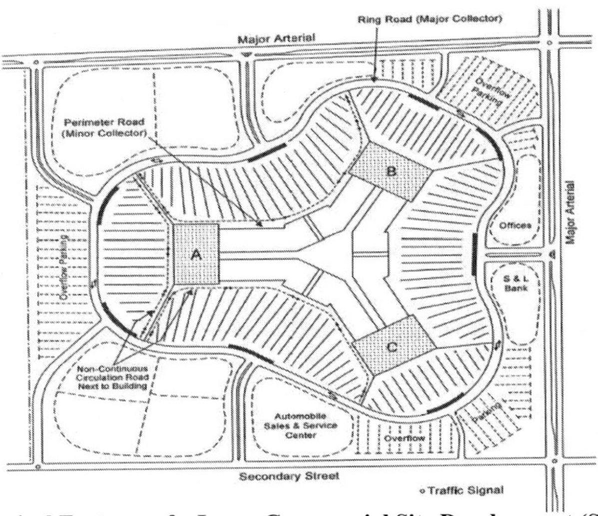

Figure 5: Typical Features of a Large Commercial Site Development (Stover and Koepke)

The main reason why this point is illustrated, is to indicate that there are practical situations which, seemingly, cannot be serviced with the standards that have been developed. The Access Management Manual's (TRB, 2003) statement that this situation can only be solved "on the basis of policy", appears to be the only way out. Can the ideal situation ever be achieved? Answer is probably NO, but that should not discourage all involved to strive for the ideal situation and to continue efforts to improve understanding and addressing of real world access management issues.

APPLICATION OF ACCESS MANAGEMENT IN PRACTICE

The real world application of a program based on policy decisions, albeit supported by scientific motivation, obviously is not simple. It could be difficult in court to defend standards that cannot be achieved in all circumstances. Is it becoming a matter of where and when are exceptions allowable? The following approach to the implementation of an Access Management program has been suggested for practical application:

i) Develop and adopt an access management policy;
ii) Classify the road network;
iii) Make road classification available to land owners/developers;
iv) Refuse rezoning of existing land use requiring access in contradiction to policy;
v) Do not allow intersection control (e.g. signalisation) at improper spacings. Rather use marginal/partial access;
vi) Retrofit where road safety/congestion is poor;
vii) Gradually upgrade mobility where possible;
viii) Use traffic calming on access/activity streets to encourage use of mobility roads.

CONCLUSIONS

This high level evaluation of the major road networks in five selected cities, indicated that the actual spacing of higher order roads in most of them is close to the "ideal" spacings as suggested in the literature. It is concluded that the standards for the ideal spacing of higher order routes are realistic.

There are practical situations for access provision which, seemingly, cannot be serviced with the standards that have been developed. The Access Management Manual's (TRB, 2003) statement that this situation can only be solved "on the basis of policy", appears to be the only way out.

Clearly, there is room for further research to understand and address real world access management issues.

REFERENCES

American Association of State Highway and Transportation Officials (AASHTO, 2004). *A Policy on Geometric Design of Highways and Streets*, Fifth Edition.

Committee of Transport Officials (COTO, 2012). *South African Road Classification and Access Management Manual (TRH26).*
Provincial Administration Western Cape, Transport Branch (PAWC, 2001). *Road Access Guidelines,* Second Edition.
Stander, H.J., Watters, M.L.(2011), *Road Access Management: Classification of Activity Routes*, First International Conference on Access Management, Athens, Greece.
Stover, Vergil G, Koepke, Frank J, Institute of Transportation Engineers, (ITE, 2006). *Transportation and Land Development*, Second Edition.
Transportation Research Board, Committee on Access Management, (TRB, 2003). *Access Management Manual.*

APPENDIX

City maps were obtained from internet and contact persons in cities.

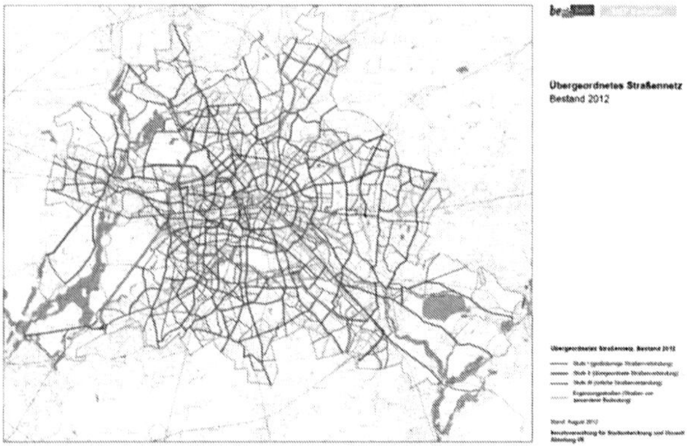

BERLIN

MELBOURNE

SHANGHAI

DALLAS

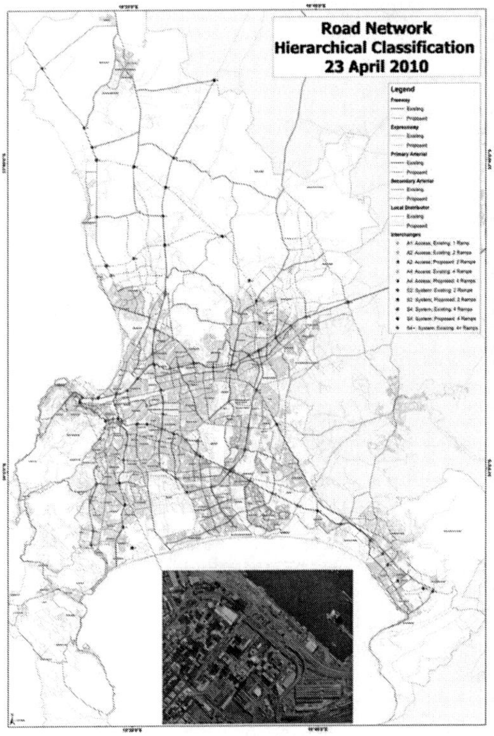

CAPE TOWN

Research on the Design of Access Road in Old City Reconstruction

Huijun Shao[1]; Yandong Fan[2]; and Junqing Li[3]

[1]The Key Laboratory of Road and Traffic Engineering, Ministry of Education, P.O. Box 201804, Shanghai, China. E-mail: shaohj2012@126.com
[2]School of Road Engineering, Changan University, P.O. Box 710064, Xi'an, China. E-mail: hdfyd2009@126.com
[3]CCCC Fourth Highway Engineering Company, P.O. Box 100024, Beijing, China. E-mail: 008ljq@163.com

Abstract
In order to improve the traffic safety and efficiency of new access roads in old city reconstruction, it is necessary to do research on site selection, layout design, and management of access road. By integrated analysis of multi-level road traffic operation, this paper determines the service level of access road using equivalent traffic capacity replacing intersection capacity as the representative value of road traffic capacity, and in according to the relationship of equivalent capacity and intersection spacing, back calculation of the optional areas of access roads is, which replaces the design method aimed at meeting the demands of the functional areas of intersection. Finally, according to this design idea, we use the road traffic simulation system to simulate the design of access road in the reconstruction of Qi county HeBi city Henan province. The results show that this method satisfies the road network service level and maximizes the system traffic capacity on the basis of traffic safety.

INTRODUCTION

In response to the problem of old city reconstruction and urban renaissance around the world, a lot of city reform and development are conducted, thus the old city reconstruction has become a hot topic around the world. Old city reconstruction refers to the partial or whole, step-by-step transformation and update of living environment in old city, in order to fundamentally improve living conditions (Carroll 1982; Breheny 1997). Generally speaking, the reconstruction covers the following: repair, renovation, duplicate, refurbishment, reconstruction and plug construction. Such as adding new urban functions within the original land; increasing buildings and improving infrastructure; expanding city afforestation, etc. The core purpose is to solve the structural imbalance in urban development, functional degradation and other issues in city development, to make cities meet the development plan. Through old city reconstruction, the living conditions of residents have could be greatly improved, and regional economic vitality is also promoted. At the same time, the rise of real estate and infrastructure construction significantly boosts the local economy and enhances city's overall economic strength, laying a solid foundation for the further development. However, there are lot of problems behind the development of old city reconstruction, like, old city reconstruction has to meet the new land use pattern of intensity and high efficiency , rural ecological environment promoting and public facilities and infrastructure co-construction and sharing encouraging. In this case, reasonable arrangement for new access roads while keeping as much original roads and infrastructure as possible is significantly important, for assuring road network reconstruction, to meet modern society requirements and provides service for cars while save more space for human.

Most domestic research on access road refers to access management technology of the United States (*Committee on Access Management*.2003; CTRE.2000; *Management Techniques*.1999). The technology mainly includes position selecting, designing, management and controlling of road intersections, median opening, interchanging. The management of the access road design chooses intersection traffic as control indicators, and realizing intersection function as goals. The standard of access road is build up with setting the size of the intersection area to control intersection spacing. There are two problems in this method (Oregon Department of Transportation.2000; Lall 1999). Firstly, the integration of old roads and new roads often occurs in old city reconstruction. Because of the different construction period, service levels are different, which causes diversity in traffic running level evaluation indicator and threshold value. However AMM doesn't give the method for co-ordination of the service level between new roads and the old; Secondly traffic capacity of intersection is calculated based on a certain model of traffic flow-density-speed, and got by reduction factor method or measurement method, which is taken as the capacity of whole road. It is suitable for roads of long intersection spacing and homogeneous section performance, but unreasonable for urban road of serried intersections. Therefore, directly using the methods of AMM cannot effectively guide the management of new-built access road in old city reconstruction in China. We need to adjust AMM methods to our local conditions.

ACCESS ROAD SERVICE LEVEL

The evaluation of transportation service level (*Transportation Research Board* 2000) is aimed at interpreting the formation mechanism of road traffic conditions, describing the feeling of travelers about traffic conditions and providing reference for management decisions. At present, the urban road is divided into four categories: fast road, arterial road, collector road and branch, which is according to the road status, traffic function and service function to buildings along the road in our country. The service level is often chosen as the indicator in classification. The geometrical target of each class of road is recommended. A lot related researches in this field are carried out by institutions of various countries. U.S, Canada, Australia, Japan, Russia, Germany have published their own capacity manual in succession, where the service level evaluation index and method are described. High way Capacity Manual (HCM) defined service level as a kind of quality standards that describes the operation conditions within traffic flow and the feeling of drivers and passengers.

However, with the development of society, especially the economic condition, the big difference in service level of roads with same grade built in different periods with grade road is appeared. The existing HCM adopts different service level evaluation indexes for different types of facilities in evaluating the multi-level urban road, and it also chooses different service level classification thresholds for different grades of the same facility, which becomes very difficult, even impossible, to achieve the system service level for the newly-built intersections, roads and other facilities in old city reconstruction.

Service Level of Access Road

For the problem of different indicators and various thresholds in different roads' traffic level evaluation in old city reconstruction, the running state of traffic can be divided into four grades: open, relatively smooth, crowd and block. Firstly, the method of expert evaluation is adopted to define the grading result of running status for each road. Then membership function is confirmed according to expert investigation statistical testing method, and the mapping relation between grade and traffic running state is established. As shown in Fig. 1. The purpose is to determine grading results of each road under the same running conditions (Shao et al 2005).

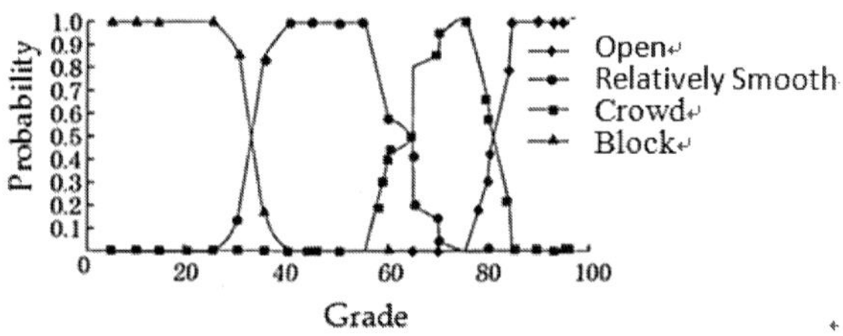

Fig. 1. Analysis of threshold values for different levels of traffic operation

Access Road Design Speed

Density ratio, which can reflect the level difference of different facilities supply, is chosen as the objective index of evaluating running level. The density ratio of each section in the system can be determined on multi-level urban road grading curve (As shown in Fig. 2.), according to the results of road grade. Then according to the density ratio, and back calculation of the design speed for each section is carried out by formula 1. Therefore the multi-facilities traffic condition evaluation is unified on system level, and reliable design speed is achieved (*Transportation Research Board* 1996).

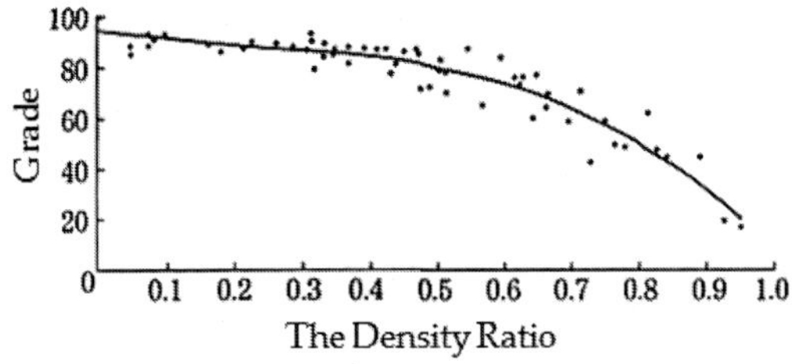

Fig. 2. Integrated grade curve for level of traffic operation

$$\beta = 1 - \frac{v}{v_f} \tag{1}$$

Where β=the density ratio, v=the actual travel speed under certain demand, and v_f= the design speed.

ACCESS ROAD LOCATION

Intersection traffic capacity is important parameters in AMM. AMM is aimed at realizing the function of intersection and the delimiting of intersection functional zone area in order to control intersection spacing (Millard 2003). HCM is usually based on a certain model of traffic flow-density-speed, and adopt reduction factor method or measurement method to obtain straight traffic capacity of the minimum point of urban road traffic capacity (usually for a particular intersection), and then take it as one-way traffic capacity for the whole city road. This method has been widely used in urban transportation planning, design and operation management nowadays. However, this method of choosing the road section capacity as the whole road traffic capacity is only suitable for those roads with bigger intersection spacing and evenly distributed section performance. It is unreasonable for those roads with serried intersection.

Equivalent Traffic Capacity

The equivalent traffic capacity (Gerbugh and Huber 1975; Zhang and Sun 2003) is defined as follows: the real road consisting of sections and intersections is abstracted as an equivalent road with no intersections. The actual road travel speed that can reflect the road conditions, intersection density, delay, interference and driving behavior is chosen as the speed of equivalent speed. The mean traffic capacity of each road is chosen as the road equivalent traffic capacity. Equivalent traffic capacity is achieved on the basis of certain traffic flow-density-speed model.

The method to obtain equivalent traffic capacity is recording the travel speed and the flow of a selected cross section for each road (taking the section of the middle of road cells) during peak hour and mean hour after choosing the road to invest. After investigation, a certain analysis period is chosen, and the average capacity of each road section and test vehicle average travel speed within the analysis period are matched to regress the relation between speed and traffic flow, in order to calculate the biggest traffic volume that can pass, which is taken as the equivalent traffic of the road. The specific procedures are shown in Fig. 3.

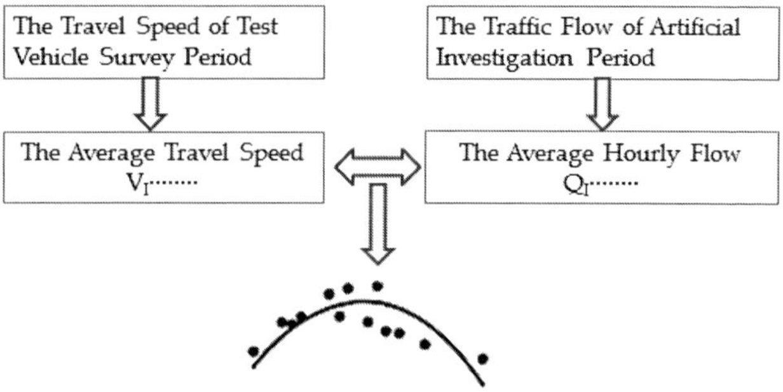

Fig. 3. Process depictions of equivalent capacity survey and analysis

Access Road Location

The greatest feature of the equivalent traffic capacity method is that urban road intersection and road section are considered as a whole, and travel speed including road conditions, signalized intersection spacing, delays and various influence factors are adopted instead of section speed in the model of classic continuous flow. The equivalent capacity of the road could be obtained by data regression. It reflects the maximum number of vehicle kilometers that the road can accommodate in time and space unit, for taking the intersection and upstream sections as an organic whole into consideration. Thus the reasonable spacing of intersection can be obtained according to the traffic capacity of road unit. As shown in Fig. 4.

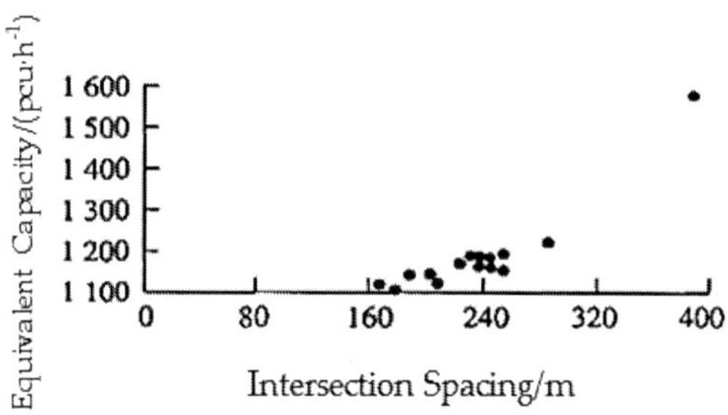

Fig. 4. Equivalent capacity changing with intersection spacing

SIMULATION TEST

Introduction of Example-Road

Qi County (HeBi City, Henan province) began old city reconstruction since March 2013. Red Flag road which is 2.6km length, two-way two-lane, 3.75m lane width, designed for speed of 40km/h, was the trunk road in the county originally. The Red Flag road mainly for light trucks, cars, electric cars, bicycles, belongs to the mixed road. In the process of old city reconstruction, there was a large number of dwelling district and commercial district on both sides of the Red Flag road. Trunk road became not only the important link among communities, but also played an important role in commerce and tourism. At the same time, there is a newly-built QiYuan road cross about 200m at the left of Red Flag road and XiaoXi street intersection, which is a total length of 2 km. It is two-way two motor vehicle lane with its lane width of 3.75 m, and non-motor vehicle lane of its width of 3m. The new road system (As shown in Fig. 5.) attracted more traffic flow by its good environment, but often led to a congested intersection and the rate of traffic accident raised significantly.

Fig. 5. The road system after old city reconstruction

The Results of Simulation Test

Road traffic simulation system has its characters of economics, security, repeatability, usability, controllability and so on. This research adopts Vissim which has various features such as model complete, strong operability, strong operability, good visibility and wide application to conduct traffic simulation. The design speed parameters of the access road were determined by rating curve determines. The traffic volume parameters were determined by the model of Underwood index equivalent traffic flow. In the simulation the type of vehicles were trucks, cars, electric cars, bicycles. The drivers' behaviors were divided into conservative strategy and adventure strategy. The model of urban traffic simulation is established for access port planning and designing and it also for the analysis of traffic manage and traffic control (Eisele and Frawley 2004).

The simulation result shows that the design speed of access road becomes lower and the equivalent capacity decreases correspondingly with the score of different road section that is obtained under the same condition decreased. At this point, the spacing between intersections should not be too close. The minimum spacing between intersections obtained by simulation is 350m, which is bigger than the actual situation (200m). Therefore, it has brought serious traffic congestions in practice.

CONCLUSION

In china, a great many urban roads lead to a village or a small business district, etc. There are entrances of individual buildings like shops, residences, hotels are directly connected to the road in old city reconstruction, which causes a lot of accident potential and traffic conflict. At present, the domestic management mechanism of access road in old city reconstruction is chaotic. Most existing researches draw on the method of AMM, instead of proposing a method suitable for small spacing intersection of access road in old city reconstruction. This study firstly gives a design speed of access road on the basis of unified urban road evaluation level. Then the traffic flow of newly molding intersection is calculated and the reasonable distance between intersections is given on the basis of equivalent capacity. The significance of the study including: make full use of the limited space of the pavement, reasonably design and set up the urban arterial road entrance, reduce interweaving, conflicts and delays of arterial road intersection function zone, improve safety and operation of arterial road traffic to reduce traffic congestion. The balance is reached between urban road traffic safety, traffic capacity, road accessibility and society, economy, environment.

REFERENCES

Breheny, M.(1997)*Urban Compaction: Feasible and Acceptable. Cities*

Carroll, G.R.(1982)*National City-Size Distribution: What Do We Know after 67 Years of Research*, Progress in Human Geography, Voll.

Committee on Access Management. Access Management Manual. Transportation Research Board. (2003)Washington D.C.

Gerbugh, D.L, Huber, M.J. (1975)*Traffic flow theory: a monograph*: special report 165[R]. Washington D C: Transportation Research Board

Lall, B.K.(1999)Analysis of Traffic Accidents within the Levinson, H.S. Access Spacing and Accidents: An Empirical Approach. Presented at theStreet Design Symposium, Dallas, Tex.

Levinson, H.S., and Gluck, J.(2000) Access Spacing and Safety: Recent Research Results. *Proc.4thNational Conference on Access Management*, Portland, Oreg.

Millard, W.(2003) *Accident Analysis Relating Crashes to Major Access Management Features-US41*,Lee County, Florida Department of Transportation

Management Techniques. (1999)TRB, National Research Council, Washington, D.C.

Shao M.H., Wang, Y., Tu, H.Z.(2005) "The urban express way evaluation based on the equivalent capacity" *Journal of Shandong Jiao tong University*, 13(4),310.

Transportation Research Board, Driveway and Street Intersection Spacing. Report NO.456(1996)

The Center for Transportation Research and Education (CTRE)(2000) *Access Management Handbook*. Iowa State University.

Transportation Research Board. (2000)Highway capacity manual: special report 209. Washington D C: National Research Council,.

Eisele, W.L., Frawley, W.E.(2004)"Estimating of The Impact of Access Management With Micro-simulation: Lessons Learned"*6th National Conference on Access Management*. Kansas City, Missouri

Zhang, J.S., Sun, L.J.(2003)"Simple measure method of the capacity of urban street intersection" *Shanghai Highway* ,(Sup),218

Spacings of Unsignalised Intersections in Urban Areas—An Empirical Approach Based on Operational and Safety Requirements

A. D. Abrahamson, F.SAICE, Pr.Eng.[1]; and C. A. Rose, F.SAICE, Pr.Eng.[2]

[1]A3 Transportation Engineers, P.O. Box 771, Rondebosch 7701, Cape Town, South Africa. E-mail: tony@a3.co.za
[2]Kantey and Templer (Pty) Ltd, P.O. Box 3132, Cape Town 8000, South Africa. E-mail: cecilr@ct.kanteys.co.za

Abstract

The paper is based on work undertaken for the 2014 edition of the "access management guidelines," developed for the Western Cape Provincial Government of South Africa, and which will replace the "road access guidelines" published by that Government in 2001. The aspect covered in the paper deals with the spacings that should be allowed between different categories of unsignalised intersections along different categories of urban at-grade roads. Firstly, the paper examines the candidate criteria for determining spacings of unsignalised public roads and private driveways, and then presents the conclusions reached on the appropriate criterion for each combination. The rationale for selecting criteria is based on one of the two presumptions: the task facing the driver negotiating the road and who must undertake processes of decision-making and actions when approaching an intersection ahead (i.e., operational requirements); or alternatively, the manoeuvre required of a driver to avoid a collision (i.e., safety requirements). The application of these principles is used to establish appropriate spacing distances along the through road, depending on the classification of the through road and the type of intersection or driveway intersecting with the through road.

INTRODUCTION

The aspect of Access Management under consideration in this paper is the appropriate frequency (spacing), in urban areas, of intersecting roads and private access driveways permitted along the length of a through route having at-grade intersection control. Some intersections may be traffic signal controlled, others may be traffic circles, while many are simply unsignalised public road intersections or private driveways. When road intersections and private driveways are permitted at spacings that are very close together, vehicular conflicts associated with these intersections and driveways may lead to unsafe or inefficient operations on the route.

Conversely, roads that have infrequent intersections and accesses may not effectively serve adjacent land use developments.

The function of a road being considered as the through route within the hierarchy of the road network – either a higher order arterial "mobility" road, a lower order collector or local "access" route – is an important element determining the degree of mobility or access appropriate to the road: on high order arterials relatively infrequent intersections and accesses should be strived for in order to preserve the mobility required; on collector and local roads more frequent access is may be permitted so as to provide for the direct access to land use developments.

While the principles of providing mobility for high-order arterials and access for low-order collectors and local roads are well established in most access management guidelines, the technical rationale for determining the appropriate spacing distances between unsignalised roads and driveway accesses is rarely clearly defined.

OBJECTIVES UNDERLYING ACCESS SPACING

One of the main objectives of good road management is to ensure that the operation of all motor vehicles and their users are safe and that drivers have sufficient time to react to events ahead in accordance with reasonable driver expectancy. Because of higher operating speeds on mobility roads, and the higher volumes of traffic using these roads, the location of driveways and unsignalized intersections are based on ensuring road safety through minimizing conflicts between vehicles.

The other important objective is to ensure mobility on high order roads, access on low order roads. Access spacing for "mobility" roads must ensure that the major junctions (signals and roundabouts), driveways and unsignalised intersections, intersecting with the through route, do not adversely impede traffic flow and capacity on the through route. This implies infrequently spaced intersections and driveway. For "access" roads the converse applies – that the intersections and driveways can be more frequent and provide high levels of access to the through routes so that adjacent property developments can easily be reached by traffic.

CLASSIFICATION OF ROADS AND INTERSECTIONS

The Access Management Guidelines rely on the classification of the roads to describe a logical hierarchy, whereby roads primarily performing a "mobility" function or primarily an "access" function. Intersecting roads or driveways with a through road are also classified based on their importance. The combination of the classification of the through road considered together with the classification of the intersecting road or driveway determines the spacing between intersecting roads and driveways.

Classification of roads

The network of roads in an urban or metropolitan area is divided into two broad categories, namely the higher order "mobility roads" and the lower order "access roads". The formal categorization and recognition of the functions of mobility and access were originally defined through work undertaken by Vergil Stover Kopke during the 1970s, and have been incorporated into the 2014 Access Management Guidelines.

The traditional S-curve diagram is shown in Fig. 1 below and the adapted "stepped" approach adopted by the Access Management Guidelines is shown in Fig. 2. The "stepped" approach is derived from the South African national road classification approach, which attributed a greater "mobility" function to minor arterials (Class 3 roads) than normally the case in the S-curve.

Classes 1, 2 and 3 roads fall under the "mobility" category of roads, with Classes 1 and 2 predominantly having a mobility function and few opportunities for vehicles to enter and leave the route. Class 3 roads are meant to allow for a balance between mobility and access, providing relatively frequent high order accesses while not severely compromising mobility along the route. Classes 4 and 5 roads are "access" roads, and are intended to allow for frequent accesses and intersections, providing access between the road and adjacent properties.

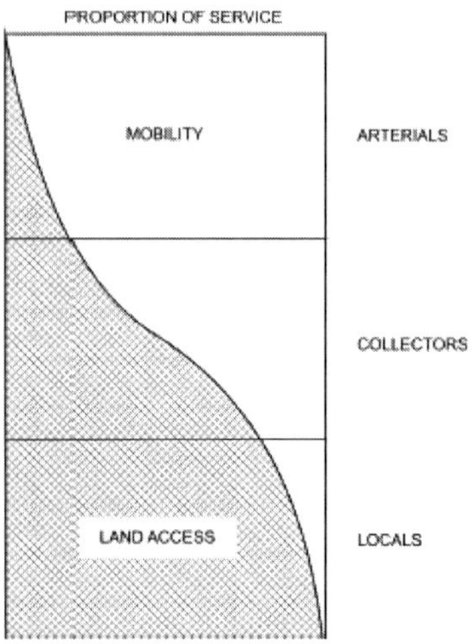

Fig. 1. Traditional S-curve

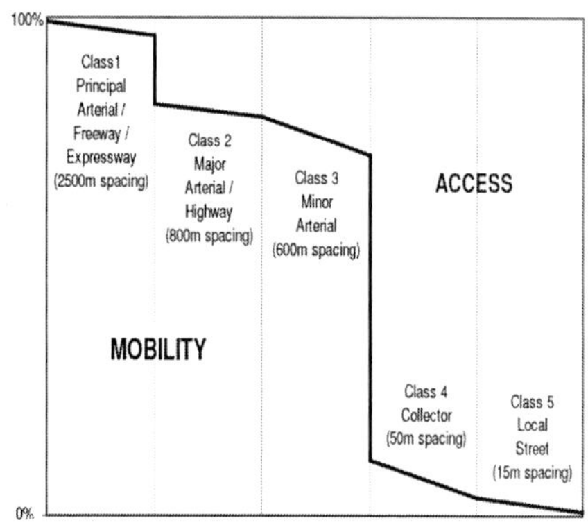

Fig.2. South African approach

Classification of intersections and driveways

Intersections and driveways are considered in their own hierarchical levels of importance, with signalized intersections at major junctions being the highest level of intersection in the hierarchy. Next are full unsignalised intersections, these allowing for all movements, and where a median exists, a break in the median for crossing and turning movements to take place. Direct access low-volume driveways to properties are the lowest level in the hierarchy.

The road network further must have logic in the manner in which connections are made between roads of different classes, and conform to the rule that a road of a particular class may connect with roads of two classes above and two classes below its class, as set out in Table 1.

Table 1. Connections to road classes.

Class of through route	Class of intersecting road	Intersecting road class permitted	Typical form of connection between through route and intersecting road
2	2	Major arterial	Signal/roundabout
	3	Minor arterial	Signal/roundabout
	4	Collector road	Full unsignalised intersection
3	3	Minor arterial	Signal/roundabout
	4	Collector road	Full unsignalised intersection
	5	Local street	Unsignalised T-intersection
4	4	Collector road	Full unsignalised intersection
	5	Local street	Full unsignalised intersection
5	5	Local street	Full unsignalised intersection

Categories of driveways

The term driveway describes the roadway intersecting a through road giving direct access to a privately owned or publicly owned property adjacent to the road, and where the roadway into the property does not usually link to other parts of the public road network. A driveway is usually the only access point to the property, although it is possible that connections could be made between properties, through connecting developments provided by agreement between property owners. Larger property developments may take access through driveways from other public roads. Three types of conventional driveways are defined:

(i) Domestic equivalent driveways that give vehicular access to private homes and micro businesses, and attract very small volumes of traffic. Domestic equivalent driveways are not permitted on Classes 2 and 3 arterials, but acceptable on certain Class 4 roads and on all Class 5 roads;

(ii) Low-volume driveways that carry larger traffic volumes than domestic equivalent driveways and are considered a class beneath that of a Class 5; and

(iii) High-volume driveways, which are equivalent to Class 5 local roads.

Equivalent driveways

Driveways that generate traffic volumes greater than high-volume driveways are accorded the status of one of the five road classes, depending on the traffic volume that is likely to be generated by the property being given access through that driveway. For the purpose of traffic control at the intersection of an equivalent driveway with the through road, the driveway is treated in exactly the same manner in which the intersection of the public road of the same class would be provided with. Thus, for example, a driveway providing access off a Class 2 major arterial into a regional shopping center may generate sufficiently high traffic volumes to justify the equivalent driveway being considered a Class 3 minor arterial. In this case the installation of traffic signals is likely to be warranted.

The categories of driveways and equivalent driveways and the traffic volumes that are related to these categories are in Table 2.

Table 2. Categories of driveways and threshold volumes

Driveway category	Class equivalent	Vehicles per hour (total in and out of driveway)
Domestic equivalent driveway	-	≤ 5
Low volume driveway	-	≤ 30
High volume driveway	5	> 30
Equivalent collector road	4	>150
Equivalent minor road	3	>750
Equivalent major arterial	2	>1500

Rules for permitting conventional driveways

Table 3 provides rules on where driveways are permitted for a given road class and the density of the urban development area. The table shows that no "conventional" driveways are permitted on Class 2 roads; only driveways that are equivalent to higher order public roads as indicated in the table above are permitted. On Class 3 roads only high volume driveways can be considered, and then only for CBD density areas (where development density is higher than 10 000sqm per hectare).

In the higher order Classes 2 and 3 roads where carriageways are usually separated by a barrier median, driveways are limited to left-in left-out movements. However, high volume driveways that are equivalent to a Class 4 side street may be treated as full unsignalised intersection where a full or partial median break is permitted.

For lower Classes 4 and 5 and where there is no plan for a future median, all movements may take place into and out of a driveway.

Table 3. Rules for permitting driveways

Development intensity	Class 2	Class 3		Class 4		Class 5
	Major arterial	Minor arterial		Collector		Local
	DED + LVD + HVD	DED + LVD	HVD	DED	LVD + HVD	DED + LVD + HVD
CBD density areas						
Suburban density areas						

	No "conventional" driveways are permitted
	"Conventional" driveways are permitted

"Conventional" driveways: DED = domestic equivalent driveways
LVD = low volume driveways
HVD = high volume driveways

CRITERIA FOR DETERMINING ACCESS SPACING

The guiding principle for determining the spacing between accesses is to provide the driver of a vehicle on the through road with an adequate distance between successive conflicts to safely negotiate the vehicle being driven along the road without colliding with other vehicles. This provides a driving environment that provides for safe "driver expectancy". The spacing must allow a typical driver traveling at the operational speed of the road sufficient time to safely undertake a variety of driver tasks in situations that may be encountered. These may include reacting to other vehicles sharing the road, to vehicles turning into and out of driveways and intersections, or for the driver to make a maneuver to turn off the through road at and intersection or driveway.

A number of candidate criteria were considered in the development of the Access Management Guidelines, and those adopted have been used to formulate the spacing distances recommended in the Guidelines.

Some criteria are based on the specific maneuver that a driver must make at the downstream driveway; others are based on the need for the driver to avoid a conflict with another vehicle – see Table 4.

Table 4. Candidate criteria to determine spacings

Candidate criterion	Based on vehicle maneuver	Based on vehicle conflict	Adopted for spacing in urban areas
Stopping sight distance (SSD)		□	□
Decision sight distance (DSD)	□		□
Upstream functional boundary distance (USFBD)	□		□
Downstream functional boundary distance (DSFBD)	□		□
Left turn conflict (LTC)		□	□
Egress capacity criteria (ECC)	□		□
Egress conflict (EC)		□	□
Communication criteria (CC)	□		□
Weaving criteria (WC)	□		□

Stopping sight distance

Stopping sight distance is the distance a driver requires to bring the vehicle to a stop in an emergency to avoid colliding with a object or vehicle in the road ahead. Stopping sight distance must always be available on any segment of a roadway. It is not regarded as a suitable distance to be used as a criterion for establishing spacing between intersections, as it relies on emergency braking in the absence of other distractions to the driver.

Decision sight distance

Decision sight distance is the distance required of a driver to detect and unexpected or otherwise difficult-to-perceive information source or hazard in the roadway ahead, and to comprehend it and act accordingly. It takes into account a roadway environment that may be visually cluttered by frequent driveways and intersections, vehicles that perform random maneuvers, regulatory and information signs and markings, and flanking developments and advertising signage. As in the case of stopping sight distance, the components of decision sight distance are:

(i) PIEV (perception-intelligence-emotion-volition) distance, also referred to as the distance covered in the perception-reaction time. This is the time that a driver needs prior to the action of braking, changing a lane or some other avoiding action. A value of 2.5 sec based on the AASHTO Green Book is assumed.

(ii) Maneuvering, lane change and braking distance is the distance covered during which the driver decelerates to a stop at the back of a queue of a signalized intersection or a right-turning lane, a lane change, or a combination of both.

Upstream functional boundary distance

Upstream functional boundary distance is the distance a driver of a motor vehicle requires on the approach to a signalized intersection or to a full unsignalised intersection, and is made up of the functional boundary distance plus the queue length that may build up in advance of the intersection. The functional boundary distance defines the minimum spacing from a driveway or intersection to a downstream signalized or full unsignalised intersection, and within this distance no intersections of driveways are generally permitted.

Fig 3 illustrates the components of functional boundary distance that are applied to the approach to a signalized or full unsignalised intersection. Of additional importance is that the spacing is considered in each direction of travel separately, as different types of driveways at different locations may be considered in the same section but in the opposite direction of travel where other criteria may apply to spacing.

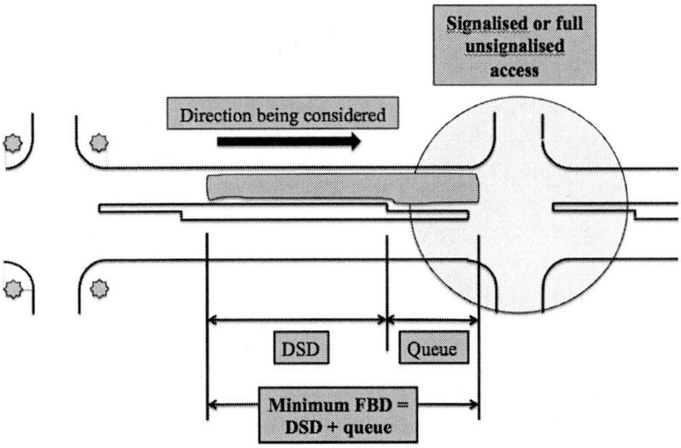

Fig. 3. Components of functional boundary distance

Left turn conflict

The left turn conflict criterion differs in principle from functional boundary distance. Left turn conflict has a similar theoretical basis to “shoulder sight distance”, the design principle on which is based a road design that ensures that a driveway or intersecting road is located in a safe location where the approaching driver has sufficient sight around or over a curve in the road to see a vehicle emerging from the side road into the path of the oncoming vehicle. Left turn conflict allows for the situation where the vehicle exiting the side road turns left and the oncoming vehicle must slow down to a degree to allow for the turning vehicle to simultaneously accelerate to operating speed.

The principle is illustrated in Figure 4.

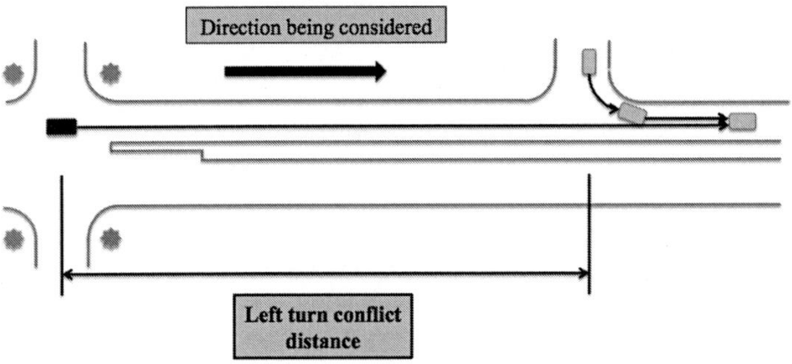

Fig. 4. Components of left turn conflict distance

Egress conflict

Egress conflict in relevant to the spacing between successive driveways. The egress conflict criterion considers vehicles entering the through road from adjacent driveways by both turning left. Drivers in this situation have equal rights of way, but frequently neither is able to predict the intended maneuver of the other. Unless sufficient spacing is provided between successive adjacent driveways the two vehicles may conflict requiring evasive action such as braking. Fig. 5 illustrates the relationship between the vehicles in this situation. The distance for the PIEV and braking distance for vehicle A to stop or slow down given the slow speeds of travel is recommended to be a minimum of 25m for this type of conflict, and is not related to the operating speed of the through road.

A separation of 25m is considered to be unacceptably short for Class 2, 3 and 4 roads where other criteria are regarded as more appropriate.

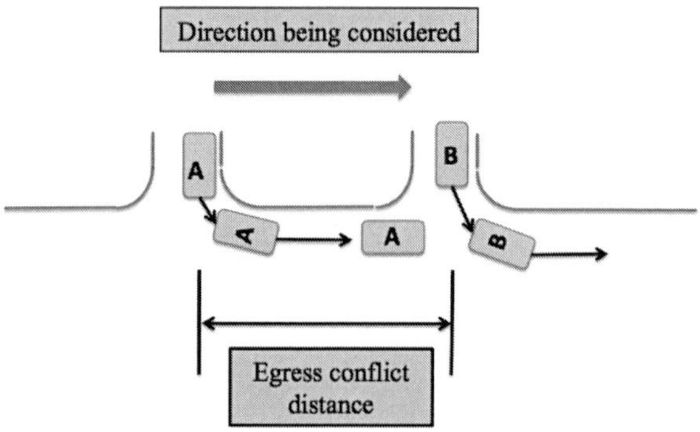

Fig. 5. Components of egress conflict

Criteria selected for application in the Guidelines

The application of the different criteria to classes of roads and for various roadside development environments to decision-making on spacing of intersections and driveways is dependent on a number of aspects.

Mobility roads – Classes 2 and 3 roads – have high operating speeds and relatively high traffic volumes and demand a spacing that gives drivers ample driver expectancy in advance of an intersection or driveway. A longer spacing is applicable and criteria that allow for generous spacing are appropriate. Classes 4 and 5 roads being predominantly for access have lower operating speeds and are expected by drivers to have frequent accesses. Consequently the criteria used for determining spacing of intersections and driveways on the lower order Classes 4 and 5 roads allow for closer spacing.

The standard criteria used to determine access spacings are functional boundary distance (FBD) where the clear distance upstream of a signalized or full unsignalised intersection (or equivalent) is concerned, and left turn conflict (LTC) in the case where the clear distance upstream of a driveway is being considered. Egress capacity (EC) is used only for spacing of driveways on Classes 4 and 5 roads.

The spacing distance corresponding to the operating speeds of the through road give the values in Table 5.

Table 5. Decision sight distance

Operating speed (km/hr)	Decision sight distance (m)	Left turn conflict (m)	Egress conflict (m)
40	120	45	25
50	160	60	25
60	205	82	25
70	240	107	
80	275	135	
90	320		
100	365		

APPLICATION OF CRITERIA TO SPACING GUIDELINES

A fundamental principle influencing the separation between intersecting roadways/driveways along a route is that a driver proceeding along the road can handle a limited amount of information while safely operating the vehicle. For safe operation, particularly on busy urban arterials, drivers can only deal with decision-making and actions related to one intersection/access at a time; thus the driver must pass an intersection/access before his or her attention can turn to the next oncoming intersection.

Establishing the positions of major junctions (signalized intersection)

A key task in the process of considering unsignalised driveways and intersections on an existing or planned road is to fix the position of the major junctions, these normally being the signalized intersections along the length of the route for as far upstream and downstream as is necessary to consider a specific series of driveways or intersections.

The task is simplified by the completion of a road network master plan (or the arterial management plan), which would have determined the locations and types of major intersections along the route, and may leave flexibility for the location of driveways.

Evaluation to be undertaken for each direction of travel

Once the major junctions have been established, the spacing of unsignalised intersections can be considered. The analysis and evaluation process is undertaken by considering the approach of a driver of a vehicle traveling along the route towards the proposed driveway or intersection is being contemplated. Where the movements at the driveway or intersection being considered are to be left-in left-out only, and other movements are prevented by a barrier median, then only one direction of travel need be considered. When the driveway or intersection involves a break in the median or no median exists, the evaluation process must consider both directions of travel along the route.

Spacing depends on the type of downstream intersection or driveway

In the case of Class 2 and 3 arterials, at least one full unsignalised intersections and several driveways can usually be fitted into the space between major signalized junctions. Where the downstream intersection relative to the access being considered is a signalized intersection or a full unsignalised intersection, the space between must consist of the functional boundary distance. Considering the space between the upstream intersection and the access being considered, the criterion for selecting the spatial separation depends on the intersection being considered: if it is a full unsignalised intersection (public road or equivalent driveway) then the criterion dictating the spacing is the functional boundary distance; if it is a driveway restricted to left-in left-out movements, then the criterion to be used in left-turn conflict.

SPACING GUIDELINE DISTANCES

The basic principles of the Access Management Guidelines outlined in the paper have been applied strictly to Class 2 and 3 arterials where mobility and safety of vehicular operation is non-negotiable. In the case of Class 4 and 5 access routes it is expected that the driver will have an expectancy imposed by multiple access and urban “clutter”, which allows the rules for Classes 2 and 3 arterials to be relaxed. The Arterial Management Guidelines contain additional rules that allow for closer access and intersection spacing on the lower order roads, not covered in detail in this paper. Table 6 summarizes the spacing distances incorporated in the Guidelines for Classes 2, 3 and 4 roads.

Table 6. Spacing distance basic guidelines (m)

Spacing		Class 2		Class 3		Class 4	
From	**To**	**CBD**	**Sub**	**CBD**	**Sub**	**CBD**	**Sub**
Signal	Unsig	235	305	180	260	90	115
Signal	Driveway	NP	NP	60	NP	90	115
Unsig	Unsig	235	305	180	260	90	115
Unsig	Driveway	NP	NA	60	NP	90	115
Unsig	Signal	235	305	180	260	140	180
Driveway	Unsig	NP	NP	180	NP	40	80
Driveway	Driveway	NP	NP	60	NP	40	80
Driveway	Signal	NP	NP	180	NP	90	115

Abbreviations:
CBD = CBD density areas
Sub = Suburban density areas
Signal = Signalised intersection
Unsig = Full unsignalised intersection
NP = Not permitted

CONCLUSIONS

The Access Management Guidelines, 2014 provide a logical and structured set of standards and guidelines to aid road authorities and planners in their management of their road networks. The Guidelines allow for the preservation of the functions of mobility on high order arterials by regulating the frequency of intersections and driveways, while also allowing for vehicular access necessary to support adjacent land use developments.

ACKNOWLEDGMENTS

The authors appreciate the support of the Provincial Government of the Western Cape, South Africa.

REFERENCES

Provincial Government of Western Cape (2001). "Road Access Guidelines." Western Cape, South Africa.

Stover, V.G. and Koepke, F.J. (1988). "Transportation and Land Development." *Institute of Transportation Engineers,* Prentiss Hall, New Jersey.

Transport Research Board (2003). "Access Management Manual." Washington DC.

American Association of State Highway Officials (2004). "A policy on Geometric Design of Highways and Streets." Washington DC.

Department of Transport (1993). "Spacing of access on major arterials." Pretoria, South Africa.

Committee of Transport Officials (2013). "TRH 26 South African Road Classification and Access Management Manual." Pretoria, South Africa.

Study on Corner Clearance of Signalized Intersection of Beijing

Xiaobo Qu[1]; Xiaokuan Yang [2]; and Yang Wang[3]

[1]Beijing University of Technology, No. 100, Pingleyuan, Chaoyang District, Beijing, China. E-mail: qu715121871@126.com
[2]Beijing University of Technology, No. 100, Pingleyuan, Chaoyang District, Beijing, China. E-mail: xiaokuan@hotmail.com
[3]Beijing University of Technology, No. 100, Pingleyuan, Chaoyang District, Beijing, China. E-mail: wangyang@126.com

Abstract
Currently, the study on access design has not yet conducted widely in large cities such as Beijing. As a result, there is no specification or criterion to dictate the layout and design of driveways connecting to roadways. Generally, these driveways are closely located to the signalized intersection. The egress and ingress of vehicles will greatly impact the operation of traffic at the intersection. In addition, this type of design brings about serious safety problems. The objective of this study is to determine the minimum spacing between access and signalized intersection of Beijing based on corner clearance technology. As is known, corner clearance technology is one of the access management strategy. By analyzing the constraints on the corner clearance of the intersection, the influencing factors are determined. On this basis, a minimum spacing model between the access and the intersection is established in the paper. In order to analyze the influencing factors on corner clearance of urban major roadways, the simulation model is used. The delay is regarded as an evaluation index. The simulation-based minimum corner clearance is determined. The result of this study can help traffic engineers to make sensible decision on the layout and design of accesses near the signalized intersections of Beijing.
Keywords: Corner clearance, Intersection, Access, Simulation

INTRODUCTION

In recent years, with the development of social economy and the general improvement of living standard, the demand for transportation of the whole society is increasing; traffic is in the increasingly tense situation, which is associated with a series of traffic problems: road congestion, vehicles disordered, frequent accidents and serious pollution etc. The city's economic development is restricted; easing traffic congestion problem is imminent. To urban road, intersection is the key nodes of the whole network. Traffic characteristic of intersection is more complex, traffic chaos and accidents of intersection area will reduce the traffic capacity of road network.

Corner clearance of signalized intersection refers to the distance between the intersection and the nearest driveway. The egress and ingress of vehicles will greatly impact the operation of traffic at the intersection. The main purpose of this study is to explore the reasonable distance between the nearest driveway and intersection, to reduce the impact of the egress and ingress of vehicles on the arterial traffic and improve the traffic operational efficiency.

Literature Review

In the other countries, especially in U.S., they carried out a series of substantive research on the access management system from the view of planning, design, operation and management. They have carried out a comprehensive systematic analysis. While Chinese research in this field is relatively small, lack of systematization, far from forming a complete system theory.

In the other countries, Texas Department of Transportation (TXDOT) put forward access management regulation and some issues that should notice in the layout and design of accesses in "Access Management Manual".[1] McCoy and Heimann assessed the impacts of corner clearances on saturation flows. [2] Long and Cheng-Tin derived formulas for estimating required corner clearance distances. [3] Kaub derived a model for estimating corner clearances. The authors found that driveway traffic can reduce the saturation flow rate on signalized intersection approaches. The amount of reduction was found to depend on the corner clearance and the proportion of curb-lane volume that enter and exit the driveway. Gluck et al. concluded that driveway obstruction is the most pervasive problem resulting from poor corner clearance, and that intersections featuring multiple, inadequate corner clearances are more likely to experience an increase in crashes. [4]

Domestic scholars have been studying access spacing these years, though they have less research on corner clearance of signalized intersection. Zhuo Xi and Zhang Ning had a research on the same-side and different-side access spacing of large public building, and then they put forward the strategy of AM which is based on access layout and traffic organization method of large public building[5,6]. Combined with practical experience, some engineering measures are given for the urban road opening design organization in "Road traffic organization optimization". [7] Deng Yajuan put forward the concepts of interchange's and off-on ramp' absolute minimum spacing, normal minimum spacing, preferable minimum spacing as well as the counting methods. [8] Zhu Shengyue studied the access settings of the urban expressway. [9] Sun Youwang had a research on the improvement of urban expressway access setting. [10] He Yulong put forward the calculation method of the minimum spacing of urban expressway interchanges, which based on the psychological characteristics of road users, the most unfavorable conditions for driving, and combined with traffic condition. [11] Cao Rongqing has studied the minim spacing model of urban road access after analyzing the relationship between access spacing and traffic conflicts, accident rate, traffic capacity. [12]

Model of minimum corner clearance

- **Intersection function area**

The intersection just act as a cross in the traditional sense, but motor vehicles need a series of complex operations while entering the intersection: reaction, deceleration, waiting in line, acceleration and steering or through, etc. The intersection function area is an area range of implementing a series of complex operations, which also can be regarded as the influence area range of intersection on the intersecting roads. The intersection function area is the area where vehicles are affected by the intersection, when vehicles going into and out of the intersection, which is different from the physical area. The intersection function area contains the upstream function area and the downstream function area.

When going into the intersection, the vehicles will slow down, change lanes and wait in line, etc. This will put a heavy psychological burden on the drivers. Driving performance is more complex than that on road section. This is an influence area that defined as upstream function area. The upstream functional area consists of three parts: □ the driving distance L1 in the response time; ②the driving distance L2 from the vehicles deceleration to stopping; ③queue length L3, which take the largest value of queue length of turning left or right and going straight. Therefore the distance L of the upstream function area can be expressed by the following formula,

$$L= L_1+ L_2+ L_3 \tag{1}$$

When leaving the intersection, the vehicles will accelerate. Compared with the vehicles entering the intersection, this area has less effect on the driver. The downstream function area consists of two parts: the length of acceleration lane and stopping sight distance. Since the acceleration lane is generally not set in the downstream of the intersection in China, distance of the downstream function area depends on stopping sight distance.

Model of corner clearance of the intersection upstream

For intersection upstream, entrance location should avoid being set in the upstream intersection function area. The upstream intersection function area constitutes the minimum corner clearance, including the distance L1+ L2+ L3 mentioned above. During the reaction time, the vehicle speed keeps constant, so L1 is calculated using the formula (2).When the driver takes a brake, vehicle slows down bound, so L1 is calculated using the formula (3).

$$L_1 = Vt/3.6 \quad (2)$$

$$L_1 = \frac{V^2}{[illegible]} \quad (3)$$

V, vehicle speed;
t, reaction time;

a, deceleration;

Signal intersection vehicle queuing model have been thoroughly studied at home and abroad, putting forward a variety of queue length calculation model. Some of the classic is MILLER, AKCELIK, SYNCHRO3, SIGNAL94, RANSYT queue length model, but it actually not necessarily completely tallies with the Chinese traffic flow. In this paper, the queue length optimization model which is suitable for the actual situation of Beijing case is used. [13] The optimization model based on SIGNAL 94, according to Beijing's traffic flow, shows the effectiveness of the optimization model using the measured data of the precision analysis. The calculation accuracy can reach about 95% in normal traffic flow situations.

$$Q_n = 2.2655 q_n R + D_{n-1} \quad (4)$$

$$D_n = \begin{cases} D_{n-1} + q_n(C - G) - S(G - L) & ([illegible]) \\ 0 & (D_{n-1} + q_n(C - L) \; [illegible]) \end{cases} \quad (5)$$

Qn, The maximum queue length of the nth cycle (veh);
Dn, The number of stranded vehicles of the nth cycle (veh);
qn, The vehicle arriving rate of the nth cycle (veh);
R, Red time (s);
G, Green time (s);
S, Saturation flow rate (veh/h);
C, Cycle length (s);
L, Start-up lost time (t);

Model of corner clearance of the intersection downstream

For corner clearance of downstream function area, the following three aspects are considered: downstream function area, influence distance of access, the impact on capacity of driveway.

- **Downstream function area**

The intersection function area is the area where vehicles are affected by the intersection. This requires access location of intersection downstream cannot be set in intersection downstream function area avoiding egress and ingress of vehicles impacting the operation of traffic at the intersection.

In China the acceleration lane is generally not set, so the distance of intersection downstream function area depends on stopping sight distance. Generally in the off-peak, speed of vehicles in the urban intersection is about 20 to 40 kilometers per hour. "City Road intersection planning norms" for intersection safety stopping sight distance gives the corresponding standards of intersection.

Table 1. Intersection stopping sight distance

design speed (km/h)	60	50	45	40	35	30	25	20
stopping sight distance (m)	75	60	50	40	35	30	25	20

- **Influence distance**

When the vehicles on the main roads turn right into the driveway, the followed straight through vehicles will take a brake. The distance from the position where the straight through vehicles take a brake to driveway entrance is called influence distance. The corner clearance of intersection must be greater than the influence distance, so as to avoid the impact of the braking vehicles on traffic operation. The distance in the table 2 can be the principles and standards of corner clearance. [1] The straight through vehicle will take a brake when the vehicle in front turns right into the driveway at the access which is called spillback. This phenomenon is called spillback. The higher the road level, the lower the acceptable spillback proportion. The acceptable spillback proportion of arterial road is 2%, while minor arterial road's is 5%.

Table 2. Influence distance (foot)

Speed	**Spillback proportion**		
(Miles / hour)	**2%**	**5%**	**10%**
30	380	335	290
35	405	355	310
40	460	400	340
45	530	450	380
50	620	520	425
55	725	590	480

- **Capacity of driveway**

The interferences of vehicles that are turning right at intersections into the main road to the vehicles running out of the driveway should be avoided when vehicles in the main road are ensured not to be affected by vehicles from the driveway at the same time. After vehicle NO.1 turning into the main road, vehicle No.2 drives out from driveway to merge with vehicle No.1. If the corner clearance is not big enough, the merging behavior cannot be coordinated, easy to lead to rear-end collisions, while the vehicles from the driveway cannot drive into the main road smoothly as shown in Figure 1. In Figure 2, corner clearance is big enough, vehicle NO.2 can merge into the traffic smoothly by the acceptance gap between vehicle NO.1 and the downstream vehicles. The following three aspects should be considered in the model: distance that No.1 vehicles merge into the main road traffic, distance that No.1 vehicles merge with No.2 vehicles and safety spacing.

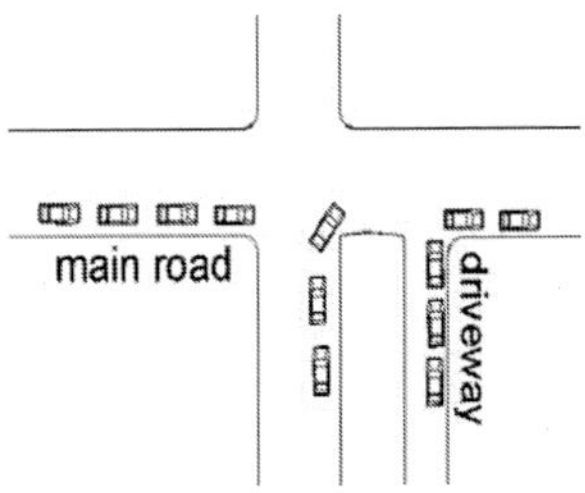

Figure. 1

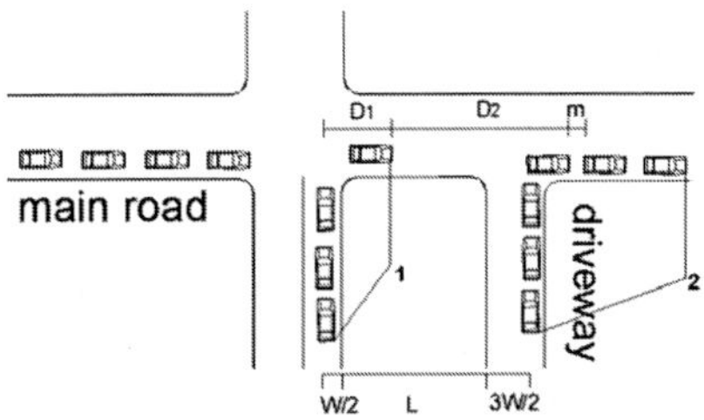

Figure. 2

$$L + \frac{W}{2} + \frac{3W}{2} \ge D_1 + D_2 + m \quad (6)$$

$$L \ge D_1 + D_2 + m - 2W \quad (7)$$

$$D_1 = \frac{V_0^2 - V_1^2}{2a_1} \quad (8)$$

$$D_2 = \frac{V_0^2 - V_1^2}{2a_2} \quad (9)$$

$$L \ge \frac{V_0^2 - V_1^2}{2a_1} + \frac{V_0^2 - V_1^2}{2a_2} + m - 2W \quad (10)$$

L, corner clearance;

D_1, distance that No.1 vehicles merge into the main road traffic;

D_2, distance that No.1 vehicles merge with No.2 vehicles;
a_1, the vehicles 'merging acceleration;
a_2, the vehicles' merging deceleration;
W, the single-lane width;
V_1, the speed the vehicles leaving driveway;
V_2, the vehicle speed on the main road;

By comparing the size of the corner clearance of the intersection under three constraint conditions, the maximum is the minimum corner clearance of the intersection downstream.

Corner clearance simulation model

Traffic simulation is a technology tracking and describing the traffic movements and space variation. Computer simulation technology is applied to conduct the research on traffic problems. Traffic simulation aims to reproduce and pre-grasp the existing and future traffic system. Without the presentation of the existing and future traffic system, simulation models can be build up to conduct the simulation tests, analyze scenarios, compare before and after result, and evaluate alternatives.

- **Corner clearance simulation design**

The vehicles from the driveway near intersection have impact on the main road traffic, while the main road traffic also has impact on the driveway traffic. Therefore, the simulation scheme considers the above two aspects. The first scheme studies the impact of different main road traffic volume on the driveway traffic under the condition of steady driveway traffic volume and different spacing. The second one studies the impact of different driveway traffic volume on the main road traffic under the condition of steady main road traffic volume and different spacing.

In the corner clearance simulation scheme, the average delay of vehicles is regard as evaluation index. The simulation parameters are calibrated by the traffic survey data of the Beijing Xi Da Wang road. The following table is the experiment scheme.

Table 3. Experiment scheme

		Design speed of main road (km/h)	Design speed of driveway (km/h)	Traffic volume of main road (veh/h)	Corner clearance (m)	Traffic volume of driveway (veh/h)	Proportion of turning traffic volume
Scheme 1	Basic conditions	50	20	A	B	200	20%
Upstream	Factor A	1500,1800,2100,2400，2700					

of intersection	Factor B	70,90,110,130,150,170,190,210					
Downstream of intersection	Factor A	1500,1800,2100,2400，2700					
	Factor B	70,90,110,130,150,170					
Scheme 2	Basic conditions	50	20	2500	C	D	20%
Upstream of intersection	Factor C	80,120,160,200,240					
	Factor D	70,90,110,130,150,170,190,210					
Downstream of intersection	Factor C	80,120,160,200,240					
	Factor D	70,90,110,130,150,170					

- **Analysis of the simulation results**

Based on the simulation results, the influence curve between the corner clearance and average delay under different traffic flow on the upstream and downstream of the main roads is plotted as indicated in figures 3 and 4. The influence curve under different traffic flow on the upstream of driveway is plotted as indicated in figures 5. To the downstream, the impact of driveway traffic on the main road traffic is little.

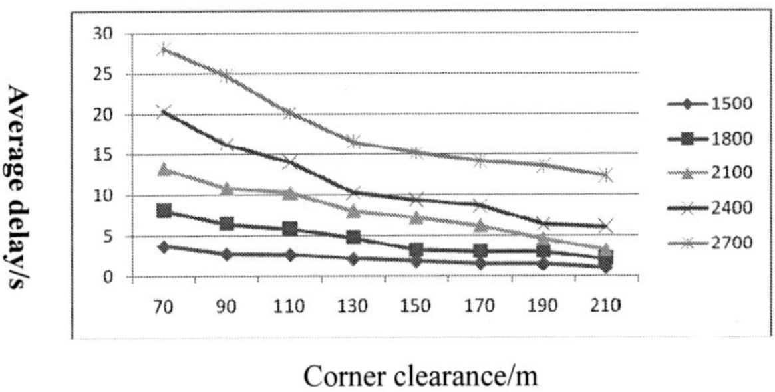

. **Figure. 3. Corner clearance delay curve of the upstream（driveway）**

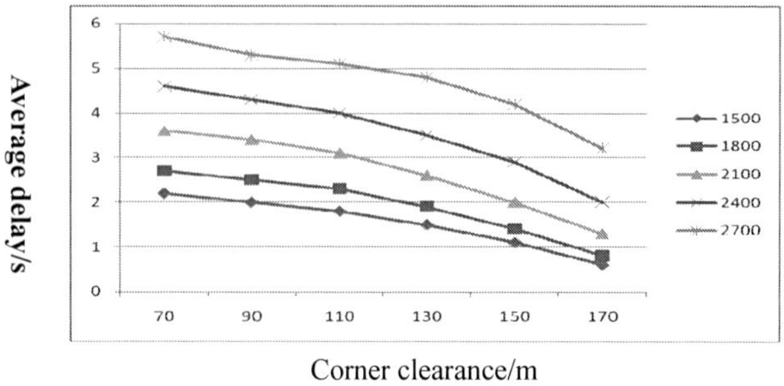

Figure. 4. Corner clearance delay curve of the downstream（driveway）

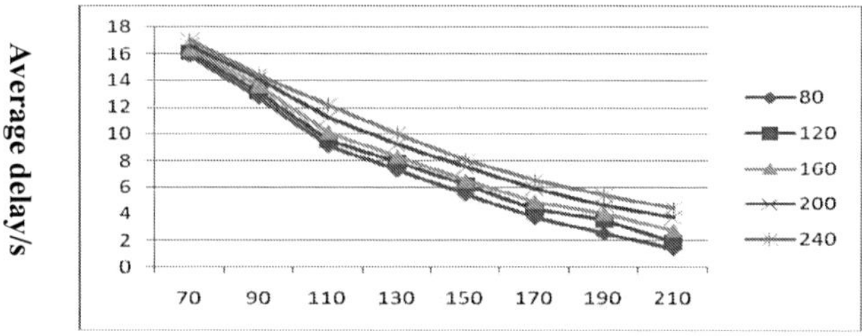

Figure. 5. Corner clearance delay curve of the upstream（main road）

By analyzing of Figure 3 and 4, we make conclusion as follows:

(1)The average delay is proportional to the main road traffic volume. The greater the traffic volume is, the bigger the average delay per vehicle is. Through comparative analysis, it can be found that the main road traffic volume has a greater influence on the upstream than the downstream of the intersection.

(2)Under the certain traffic volume, the average delay per vehicle is inversely proportional to the corner clearance. The longer the clearance is, the smaller the average delay per vehicle is, and vice verse.

(3)Under the certain spacing standard, different traffic volume has different influence on the average delay per vehicle. It can be seen from the change in slope of the curve that the impact of different spacing on delay does not vary much if traffic volume is heavy. This impact on delays varies greatly if traffic volume is little.

By analyzing of Figure 5, we make conclusion as follows:

(1)Under the certain driveway traffic volume, the average delay per vehicle is inversely proportional to the corner clearance. The longer the clearance is, the smaller the average delay per vehicle is, and vice verse.

(2)Under the certain spacing standard, different driveway traffic volume has different influence on the average delay per vehicle. But the influence is little.

CONCLUSIONS

Through analyzing constrains of the corner clearance, the influence factors are determined in this paper. Based on these, the study develop corner clearance model of the signal intersection for the city of Beijing. After analyzing the corner clearance influence factors, Vissim simulation model is then established. Based on evaluation of the average delay per vehicle produced from Vissim outputs, the impact on the average delay per vehicle under different traffic volume and corner clearance is determined. It should be noted that the model doesn't take the factor of traffic accident into account, because the traffic accident data is not readily available. As a result, the further improvements of corner clearance model combined with the traffic accident data are needed in the further study.

REFERENCES

Committee on access management (2003). Access Management Manual[R]. Washington D.C. Transportation Research Board.
McCoy p, HeimannJ (1990). Effect of driveway traffic on saturation flow rates at signalized intersections. Institute of Transportation Engineers J. 60(2):12-15.
Gary Long, Cheng-Tin Gan (1997). Model for Minimum Driveway Corner Clearances at Signalized Intersections. Journal of the Transportation Research Board, Volume 1579, 53-62.
Gluck j, H S Levinson, V Stover (1999). NCHRP Report 420: Impacts of Access Management Techniques [r]. Washington D.C. Transportation Research Board.
Zhuo Xi, Qian Zhendong, Zhang Ning (2012). Spacing calculation for same-side vehicle access of large public building [J]. Journal of southeast university, 42(3).
Zhuo Xi, Zhang Ning, Qian Zhendong (2010). Spacing calculation for same-side vehicle access of large public building [J]. Journal of Traffic and Transportation Engineering, 10(4).
Zhuo Zhongmin (2004). Road traffic organization optimization [M]. Beijing: People's Commumication Press.
Deng Yajuan (2004). Research on the linking problem of Expressway and conventionnal road [D]. Chang'an University.
Zhu Shengyue (2004). Study on the setting of expressway access [J]. Urban Transportation of China, 2(4).
Su Youwang (1997). Study on improvement program of expressway access [J]. Journal of Road Construction, 1997(1).
He Yulong, Liu Xiaoming, Ren Futian (2001). The minimum distance between interchanges on urban expressway [J]. Journal of Beijing Polytechnic University, 2001(01).
Cao Rongqing (2007). The research on the theory of urban road access spacing[D]. Chang'an University.
Rong Jian, He Min, Chen Chunmei (2002). Dynamic model of calculating queue length at signalized intersection [J]. China Journal of Highway and Transport, 15(3).

Study on the Effect of Access on the Capacity of Signalized Intersection

Wenbin Li[1]; Yu Bai[2]; Siqi Li[3]; and Guanzhe Huo[4]

[1]Key Laboratory of Road and Traffic Engineering of Ministry of Education, Tongji University, 4800 Cao'an Road, Shanghai 201804, P. R. China.
E-mail: 07liwenbin@tongji.edu.cn
[2] Key Laboratory of Road and Traffic Engineering of Ministry of Education, Tongji University, 4800 Cao'an Road, Shanghai 201804, P. R. China.
E-mail: baiyu@tongji.edu.cn
[3] Key Laboratory of Road and Traffic Engineering of Ministry of Education, Tongji University, 4800 Cao'an Road, Shanghai 201804, P. R. China.
E-mail: tj_lisiqi@163.com
[4] Key Laboratory of Road and Traffic Engineering of Ministry of Education, Tongji University, 4800 Cao'an Road, Shanghai 201804, P. R. China.
E-mail: 578342368@qq.com

Abstract

In the calculation of urban signalized intersection capacity, the effects of road and traffic conditions, which includes on-street parking maneuver rate, local bus stopping rate but not the access. Traffic that drives in and out of the accesses have a negative impact on both the smoothness of the main road traffic and the capacity of the downstream intersection. An analysis of how the assess traffic affects the downstream signalized intersection shows that the merging or passing through of the access traffic leads to green time loss of the intersection. The green time loss can be expressed by the merging time of the access. A regression analysis reveals that the merging time of the passing through traffic have a correlation with the distribution of the main road, but not strong. Then, the capacity calculation model, which consider both the access traffic and the main road traffic arriving is formed. The model shows that accesses lead to the capacity loss of the signalized intersection. At last, the result of the model is demonstrated with a numerical example.

INTRODUCTION

To begin with, a brief introduction of the definition of access and access traffic are provided. Then the effects of the access on the downstream intersection capacity are discussed in detail.

There are two different types of definitions of access. The macroscopic one includes the signalized and the un-signalized intersections, the openings of the median separators of the main road and the openings of the neighborhoods between two signalized intersections. The microscopic one is the openings of the neighborhoods and the un-signalized T intersections between two signalized intersections. This paper only focus on the microscopic type.

The access traffic refers to the traffic that comes in and out of the access which includes 3 types of traffic flow: separation flow, merging flow and passing through flow. The separation flow refers to the flow that separates from the flow of the main road while the merging flow is the flow that merges into the outermost lane of the main road. And the passing through flows are those who must pass through the straight lane, the right turn lane or the left turn lane while they get in or out of the accesses. (see Fig. 1).

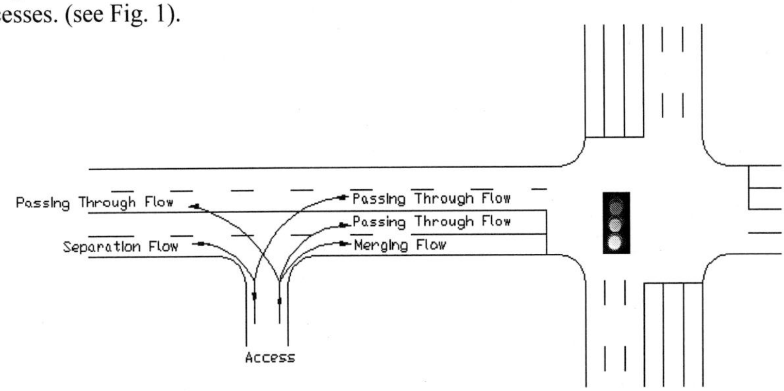

Fig. 1. Microscopic Type of Access at the Intersection.

The access traffic leads to lane and speed change of the main road and queuing and green time loss at the downstream intersections. A floating car experiment was done in Cixi, a medium-sized city in the southeast of China. Five float cars were running on the 240-meter road that has 2 accesses for a half day on a Wednesday morning. In order to get more data, the section was selected near a hospital which has a large number of access traffic. The passive lane change means that the floating car was forced to change lane by the road environment, esp. the access traffic. The active lane change means that the floating car overtake the fronting car. The passive and active speed reduction have the similar meaning. The data shows that there were 39 times of passive lane change while only 18 times of active lane change and 62 times of passive speed reduction and 4 times of active speed reduction. The main reason for this phenomenon is the access traffic leads to the passive lane change and speed reduction.

The separation flow needs to wait for the critical gap of the bicycle or pedestrian flow to get into the access. The influence factors are the distributions of the separation flow and the bicycles and pedestrians, the signal control of the outermost lane at the downstream intersection. The merging flow leads to speed reduction and lane changes of the outer most lane and even the inner lanes. The influence factors are the distributions of the merging flow and outermost lane traffic flow. The passing through flow often refers to the flow that needs a left turn. On their way to the exclusive lane, they need to cross all the straight lanes. That means the influence factors contain the distributions of the main road and the passing through flow, the signal control of the downstream intersection.

In this paper, we focus on the intersection that has an exclusive left turn phrase and lane. (see Fig. 2). During the green time of the straight lane, the left turn vehicles are waiting at the left turn lane. But the passing through flow is waiting at the straight lane which leads to green time loss, thus results in the reduction of capacity at the intersection.

The remainder of this paper is organized as follows. In the Literature review section, a brief review the literature is given. In the Methodology section, we develop the model of signalized capacity involving distribution of the access traffic and the location of the access. In the numerical example section, an intersection of Shanghai is provided to illustrate the results. The conclusion section summarizes the work done in this paper and provides the conclusion.

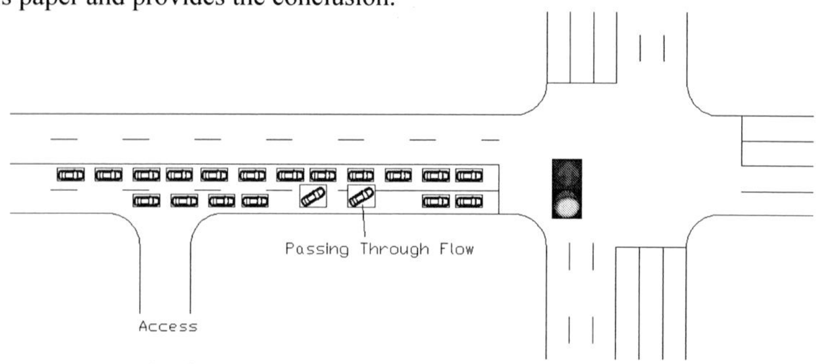

Fig. 2. Passing Through Flow at the intersection.

LITERATURE REVIEW

The influence factors of the capacity include road, traffic and environmental factors. The road factors include the width, number, designing speed of the lane, alignment and grade of the road, range of visibility, etc. The traffic factors include the volume, composition and direction distribution of the traffic; traffic control mode and traffic management, etc. As for the environment factors, the main elements are commercialization of the street, bus station and curb parking and the weather conditions, etc. (Yan 2003). However, the accesses at the upstream or downstream of the intersection are barely considered.

The main calculation method of capacity at the signalized intersection is Saturation Method. The capacity at the intersection is the product of saturation and split ratio. There are some other similar methods like ARRB (Australian Road Research Board), Webster Method. (HCM 2010; HCM 2013). Considering the driving behavior of Chinese drivers which means the drivers seldom obey the rules of stop sign, some local calculation methods like Stop Line Method, Conflict Zone Method. These methods take the driving behavior, composition of traffic into consideration. All in all, these methods are all on the basis of the Saturation Method. (Zhou 1987). Jing Zhao formed a capacity calculation method that divide the access traffic into 6 types according to the flow directions (Zhao 2008). But he did not analysis the traffic flow that affects the capacity most.

METHODOLOGY

Driving Behaviors in China are quite different from the western countries. At the two-way stop-controlled intersections, Chinese drivers on the secondary road seldom give way to the arterial roads. The critical gap in China is much lower than the U.S. Vehicles at the secondary roads often squeeze into the main road. That means the traffic flow characteristics at the two-way stop intersections are more likely to be an all-way stop-controlled intersections or the traffic actualized signal intersections. (Zhao 2008). Assuming that whenever an access traffic appears, the green time of the main road will suffer a loss and the access traffic doesn't have any impact on the main road capacity during the red lights.

There are 3 kinds of passing through flows, as we can see in Fig. 1. In this paper, we only focus on the left turn passing through flow, which is shown in Fig. 2. The passing through traffic must drive across all the lanes except the target lane. It has 3 kinds of impacts on the intersection arm.

The first impact is disturbing the outermost lane and the straight lanes. The vehicles were forced to slow down or change lane while the passing through cross it. This can be shown on Fig. 3. The second impact is stopping the vehicle on the straight lane which is adjacent to the left turn lane. As there is an exclusive left turn phase, the left turn passing through vehicle must wait on the straight lane until the left turn lane get an green light and then a gap on the left turn lane. The third impact is disturbing the vehicle on the left turn lane. The left turn passing through vehicle must wait a gap on the left turn lane. Green time loss on the outermost lane, straight lane and left exclusive lane are shown as equation 1 to equation 3.

$$g_{rs} = kt_1 \tag{1}$$

$$g_s = kt_2 \tag{2}$$

$$g_l = kt_3 \tag{3}$$

Where g_{rs} = Green time loss at the outermost lane and the straight lane which is not adjacent to the left turn lane, g_r = Green time loss at the straight lane that is adjacent to the left turn lane, g_l = Green time loss at the left turn lane, k = the number of the left turn passing through vehicles, t_1 = merging time at the outermost lane and the straight lane, t_2 = merging time at the straight lane which is adjacent to the left turn lane. t_3 = merging time at the left exclusive lane

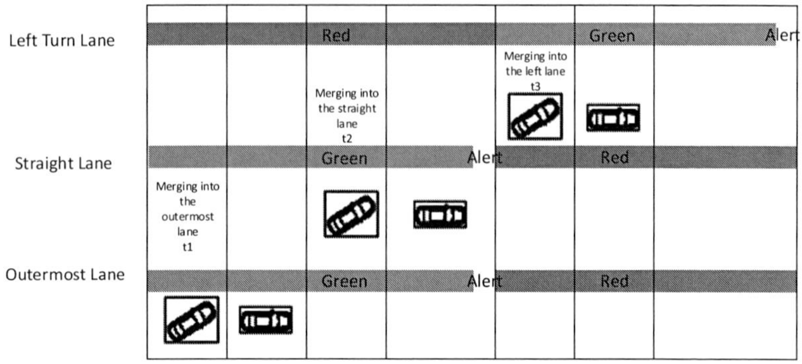

Fig. 3. Passing through flow at the intersection arm.

As we get the green time loss, then the number of left turn passing through vehicles must be known, which means we need to know k.

So the possibility of k vehicles that come out of the access is P_k.

$$G_{rs} = kt_1P_k \tag{4}$$

$$G_s = kt_2P_k \tag{5}$$

$$G_l = kt_3P_k \tag{6}$$

Where G_{rs} = Green time loss at the outermost lane and the straight lane which is not adjacent to the left turn lane at the downstream intersection, G_r = Green time loss at the straight lane that is adjacent to the left turn lane at the downstream intersection, G_l = Green time loss at the left turn lane at the downstream intersection, P_k = the possibility of k passing through vehicles that comes out of the access.

Then the capacity of the intersection arm of each lane are:

$$C_{rs} = S_{rs}\frac{G - G_{rs}}{C} + k \tag{7}$$

$$C_s = S_s\frac{G - G_s}{C} + k \tag{8}$$

$$C_l = S_l\frac{G - G_l}{C} + k \tag{9}$$

Where C_{rs} = the capacity of the outermost lane and the straight lane that is not adjacent to the left turn lane, C_s = the capacity of the straight lane that is adjacent to the left turn lane, C_l = the capacity of the left lane, S_{rs} = the saturation flow rate of the outermost lane and the straight lane that is not adjacent to the left turn lane, S_s = the saturation flow rate of the straight lane that is adjacent to the left turn lane, S_l = the saturation flow rate of the left turn lane, G= effective green time at the intersection arm, and C= the cycle of the intersection. k= the of the left turn access traffic.

NUMERICAL EXAMPLE

Access Traffic Distribution

According to the traffic survey of Shanghai in 2013, the distributions of the access traffic vary from place to place. Ten intersections of Shanghai show that the access traffic have different distributions. Let's take Wusong Road ~ Haining Road intersection as an example. This intersection is a four-leg intersection at the central area of Shanghai. The east approach has 5 lanes. An access is 80 m away from the stop line. The distribution of the passing through flow of this access can be shown in Fig. 4.

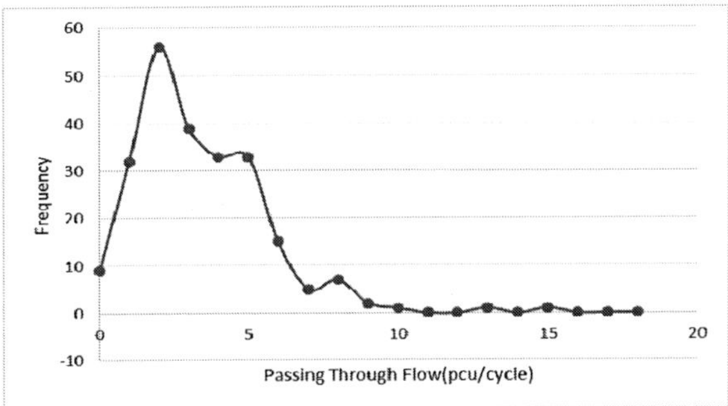

Fig. 4. Passing through flow distribution.

As we can see, it seems to be a negative binomial distribution. A nonparametric test was done. The expectation is 3.4 and the variance is 5.

According to the chi-square test,

$$\chi^2 = 1.90 < \chi_5^2(0.05) = 11.07 \tag{10}$$

The distribution of this access traffic flow is submitted to the negative binomial distribution. Thus we can get the probability of the left turn vehicles' number.

The distribution function of negative binomial distribution is

$$P_0 = (1 + \lambda t/\beta)^{-\beta} \tag{11}$$

$$P_k = \frac{(k + \beta - 1)}{k} \frac{\lambda t}{\beta + \lambda t} P_{k-1} \tag{12}$$

Where P_0 = the possibility of no passing through vehicle, P_k = the possibility of k passing through vehicles that come out of the access, and k= the number of vehicles that comes out of the access, λ= the arriving rate of the passing through vehicles, t= the counting interval, β= the parameter of the distribution.

As the expectation and the variance have been given, the distribution function is shown as follows:

$$P_0 = 0.06 \tag{13}$$

$$P_k = 0.405\frac{k+4}{k}P_{k-1} \tag{14}$$

Green Time Loss

The merging time of the passing through traffic is the interval between the time when the passing through traffic touches the lane line and the time when it leaves the lane or drive smoothly on this lane. (Fig. 3). The traffic arriving rate is the number of vehicles shown 10 seconds before the access shown at the main road.

A regression of green time loss and the traffic arriving rate at the main road was done. The results of the access at the Wusong Road ~ Haining Road intersection can be shown as below:

$$t_2 = 0.0024Q + 3.704\text{, } R^2\text{=0.1531} \tag{15}$$

Where Q= the arriving rate of the main road traffic.

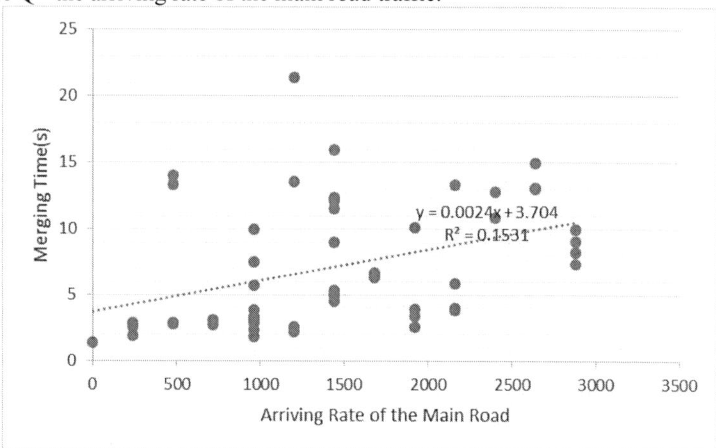

Fig. 5. The regression of main road arriving rate and merging time.

But the as we make the regression of the average value of merging time of each volume of the main road, we get an improved regression. The results can be seen as Fig. 6.

$$t_2 = 0.0024Q + 2.6086\text{, } R^2\text{=0.751} \tag{16}$$

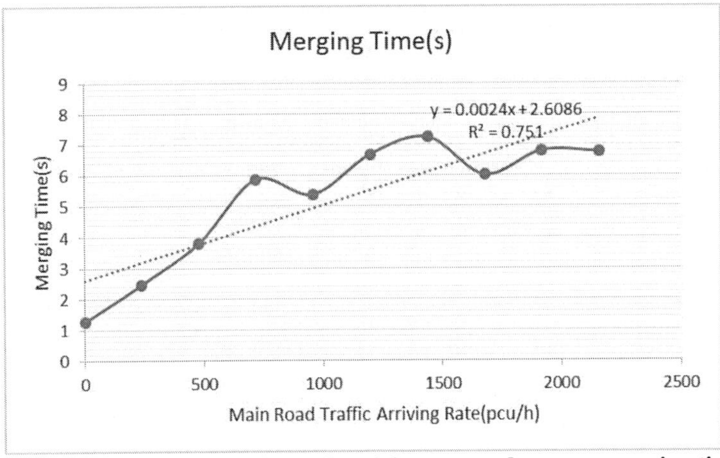

Fig. 6. The regression of main road arriving rate and average merging time.

Since the merging time (green time loss) has a strong connection with the main road traffic flow, we can get the green time loss of each lane through the arriving rate of the main road traffic.

Capacity of the intersection approach

As we get the green time loss and the probability of the access traffic number, the capacity of the intersection approach can be get. The cycle of this intersection is 210s, the green time of the left turn phrase is 70 s and the green time of the straight phrase is 90 s. The average saturated headway is as shown in Table 1. The capacity of the approach is as shown in Table 2.

Table 1. Average saturated headway

	Lane 1	Lane 2	Lane 3	Lane 4	Lane5
Headway	2.02	2.02	2.05	2.14	2.24

Table 2. Capacity of each lane

	Q (pcu/h)	Calculated Capacity (pcu/h)	Actual Capacity (pcu/h)	Difference (pcu/h)	Percentage
Lane 1	1680	732	716	16	2%
Lane 2	1680	729	744	15	2%
Lane 3	1440	759	744	15	2%
Lane 4	1680	561	514	47	9%

The difference between these capacities is quite small. So, this calculation method can be accepted.

CONCLUSIONS

First, this paper give a detailed introduction of how the access affect the main road, including the speed and lane change, and capacity of the intersection. Then a model was built based on the analysis of the merging of the access traffic to identify the green time loss and capacity loss of the intersection. At last, the result of the model is demonstrated with a numerical example.

ACKNOWLEDGMENTS

The authors appreciate the support of the Shanghai Highway Capacity Manual Project (12231201500).

REFERENCES

Chen, K. M., Yan, B. J. (2003). "Highway Capacity Manual (in Chinese)." *Renmin Jiaotong Press.* Beijing.

Transportation Research Board (2000). "HCM 2000." Washington, DC.

Transportation Research Board (2003). "Access Management Manual." Washington, DC.

Wang, D., QI H., and Xu, C. (2010). "Reviewing traffic reliability research." Journal of Transportation Systems Engineering and Information Technology(in Chinese)

Zhao, J. (2008). "Impact of Upstream Weaving Segment on Signalized Intersection Capacity." 2008 International Conference on Intelligent Computation Technology and Automation.

Zhou, S. W. (1987). "Traffic Engineering." Tongji University Press. Shanghai. (in Chinese)